Medicine and Biomedical Sciences in Modern History

The aim of this series is to illuminate the development and impact of medicine and the biomedical sciences in the modern era. The series was founded by the late Professor John Pickstone, and its ambitions reflect his commitment to the integrated study of medicine, science and technology in their contexts. He repeatedly commented that it was a pity that the foundation discipline of the field, for which he popularized the acronym 'HSTM' (History of Science, Technology and Medicine) had been the history of science rather than the history of medicine. His point was that historians of science had too often focused just on scientific ideas and institutions, while historians of medicine always had to consider the understanding, management and meanings of diseases in their socio-economic, cultural, technological and political contexts. In the event, most of the books in the series dealt with medicine and the biomedical sciences, and the changed series title reflects this. However, as the new editors we share Professor Pickstone's enthusiasm for the integrated study of medicine, science and technology, encouraging studies on biomedical science, translational medicine, clinical practice, disease histories, medical technologies, medical specialisms and health policies.

The books in this series will present medicine and biomedical science as crucial features of modern culture, analysing their economic, social and political aspects, while not neglecting their expert content and context. Our authors investigate the uses and consequences of technical knowledge, and how it shaped, and was shaped by, particular economic, social and political structures. In re-launching the Series, we hope to build on its strengths but extend its geographical range beyond Western Europe and North America.

Medicine and Biomedical Sciences in Modern History is intended to supply analysis and stimulate debate. All books are based on searching historical study of topics which are important, not least because they cut across conventional academic boundaries. They should appeal not just to historians, nor just to medical practitioners, scientists and engineers, but to all who are interested in the place of medicine and biomedical sciences in modern history.

More information about this series at
http://www.palgrave.com/series/15183

Alan Mackintosh

The Patent Medicines Industry in Georgian England

Constructing the Market by the Potency of Print

palgrave
macmillan

Alan Mackintosh
Centre for History and Philosophy
of Science
University of Leeds
Leeds, UK

Medicine and Biomedical Sciences in Modern History
ISBN 978-3-319-69777-2 ISBN 978-3-319-69778-9 (eBook)
https://doi.org/10.1007/978-3-319-69778-9

Library of Congress Control Number: 2017958013

Printed on acid-free paper

This Palgrave Macmillan imprint is published by Springer Nature
The registered company is Springer International Publishing AG
The registered company address is: Gewerbestrasse 11, 6330 Cham, Switzerland

Preface

The research into patent medicines did not turn out as originally intended. My initial aim was to explore the use of the printed word in the promotion and effectiveness of Georgian patent medicines, and to see if the booksellers and newspaper printers, who were often selling the medicines, exhibited specific skills or experience in the use of print. First, I had to identify the owners, wholesalers and retailers across England, rather than rely on earlier accounts which were largely anecdotal or confined to a small area. It quickly became clear that most of those involved with patent medicines were not the 'quacks' of previous reports, but tradesmen and others who were engaged in making and distributing a popular consumer product—just like the many other Georgian tradesmen who produced a growing range of consumer products for sale across England. Thanks to the prejudices and censorship of Victorian and later writers, the few irregular, often itinerant and very colourful practitioners ('quacks') who did produce patent medicines had stolen the show, hiding the greater number of reputable tradesmen and their techniques from historical inspection. This bias was shaped by the fully trained, registered medical practitioners not only capturing much of the medical market by the end of the nineteenth century, but also monopolising the discussion on its history. Quickly, the research broadened into a larger revisionist study of these medicines within the medical market, rather than a fine-tuned investigation of the use of the printed word.

Simultaneously, realisation dawned that perspective was crucial, and my 30 years as a cardiologist became an important collaborator. Historians have moved a long away from the great-men-doing-great-things vision of medical history, but some of this approach persists: officially approved activity is placed well ahead of the empirical, and the uneducated are often assumed to be fraudulent while the erroneous educated are merely ill informed. Avoiding these assumptions is difficult because our familiarity is confined to regulated twentieth and twenty-first century medicine which appears to hold the answers. However, my experience as a doctor reveals that this is not the case: much of modern medicine is based on flimsy foundations without strong scientific evidence. Even when good randomised trials of therapy have been performed, current medical science often reveals more about the optimum treatment of a population, for example the hundred patients in a selected trial group, than it does about the best treatment for an individual. Sometimes, modern medicine is completely wrong: for example, when I was a newly qualified doctor in the 1970s, the modern treatment for resistant stomach ulcers was a major operation which involved reshaping the stomach and severing some of its nerve supply, in order to reduce the acid in the stomach and duodenum. This irreversible operation was associated with a small mortality rate and a greater risk of disabling side effects: theses and papers were written about it and its benefits, and academic promotions were obtained by the experts in the procedure. But the underlying assumption that the ulcers were secondary to excess acid production was incorrect. Later investigation revealed that the ulcers were produced by a chronic bacterial infection, and the treatment quickly became a simple two-week course of antibiotics. The operation disappeared, but its advocates had been far from dishonest or ignorant: they were correctly exploring the best available treatment within the knowledge of the time.

In every period, many medical practitioners and others strive to produce the best medical treatment within the knowledge of the time. No era has all the answers, but it is difficult for the historian to exclude a bias secondary to the use of current medicine as a benchmark. In this research, the attempted solution to this problem was to approach patent medicines as an industry which supplied a form of healthcare, rather than a type of medical care accompanied by practical details. For reasons which can be debated, greater objectivity can be applied to the study of past industry and commerce than it can be to exploring medicine, and by taking this approach we can escape the Victorian blanket of prejudice against any healthcare which was not sanctioned by registered medical practitioners.

With this new attitude to patent medicines, students of the history of medicine and pharmacy will find that this book delivers a fresh perception of the medical market, and that it provides a rare insight into the *market* aspect of the medical market. Also, patent medicines had to be easily identified and extensively promoted, and so their distribution and sale are simpler to explore in comparison with other Georgian consumer goods. Business historians should be interested in the methods employed: these methods do not demonstrate how other goods of the time were sold, but they do reveal how other goods *could* have been sold. Patent medicines were largely promoted and taken with the help of the printed word, which also made a substantial contribution to their effectiveness (and they were effective, or they would not have been sold in large quantities). In this book, historians of print can discover the therapeutic potency of the printed word.

All research must be built on sound foundations, and I am grateful to John Chartres for my education on eighteenth-century commerce and industry, to Adrian Wilson for his knowledge of the medical history, to Jonathan Topham for his understanding of print history and culture, to Christine MacLeod for steering me into the difficult domain of eighteenth-century patents, and to Malcolm Chase for the tradesmen's perspective of the industry of the time. Alan Humphries at the Thackray Medical Museum in Leeds uncovered some unusual sources, and much would have been omitted without the help of the staff of Leeds Central Library, the Library of Birmingham, the National Archives, and the county archives in Leeds, Preston, Carlisle, Chester and Chippenham. Jonathan Topham, Adrian Wilson, Malcom Chase and Michael Brown read many early drafts and interim chapters, and I am indebted both to their many critical but encouraging, remarks, and also to their consistent advice that a book should be written. The staff and students of the Centre for History and Philosophy of Science at the University of Leeds and the Leeds Humanities Research Institute provided many comments and challenging questions, as well as unswerving support, and I am particularly grateful to Becky Bowd and Jo Elcoat for their prolonged help, companionship and the important coffee breaks.

Finally, I would like to thank my wife, Susan, for not just tolerating but encouraging my metamorphosis from a doctor into a historian. Readers will judge whether it was successful.

Leeds, UK — Alan Mackintosh

Contents

Abbreviations

ABG	Aris's Birmingham Gazette
BMA	British Medical Association
CP	Cumberland Pacquet
ECCO	Eighteenth Century Collections Online
GM	Gentleman's Magazine
LI	Leeds Intelligencer
LM	Leeds Mercury
LRO	Lancashire Records Office
MCR	Medical and Chirurgical Review
MPJ	Medical and Physical Journal
ODNB	Oxford Dictionary of National Biography Online, January 2008, Oxford University Press, http://0-www.oxforddnb.com
OED	Oxford English Dictionary Online, March 2017, Oxford University Press, http://0-www.oed.com
SJ	Salisbury Journal
SWJ	Salisbury and Winchester Journal
d	pennies (12 pennies in a shilling)
s	shillings (20 shillings in a pound sterling)
£	pounds sterling

List of Figures

List of Tables

CHAPTER 1

Introduction

Prologue: An Obscured History

The imposing tomb of John Newbery, far larger than other graves and stones nearby, dominates the picturesque churchyard of Waltham St. Lawrence, Berkshire (Fig. 1.1). As a successful London publishing bookseller, John Newbery (1713–1767) was a close friend of Oliver Goldsmith, Samuel Johnson and many other literary figures, as well as an employer of Tobias Smollett. Today he is well known as the father of children's publishing, with the John Newbery Medal being awarded each year by the American Library Association for the most distinguished US children's book. However, the tomb reveals that Newbery was more than just an innovative London bookseller.

As might be expected, the south and north faces of the tomb commemorate his publishing activities, with some of the inscriptions probably added at a later date. But examination of the smaller west face reveals a different facet of John Newbery, and also a great deal about the prejudices of later generations. The west face, unlike the north and south faces, has not been recently cleaned, leaving its inscription partly obscured by lichen. Closer inspection shows that the wording is a sixteen-line eulogy extolling Newbery's virtues. The middle seven lines read:

A. Mackintosh, *The Patent Medicines Industry in Georgian England*, Medicine and Biomedical Sciences in Modern History,
https://doi.org/10.1007/978-3-319-69778-9_1

Fig. 1.1 The tomb of John Newbery at Waltham St. Lawrence, Berkshire (south-west aspect)

> Sagacity, that discerned, and
> Skill, that introduced
> The most powerful discovery
> In the annals of medicine.
> The humble wisdom that taught,
> And still teaches, moral lessons
> To the rising generation.

'The most powerful discovery in the annals of medicine', a truly breathtaking claim, refers to the most successful patent medicine in Georgian England, Dr. James's Fever Powder. Newbery was the wholesaler and part-owner of this powder. On the tomb, this activity is placed ahead of, and expressed in more superlative terms than, Newbery's children's publishing, which taught 'moral lessons to the rising generation'. The words on any tomb are not a spur-of-the-moment impression of little consequence; they reflect the considered opinion of at least some family members or contemporaries on the life of the deceased. And for John Newbery, this opinion was that selling a patent medicine was as important as, or perhaps more important than, publishing children's books.

Newbery's will, written just three months before his death, suggests that the medicines were the most profitable aspect of his business.[1] An indication from his will of the importance of patent medicines to Newbery and his associates is the three friends who were left money for a commemorative ring. They were: Dr. Robert James, a physician, writer of a well-known medical dictionary and inventor of the eponymous fever powder, Thomas Greenough, an apothecary and business collaborator, and Benjamin Collins, publishing bookseller, banker and printer of the *Salisbury Journal*. The physician, apothecary and bookseller, all prominent in their vocations, had one thing in common: they were medicine owners who had patented at least one medicine each and seven between them. The inscription on the tombstone was not a mistake: patent medicines were a substantial part of Newbery's business.

So why has this inscription, which tells us so much about John Newbery, escaped the scrubbing brush in contrast to the wording on the rest of the tomb? The probable answer reflects a common modern attitude to Georgian patent medicines, an attitude which is so powerful that it can trump the available evidence. It seems that the restorers wanted to reduce awareness of Newbery's involvement with patent medicines, even though his medicine interests have been well documented, but underestimated, by biographers.[2] They did not regard it as a respectable activity for a revered eighteenth-century publishing bookseller, especially one associated with children's books. A blind eye had to be turned. By contrast, John Newbery's relatives and friends two-and-a-half centuries earlier had wished to herald that involvement.

The inscription provides a brief glimpse of an organised Georgian industry which was regarded as being perfectly respectable by many of its participants, and by their friends and relatives, despite the opposition from some members of the medical professions. But later generations hid Georgian patent medicines away or portrayed them with condescending humour. Like deciphering the words obscured by lichen on the west face of Newbery's tomb, we need to find the authentic picture of Georgian patent medicines underneath the prejudices of later centuries. Only then can the position of these owned, secret, medicines in the medical market be properly assessed, and the full structure of Georgian healthcare be allowed to emerge. That is the object of this book.

Patent Medicines in the Medical Market

Few topics of the eighteenth century have been so vulnerable as patent medicines to the hostile judgements of later generations. The supposed dishonesty and avarice of their producers and the assumed ignorance of their consumers can still provide amusement today, with words such as 'crooks', 'quacks' and 'fraud' being frequently employed.[3] The very success of Georgian patent medicines led to this historical aberration: a minor role could have been ignored, but their success was a threat. Many orthodox medical practitioners of that period sought to diminish this commercial and professional challenge in the crowded medical market by questioning the safety of the medicines, and by describing any apparent benefits as a delusion. Then the later medical reformers of the mid-nineteenth century needed to brand them as a menace to the community in order to establish a separate and unified medical profession, the sole guardian of the public interest. Medical care that was not under the control of the profession had to be immoral and inconsequential.

In recent decades, historians have adopted a more balanced approach to alternative Georgian healthcare, with the medical market as the preferred framework. Orthodox medicine was only one of the market's components, and the consumer could choose freely between them in the largely unregulated medical world of the British Isles. Patent medicines are readily observable in this market, but they have been largely conceived as a constituent, and indeed an accessible illustration, of irregular practice.[4] They provided historians with access to 'quackery', but have rarely been considered as a separate entity. This study of patent medicines demonstrates, in contrast, that the ownership, distribution, retailing and promotion of patent medicines was a stable, successful, and mostly honest industry, which can be recognised as such from the middle of the eighteenth century. Much of this industry was run by tradesmen, specialised medicine wholesalers, chemists, regular medical practitioners, newspaper printers, booksellers and druggists, and it was distinct both from irregular practice and from orthodox medicine. As a result, the structure of late Georgian healthcare needs to be redrawn with the patent medicines industry regarded as a separate entity which overlapped with the care provided by both regular and irregular practitioners, but had its own specific character.

This exploration of patent medicines goes deeper than a revision of the overall configuration of the medical market: it provides the functional detail for an important section of that market. Jenner and Wallis have pointed out that the term 'medical marketplace' has been used to avoid precise investigation of the market.[5] This book brings the market back into the medical market by approaching the ownership and sale of patent medicines as a commercial activity which supplied a form of healthcare. Commercial success depended on satisfied consumers, so this approach then directs attention to the health benefits of patent medicines. In contrast, many histories of Georgian medicine have come from the opposite direction, exploring some aspect of healthcare and then adding in the commercial background if possible. Thus, we have surprisingly little detail on how the medical market, a commercially driven enterprise, functioned across the country. This book exposes, for the first time, some of the business principles and practices throughout England of a major section of the medical market.

Revealing these business practices shines a spotlight on the use of the printed word in promoting, and ensuring the effectiveness, of patent medicines. With a few exceptions, both regular and irregular medical practice depended on direct, verbal, communication, however perfunctory. In contrast, the large majority of consumers of patent medicines never met the producers of their therapy, and the printed word was the essential vehicle for publicising and explaining the medicines. The use of the printed word involved trust, authority and repetition, and it was more nuanced than the isolated examples of hyperbole which have often been quoted for entertainment as well as information. The printed word had an important additional role: a direct effect on the well-being of the consumers. As contemporaries were aware, reading about patent medicines increased their efficacy over and above the pharmaceutical actions of their components. Print historians are constantly exploring the impact, whether rational, irrational or unconscious, of the printed word on human ideas and behaviour, but investigating the patent medicines industry uncovers the direct benefits of the printed word on human health.

Having stated some of its aims, the boundaries of the book should also be made clear. As it is not a history of pharmacy, it does not attempt to reveal the constituents of the medicines, except for the information which the consumers of the medicines could easily obtain themselves. The secrecy of the recipe was essential for the commercial success of patent medicines, and as we shall see, the owners were reluctant to reveal even the most general description of the contents. In line with the commercial approach, the level of knowledge of the consumers who were contemplating a purchase is more relevant than later exposures of the constituents.[6] Similarly, the medical efficacy of patent medicines will not be assessed by twenty-first century medical science: such an approach would be anachronistic as it would test this Georgian therapy by today's assumptions and knowledge, something which is not required for other historical topics. For example, we do not study the ability of the hand press to transfer knowledge in comparison with the Internet, neither do we ask whether the horse-drawn coach was as comfortable and fast as a modern vehicle. The effectiveness of Georgian medicines by today's standards is irrelevant. The question should be: Did the Georgian consumers benefit from these medicines? The large volume of sales over many decades shows that many of them clearly did.

The temporal and geographical boundaries of the book also need clarification. The whole of the Georgian period is included, but the patent medicines industry became a distinct entity in the middle of the eighteenth century, and so this book inevitably has more to say from that period onwards. The exploration ends around 1830 when the nature of the industry was changed by the introduction of radical alternatives to orthodox medicine, by alterations in wholesaling and retailing, and by the growing demand for the exclusion of all forms of unregulated medical activity. A particular virtue of the book, which sets it apart from most studies of medicine and commerce in the period, is that the whole of England is actively included: most consumers lived outside London and this book is unashamedly provincial in many of its sources. Scotland and Wales are not excluded on any point of principle, but sales of patent medicines were modest there in comparison to England, even allowing for smaller populations, and a more coherent account emerged by concentrating on their larger neighbour.

Exploring Patent Medicines

The growing interest from the 1980s in Georgian consumption and its medical market sparked a new awareness of all forms of alternative medicine, led by the research, energy and memorable phrases of Roy Porter. Overall, Porter brought a fresh, open-minded, approach to the topic of Georgian alternative medicine in many publications, which substantially altered the previous views and brought this type of therapy out of the shadows.[7] His main contribution was to demonstrate that the Georgian medical market was driven by consumers who sought out both the diverse forms of regular medicine and a wide variety of alternative practitioners. To fulfil this demand, the irregulars imitated the orthodox, and the boundary between the two was indistinct. In contrast to earlier writers, he largely considered alternative therapies, including patent medicines, to be as effective as regular treatment, and so they should not be considered as a trick on the gullible on most occasions. Porter's emphasis was on the wide range of alternative practitioners and therapies which had escaped the attention of historians, but he had little engagement in the overall structures, mechanisms, geographical patterns or temporal changes. As Harold Cook expressed it, 'the ways in which medical buying and selling occurred did not occupy his attention for long'.[8] In particular, he did not analyse patent medicines as a distinct entity: rather, he described them as one of the tools of 'quackery', often using their promotional wording to illustrate the aims and activities of irregulars in general.

Several historians have joined Porter in exploring Georgian alternative medicine and the patent medicines available within it. Some have investigated patent medicines in a particular geographical area, especially Brown in his often-quoted studies of patent medicines in Bath, or from a single source.[9] Helfand has provided an illustrated, transatlantic, perspective.[10] More investigators have focussed on aspects of their retailing, often being concerned with their promotion and advertising. In studies of the development of national markets, patent medicines have provided an easily identified product for research.[11] These medicines can open up our understanding of how non-essential goods in general were sold.[12] In addition, patent medicines have appeared in the early sections of histories of the pharmaceutical industry.[13] However, many of these accounts of Georgian pharmacy only explore patent medicines by anecdotes, with the notable exception of two papers by Burnby which are rare attempts to investigate their ownership and distribution outside a study of irregular medicine.[14] Thus, patent medicines have not escaped the attention of recent historians, but only minimal efforts have been made to analyse the ownership, distribution and sale of patent medicines as a whole industry, spread across the country. As a result, their distinct role in the inseparable healthcare and commerce of late Georgian England has not emerged.

A lack of available records from the businesses involved explains some of the reluctance to undertake this analysis.[15] Eighteenth-century businesses rarely retained records once they ceased to have a practical use. However, there was also a degree of censorship which probably discouraged the retention of relevant documents. The activities of medicine owners, distributors and retailers were readily discussed by their contemporaries, but their descendants preferred to ignore any details or even not to mention the business at all. For example, the patent medicines business started by John Newbery remained within the family for over 150 years after his death, and a family memoir in 1911 described the succession of Newbery directors, while printing a photograph of the current large premises in central London.[16] The author of the memoir mentioned the medicine-related activities of John and his son Francis: he could hardly do otherwise. But he was then unable to describe at all what the business had been doing for the subsequent hundred years, just referring to it as 'the company', 'the house' or 'the business'. Later chapters in this book contain further examples of the frequent suppression of patent medicines from family memoirs and biographies.

Another reason for the hesitancy in studying patent medicines as a distinct entity is that many writers have still found it difficult to avoid taking a moral position on these medicines: patent medicines were perhaps not worthy of their attention. Cody provided a blanket denunciation when she described 'quack medicines' as 'perhaps the most worthless of consumer goods in the eighteenth-century marketplace'.[17] Most recent accounts have avoided such overt condemnation, but some of these earlier attitudes are still evident. Barker described them as '"quack" cures, which were likely to have produced little benefit for those who took them in physiological or pharmacological terms', and Gardner has written of 'products of dubious effectiveness'.[18] Many writers implicitly or explicitly regard patent medicines as a confidence trick, while ignoring the strong possibility that the medicines prescribed by regular practitioners may have been no better, or perhaps worse. Even an apparently balanced account can have an intrinsic bias. For example, Porter, who tried harder than most to be open-minded about patent medicines, stated that taking them was not a mass delusion as they were probably as effective as orthodox therapy:[19] but this approach was undermined by his choice of words which tended to diminish patent medicines as a genuine healthcare provision. For example, his common use of the description 'quack medicines' linked them to colourful irregulars, while his alternative term 'nostrum' carried an implication of inefficacy. Similarly, his description of medicine wholesalers and some of the newspaper printers who sold the medicines as 'cronies' does not encourage a careful assessment of their true relationship.[20] We need to assess the production and selling of patent medicines by the normal healthcare and commercial standards of an era largely untroubled by medical and pharmaceutical regulation.

What Was a 'Patent Medicine'?

The meaning of 'patent medicine' requires clarification, given that only a minority had ever received a royal patent, and even fewer had been granted one within the previous fourteen years, the legal duration of a patent. In 1830, over 1300 medicines of this type were listed for taxation by the excise stamp, while only a total of 118 medicines had been patented up to that year (Appendix B), and the number of medicines with a current patent never exceeded 31.[21] None of the alternative terms used to describe these owned medicines provides an entirely satisfactory use of Georgian terminology that is also easily understood today. In recent years, they have often been described as proprietory medicines, but this term was rarely used for medicines in the eighteenth century. At that time, 'proprietory' referred more often to land possession, or associated legal rights, than to commercial ownership.[22] 'Public medicine' was used in the period, but this term is vague, and also potentially confusing for today's readers who might regard these secret owned products as being private. Regular practitioners and other critical commentators of the period often referred to them as nostrums, quack medicines, or empirical medicines, but a more neutral term is preferable. So the predominant eighteenth-century usage will be continued, with all these owned medicines described as 'patent medicines'.

The consumers of patent medicines appear to have had no difficulty in identifying patent medicines as a coherent category. Nevertheless, the 1783 Medicines Act, which attempted to tax patent medicines for the first time, immediately ran into difficulties because the government assumed that identifying the medicines would be easy and then found that this was not legally straightforward.[23] The Act was soon replaced by the 1785 Medicines Act, which described the taxable medicines as all medicines which had been patented at any time, together with any other medicines which had an owner and a secret recipe, and were publicised 'by any public notice, advertisement, or by written or printed papers or hand bills'.[24] This book will refer to all medicines which satisfied these three criteria as patent medicines and it will also include a few more which fell just short because their recipes had been divulged, because the intended advertising had not yet taken place, or because their owners in the upper reaches of society were reluctant to advertise. It is important to realise that some medicines referred to here as patent medicines were not described as such by their owners, especially if the owners were regular practitioners. Fulfilling these three criteria is more important than any gloss put on the medicine by owners seeking to maintain a professional reputation or a social position.

So how did these owned medicines become known as 'patent medicines' in late Georgian England, regardless of whether they actually had received a patent? The answer lies in the proprietors' frequent desire to extend the authority of a genuine patent to their other, unpatented, medicines, and also in the lack of a memorable alternative name to describe these owned, but unpatented, medicines. The authority of the royal patent could be unofficially widened to unpatented medicines by a number of means. One method was the ambiguous use of a heading: for example, a handbill from Francis Newbery, John Newbery's son, describing a mixture of patented and unpatented medicines, was headed 'By Virtue of the King's Patent', thereby extending this royal authority to all the medicines.[25]

Another technique for spreading the authority of the patent was to publicise wholesale or retail premises as a patent medicine warehouse. Thus, a 1781 newspaper advertisement from Jackson, Warter and Co. had two headings. The first was 'By the authority of the King's Patent granted to Jackson, Warter and Co', and the second was 'At their royal patent wholesale medicinal warehouse, Fleet Market, London'.[26] Fifteen medicines were then mentioned in the advertisement, only three of which can be clearly identified as having ever received a patent. This advertisement is also an example of the extended description of a patent for publicity purposes. Jackson, Warter and Co. could not have been granted a medicine patent, as all medicine patents in this period were granted to individuals, or a small group of individuals, not to companies (Appendix B). The senior partner, Thomas Jackson, had obtained one in the past, but this was granted in 1761 which meant that it had expired by the time of the advertisement. It suited many vendors to be vague about which of their medicines were genuinely patented and to imply that patented and unpatented medicines were similar. As a result they could all be referred to as patent medicines.

Wholesalers and retailers sometimes did provide a separate descriptive term for their unpatented medicines, but this was a variable, and less memorable, word than 'patent'. For example, a catalogue from William Bacon in Oxford Street used 'patent and public medicines', a handbill from Pearson and Rollason in Birmingham headed a section as 'genuine and patent medicines', a catalogue for Shiercliff's Circulating Library in

Bristol included 'patent and other medicines', and a newspaper advertisement for Shaw and Edwards of St. Paul's Churchyard was content with just 'patent medicines, etc.'.[27] In the absence of single clear word to describe the unpatented medicines, *all* the owned medicines sold by a medicine vendor were commonly referred to simply as patent medicines with any additional description omitted.

Provision of Medicines

Within the economic commotion of the Georgian medical market, even the term 'medicine' lacked precision. Today, a medicine can be defined easily in terms of some form of government regulation, or a list created by a professional body. By contrast, the eighteenth-century English medical world was almost entirely unregulated, and any earlier supervision of medical therapy, such as the right of the London College of Physicians to inspect apothecaries' shops, had fallen into disuse.[28] The result of this lack of definition was that some items were simultaneously medicines and consumer products with non-medical benefits. Sometimes this uncertainty was intrinsic to the type of product; for example, a toothpowder could be promoted as both a cleaner of the teeth and a means of preventing mouth scurvy, or a skin preparation could have cosmetic effects and cure 'eruptions' of the face.[29] A health benefit could also be claimed for items which were normally ingested for refreshment or nutrition. The well-known examples at the beginning of the eighteenth century were tea and coffee. Their promotion by promised health benefits was less obvious later in the century, although a medicine for consumption and other complaints in the 1780s was named 'English Coffee'.[30] A more precise example was the promotion of chocolate as a medicinal product by Fry's of Bristol. The founder, Joseph Fry, originally sold chocolate for health reasons in his apothecary's shop before starting a chocolate factory in 1761, and as late as 1843 the company chose to name its new, very popular, drink 'Homeopathic Cocoa'.[31]

The relative importance of the medicinal element of a product diminished in some cases over our period. Thus, in the mid-eighteenth century, mineral waters were often ingested for health reasons and were sold by specialist water dealers in a similar manner to medicine distribution: but Jacob Schweppe's new artificial mineral waters were designed to be drunk at table in the 1800s, although they did not escape taxation as patent medicines until 1840.[32] So the concept of a 'medicine' was not ring-fenced, and medical treatment spread out from agents which were exclusively used for health purposes into a broader range of consumer products which also fulfilled other purposes. For clarity, this book restricts the word 'medicine' to products which seem to have been devised primarily for healthcare, and it excludes items created for other purposes.

So how could people acquire their medicines? Physicians, surgeons and apothecaries could prescribe them, but medicines were also bought from chemists and druggists without a prescription, and, as we shall see, patent medicines were often acquired from other tradesmen, especially printers, booksellers and stationers. In addition, many medicines were made at home or within a small community, usually for immediate consumption. These different origins will be briefly reviewed, though many sources did not fit conveniently into these categories as Georgian healthcare was largely free of any boundaries, whether regulatory, medical, social or economic. Also, a patient often used more than one source of medicines for a single illness.

Medicines could be prescribed by the regular practitioners as physicians, surgeons, apothecaries and surgeon-apothecaries, the overlapping branches of what later became the united medical profession in the mid to late nineteenth century. All varieties of regular medical practitioners prescribed and also supplied medicines, usually after seeing the patient, but sometimes after a postal consultation.[33] Some physicians restricted themselves to writing a prescription for the apothecary to dispense, but outside London many physicians and surgeons prepared and supplied the required medicines themselves, thereby increasing their incomes.[34] For the apothecary, his income was often reliant on supplying medicines. A common feature of all the dispensing by the regular professions is that the patient did not necessarily know what was being taken. Some prescribed medicines had simple names, but many were compounds written by hand in abbreviated Latin which was incomprehensible to most patients. This would have made it more difficult for patients to decide if their treatment was appropriate, possibly increasing any suspicions about the regulars' true motives. By contrast, the names of patent medicines were written on all the packaging, and they were easy to remember.

As the apothecaries increasingly provided medical advice in the patients' homes and were less concerned with their shops, their place as day-to-day medicine suppliers was often taken up by chemists and druggists, a controversial group.[35] Originally, chemists could compound drugs and might be wholesalers, while druggists only supplied ready-made or simple preparations: but in practice the absence of regulation meant that the two terms were used interchangeably, and often together, especially outside London. This resemblance is shown in a book offering practical advice to the parents of future apprentices which listed chemists and druggists separately, but with similar apprentice fees, similar sums to set up in business, and similar annual incomes for journeymen.[36] Chemists and druggists sometimes had completed a period of apprenticeship with another chemist or druggist, but many seem to have had no organised training. Thus, the degree of expertise of the average druggist was disputed in this period, and remains so amongst modern historians. One Georgian polemicist depicted them as being poorly trained, careless and regularly adulterating medicines to increase their profit[37]; but the *Universal British Directory* entry for Leeds confidently listed them in a separate section entitled 'Physic', together with all the physicians, surgeons and apothecaries.[38] More recently, Loudon has regarded druggists as being more akin to grocers than part of the healthcare system; whereas Holloway has suggested that they were very similar to apothecaries, except that they remained in their shops.[39] Probably, all these points of view have some merit as the lack of a monitored apprenticeship system would result in a variable degree of expertise.

Whatever their level of skill, druggists had a key role in the provision of medicines, responding quickly to consumer pressure and often undercutting the apothecaries' prices.[40] Present in nearly every town by the end of the eighteenth century, they ensured a local supply of a wide range of products. Their surviving handbills reveal hundreds of conventional stock items and they could compound others.[41] Also, as we shall see, they increasingly sold patent medicines during the late Georgian period. Some, at least, did provide professional advice by choosing medicines on behalf of the consumer, as illustrated by this succinct letter to Samuel Glover, a Leeds druggist, which is reproduced in full:

> I will thank you to send a few pills for my wife, such as you think will suit her, she got her bed about three weeks ago, also ½ lb of the best salts. Your obedient servant J Barstow. June 25th 1822.[42]

Even the poor had access to the druggists' medicines, as many were cheap and the druggist could also provide small quantities out of a bottle or box of an expensive medicine.[43]

In this unregulated period, anybody could sell medicines. In addition to their normal business, many tradesmen sold small quantities of pre-prepared medicines as a sideline, particularly patent medicines, and a few sold them in larger quantities. When these non-medical sources are mentioned, the conventional story is that a wide range of shopkeepers can be found amongst these major vendors[44]; but this book will show that, in *late* Georgian England, the sale of patent medicines in large quantities outside druggists' shops was concentrated amongst printers, booksellers and stationers. Whether the consumers arrived at the shops knowing which medicines they needed, or whether the members of these print trades selected the medicine on behalf of consumers is a question which will be considered in Chapter 5.

Many medicines were not purchased, but were created at home and taken without any professional advice. John Wesley's best-selling book on medical self-help, *Primitive Physic*, aimed to provide a choice of suitable home-produced medicines for all common conditions.[45] The ability to prepare medicines was regarded as part of good housekeeping, just as much as a knowledge of food preparation or the management of the linen. For instance, when Sabine Winn of Nostell Priory in Yorkshire died in 1798, she possessed eleven volumes of handwritten, mainly medicinal, 'receipts' handed down from her mother for the benefit of her gentry family and household, together with articles on medical topics copied from the *York Chronicle*.[46] Lower down the social scale, *The Compleat Housewife* of 1753 provided over 300 medical receipts, and much other practical information, for five shillings.[47] Jane Austen provided a vignette of this frequent recommendation of domestic medical preparations in *Sense and Sensibility*, when one of the characters, Marianne, developed a cold which then deteriorated and, in advance of any trained medical help, 'prescriptions poured in from all quarters, and as usual, were all declined'.[48] Domestically prepared medicines were often the initial, and sometimes the only, treatment for many conditions.

So medicines in this period were obtained from many and variable sources, allowing consumers not only to choose their supplier but also to decide how much accompanying medical advice they wanted, if any. The provision was also flexible, enabling different sources to be used in the same illness and ensuring that all, with the financial assistance of the Poor Laws if necessary, could have access to medicines.[49] On the whole, all but the poorest late Georgian consumers could quickly buy or make most widely known medicines, with or without medical advice.

The Patent Medicines Industry

Exploring Georgian patent medicines necessitates getting underneath the later critical assumptions and censorship which have distorted the understanding of their role in the medical market. Good primary sources on events, not comments on intentions or later memoirs, are the key to finding out what really happened. Such sources should be, as far as possible, resistant to tampering by later generations, and many should be provincial to capture the national market, not just the different commercial and healthcare circumstances of London. Newspapers fulfil these criteria admirably: they are exactly the same now as the day they were printed and then read by local consumers, and they were published all over England from the mid-eighteenth century. In particular, the texts of the newspaper advertisements expose much of the core ownership and wholesaling of the industry, together with some of the retailing arrangements, and so runs of newspapers from the towns and city of Leeds, Birmingham and Salisbury, chosen to provide a range of urban and more rural locations from across the country, have been analysed in a systematic fashion. They provide a solid spine for studying patent medicines, especially when they link up incomplete information from other sources. These runs of the *Leeds Intelligencer*, *Leeds Mercury*, *Aris's Birmingham Gazette* and *Salisbury and Winchester Journal* are referred to throughout this book as the 'studied newspapers'. All the issues of each newspaper were inspected for the first six months of five years, each at least twelve years apart, namely 1769, 1781, 1794, 1807 and 1822 (the 'studied periods').[50] The initial year was chosen as the first year when complete runs were available for all four newspapers, and the other years were selected to coincide as far as possible with local trade directories. Many of the medicine advertisements in the second half of each year had been printed in the first half, so the analysis was confined to the first six months.

Other good primary sources of this type are the few surviving account or day books of newspaper printers, and books from Winchester, Whitehaven and Chester have been inspected. Government, parliamentary and other official or semi-official records have also proved to be revealing sources when they are available. Some are unaltered, such as the records of the Old Bailey and the wills of those rich enough to require one. Others are edited to some extent: for example, the early accounts of Parliament initiated in this period by William Cobbett and Curson Hansard were largely summaries, with *Hansard* only becoming a true verbatim report in 1907.[51] However, the selection in this type of record was usually based on criteria which were unrelated to the patent medicines themselves. Thus, Bennet Woodcroft's monumental work, published in 1872, on the abridgements of patent specifications does not provide the original specifications, but it includes similar information for medicines as it does for other types of eighteenth-century patents.[52]

Away from the newspapers and official documents, Edward Harrison's attempt at medical reform in 1806 has provided a window on the sale of patent medicines across the country and the reactions of the regular medical practitioners. Harrison, a Lincolnshire physician, was leading an unsuccessful effort to bring in compulsory qualifications for practitioners, and, as part of the evidence gathering, he circulated a letter to practitioners across England asking for information on irregular practice, the activities of druggists, the sale of medicines and other issues in their locality.[53] The replies, published in the *Medical and Chirurgical Review*, provide both eye-witness accounts and opinions from across the country.

The major participants in the patent medicines industry have left only occasional opinions or explanations, and any details of their day-to-day activities, such as account or day books, are patchy; but, paradoxically, the shortage of documents from participants has a methodological advantage. A wide range of imperfect sources predominately describing events, from across the country, had to be used, and this encouraged an emphasis on the *actions* of many participants, rather than on the *aspirations* of a few, which may not have been accomplished. The research was not dominated by one or two extensive archives, often preserved because a business was unusually successful and therefore atypical: the structure of the industry was derived from modest amounts of information from many sources.

One of the problems in exploring the actions, triumphs and failures of the patent medicines industry is that many of these sources were mediated by either sympathetic or critical opinions. Several sources are directly or indirectly promotional in nature, and so they may be incomplete, or just made up. Cody concluded, without much accompanying evidence, that many of the testimonials in the newspaper advertisements were complete fabrications.[54] However, each source should be assessed on its own merits: truthful, favourable, news may have been the best form of promotion. If we are aware of the character of these mediated sources and their relationship to other sources, then they can be very revealing. On the whole, practical information, such as the recommended indications and the names and addresses of the owners, distributors and retailers, can be assumed to be correct, as there would seem to have been no purpose in misleading potential customers on this type of factual information. In assessing the information contained in the sources, we must avoid the later bias against patent medicines: the promotional material for medicines was not automatically less trustworthy than similar material for other Georgian goods.

Whatever their views on the merits of patent medicines, critics and sympathisers agreed that sales were substantial in England, though less so in Scotland and Wales. The general comments of contemporaries on the ubiquity of patent medicines, and the indirect evidence from created fortunes, will be solidified in Chapter 2 by the direct evidence from excise returns and other records. But sales were more limited in the early eighteenth century: only from the mid-century can we see a true national market which partly replaced the earlier role of patent medicines as a distant extension of an individual's practice. The chapter will go on to explore the fluid and porous boundary between regular and irregular practitioners, and also between regular and irregular practice. The views of regular practitioners on these medicines were more complex than the received impression of blanket condemnation, and indeed some of them participated in the industry. Amongst the public, the status of patent medicines was boosted by the apparent inadvertent recognition by the state through the patent system and later through the excise stamp, at a time when prescribed medicines received no equivalent acknowledgement. Lastly, Chapter 2 will discuss who took the medicines and explore the range of conditions they were recommended for.

Patent medicines have often been associated with itinerant irregular practitioners ('quacks'), but these people only owned a minority of the medicines advertised in newspapers. Colourful dishonest mountebanks selling medicines on a stage linger in the popular imagination, but they were disappearing by the late eighteenth century, and they played only a small role in the industry before that. Chapter 3 will show that most of the nationally available medicines were owned and produced by medicine specialists, by other tradesmen, and by medical practitioners who did not practise irregularly. Many of these owners were reputable, operated from fixed premises for substantial periods of time, and passed their profitable businesses on to their families or business partners. Members of the higher ranks of society could also own patent medicines for a mixture of philanthropic and commercial reasons.

Individual Georgian patent medicines were readily identifiable, unlike most other products of the period; so investigating their distribution across England in Chapter 4 reveals not only some of the detailed mechanisms of the medical market, but also the available methods for the wholesaling of consumer goods in general. The major wholesalers in London dealt directly with retailers all over the country, even in Cumberland, sometimes spending hundreds of pounds a year on advertising in regional newspapers. Booksellers initially played a large part in this process, but by the beginning of the nineteenth century, wholesaling was in the hands of medicine specialists and chemists. Chapter 5 confirms the previous reports of the prominence of local booksellers and printers in retailing patent medicines: they had an even more dominant role than previously supposed. Why did so many booksellers sell a product that was physically unrelated to their books, other printed matter and stationery, and which normally came from different sources? Should these booksellers be regarded as irregular medical practitioners?

The industry was largely constructed by the printed word and the next three chapters explore the potency of print both in promoting the medicines and in increasing their health benefits to the consumers through the modulation of the 'imagination', as Georgians described it. Newspaper account books, and the newspapers themselves, will be utilised in Chapter 6 to describe how the many advertisements for the medicines were ordered and paid for. The medicine advertisements were all charged to wholesalers and local retailers, and the newspaper printer had to pay a tax on each one: he did not just insert them at a whim to fill up space. The medicine advertisements made a substantial contribution to the overall finances of a regional newspaper. Medicines were also promoted in print by bills passed by hand or fixed to the wall. These bills could be lengthy and the large number printed reflects their importance; unfortunately, little can be discovered on how they were used.

Chapter 7 moves away from commerce and analyses some contemporary views on how a patent medicine helped the consumer. For many Georgian physicians, changes in the 'imagination' secondary to the 'passions' enabled medicines, including patent medicines, to achieve benefits which were greater than the sum of their pharmaceutical ingredients. The physicians demonstrated the effect of the imagination on patients, and showed that unusual and unorthodox therapies could have therapeutic benefits, but not by the mechanisms advocated by their proponents. In other words, these proponents could be perfectly honest about the benefits of their therapies, though misguided on the means by which the benefits were obtained. This conclusion should alter our attitude to the many therapies, orthodox or irregular, which have been regarded as ineffective, and perhaps humorous, by the scientific standards of later generations. It is tempting to link the altering of the imagination with the twentieth-century placebo effect. But this should be resisted as a historical discontinuity exists between them, and both are heavily dependent on their contemporary cultures which are very different across a gap of two centuries.

For patent medicines, the imagination was largely modulated by the printed word, and Chapter 8 describes how this was done in advertisements, bills and treatises. The confidence of the consumer, accompanied by a degree of mystery, was the key to success; and the systematic analysis of continuous runs of provincial newspapers provides a Georgian consumers' impression of the information received. Contrary to earlier notions, most advertisements avoided hyperbole and adopted a relatively low-key approach, which mimicked regular medicine and was reinforced by frequent repetition. The advertisements predominately sought to build up trust by linking the producers' straightforwardness with authority, which often came from the state through the patent system and the medicine excise stamp. Chapter 9 summarises the historical legacy of the Georgian patent medicines industry, and it surveys the changing status of patent medicines in the Victorian medical market. Patent medicines were legally prohibited in 1941, but many of the medicines on today's supermarket shelves still reflect their styles of promotion and retailing.

Notes

1. National Archives, Will of John Newbery.
2. Welsh, *Bookseller*; Roscoe, *John Newbery*.
3. Recent examples are Rance, *Quack Doctor* and Strathern, *Quacks*.
4. Especially Roy Porter in his influential book, *Health for Sale*.
5. Jenner and Wallis, 'Medical Marketplace', 2.
6. Twenty-four 'compositions of quack medicines' were published in the first four issues of the *Lancet* (*Lancet*, 1 (1823), 30, 62, 89 and 138).
7. Amongst many publications, two important books were Porter and Porter, *Patient's Progress* and Porter, *Health*.
8. Cook, 'Roy Porter', 20.

9. Barry, 'Publicity'; Brown, 'Medicines Advertised'; Brown, 'Venders'; Rousseau, *John Hill*; Doherty, 'Anodyne Necklace'.
10. Helfand, *Quack*.
11. Cody, 'No Cure'; Rousseau, 'Stung into Action'; Wallis, 'Consumption'; Cox, *Complete Tradesman*, 103–110; Mui and Mui, *Shops*, 228–231.
12. Isaac, 'Pills'; Barker, 'Medical Advertising'; Strachan, *Advertising*, 4–96.
13. Holloway, *Pharmaceutical Society*; Tweedale, *Sign*; Corley, 'Pharmaceutical Industry'.
14. Burnby, 'Preparers'; Burnby, 'Early Years'.
15. Tweedale, 'Archives', 37; Helfand, *Quack*, 8.
16. Newbery, *Records*, 59–62, 82.
17. Cody, 'No Cure', 103.
18. Barker, 'Medical Advertising', 379; Gardner, *Business*, 58.
19. Porter, *Health*, 17, 141.
20. Porter, *Health*, 116.
21. *Journal of the House of Commons*, 85 (1830), 313–319; Mackintosh, 'Authority'.
22. 'Proprietory', *OED*, accessed 30 May 2017.
23. Spilsbury, *Discursory Thoughts*, 3–7.
24. Kearsley, *Tax Tables 1786*, 89.
25. John Johnson Collection, Patent Medicines, 14 (41).
26. *LM*, 23 January 1781.
27. Bacon, *Medicines*; *Pearson and Rollason*; *Shiercliff's Library*; *LI*, Messrs. Shaw and Edwards, 21 December 1807.
28. Medicus, 'Adulteration' 30.
29. *LM*, Dickinson's Gowland's Lotion, 4 January 1794; *ABG*, Jackson's British Powder for Teeth, 8 January 1781.
30. *ABG*, English Coffee, 1 January 1781.
31. Diaper, 'Fry & Sons', 34, 38.
32. McIntyre, 'Mineral Water', 3–4, 10; Burnett, *Liquid Pleasures*, 98.
33. The elite form of postal consultation is well described in Wild, *Medicine-by-Post*.
34. Holloway, 'Orthodox Fringe', 154.
35. Holloway, *Pharmaceutical Society*, 32–42; Burnby, *English Apothecary*, 22.
36. Kearsley, *Trades*, 5–24.
37. Medicus, 'Adulteration'.
38. Barfoot and Wilkes, 536.
39. Loudon, 'Vile Race', 109–111; Holloway, *Pharmaceutical Society*, 38.
40. Holloway, *Pharmaceutical Society*, 42.
41. A typical example is an 1806 handbill from Robert Spencer & Sons in Leeds listing over 400 items (Yorkshire Archaeological Society,

Small Accounts and Vouchers for Lord Ribblesdale 1806, MD 335/1/7/2/56).

42. West Yorkshire Archives, Leeds, Business Papers of Samuel Glover, WYL 33/GL/A 18.
43. Mui and Mui, *Shops*, 231.
44. Porter, *Health*, 113.
45. Wesley, *Primitive Physic*, x.
46. Day, 'Household Management', 210.
47. Smith, *Compleat Housewife.*
48. Austen, *Sense and Sensibility*, 302.
49. By the late eighteenth century, the Poor Law surgeon of the parish was often paid to provide medical services, including medicines, at the same cost as private practice (Loudon, *Medical Care*, 231).
50. Out of a total of 520 issues of all four newspapers published in these five periods, ten issues were missing from the archives (seven from the *Leeds Mercury*, three from the *Salisbury and Winchester Journal*).
51. Jones, *Debates*, 73.
52. Woodcroft, *Abridgements.*
53. For a summary of Harrison's attempts at reform see Loudon, *Medical Care*, 140–147.
54. Cody, 'No Cure', 123.

Bibliography

Archives

John Johnson Collection, Bodleian Library, Oxford.
National Archives, Kew, Will of John Newbery, Prob. 11/935.
West Yorkshire Archives, Leeds, Business Papers of Samuel Glover, WYL 33/GL/A18.
Yorkshire Archaeological Society, Leeds, Small Accounts and Vouchers for Lord Ribblesdale 1806, MD 335/1/7/2/56.

Newspapers and Journals

Aris's Birmingham Gazette (ABG).
Journal of the House of Commons.
Lancet.
Leeds Intelligencer (LI).
Leeds Mercury (LM).

Printed Primary Sources

Austen, Jane, *Sense and Sensibility* (First published 1811. Harmsworth, Middlesex: Penguin Books, 1986).
Bacon, W., *An Account of Several Valuable and Excellent Genuine Patent and Public Medicines* (London, 1790?).
Barfoot, Peter, and John Wilkes, *Universal British Directory, 1793–1798* (London, 1793–1798).
Kearsley, George, *Kearsley's Tax Tables 1786*, (London, 1786).
Kearsley, George, *Kearsley's Table of Trades* (London, 1787).
Medicus, 'On the Adulteration of Medicines, and the Religion of Quacks', *The Scourge*, 2 (1811): 29–35.
Pearson and Rollason, Printers, Booksellers, and Stationers (Birmingham, 1782?).
Shiercliff's Circulating Library, The Following Articles are Sold (Bristol?, 1779?).
Smith, Eliza, *The Compleat Housewife*, 15th edn (London, 1753).
Spilsbury, Francis, *Discursory Thoughts*, 2nd edn (London, 1785).
Wesley, John, *Primitive Physic*, 24th edn (London, 1792).

Secondary Sources

Barker, Hannah, 'Medical Advertising and Trust in Late Georgian England', *Urban History*, 36 (2009): 379–398.
Barry, Jonathan, 'Publicity and the Public Good: Presenting Medicine in Eighteenth-Century Bristol', in *Medical Fringe and Medical Orthodoxy 1750–1850*, edited by W. F. Bynum and Roy Porter (London: Croom Helm, 1987), 29–39.
Brown, P. S., 'The Venders of Medicines Advertised in Eighteenth-Century Bath Newspapers', *Medical History*, 19 (1975): 352–369.
Brown, P. S., 'Medicines Advertised in Eighteenth-Century Bath Newspapers', *Medical History*, 20 (1976): 152–168.
Burnby, Juanita G. L., *A Study of the English Apothecary from 1660 to 1760* (London: Wellcome Institute, 1983).
Burnby, J. G. L., 'The Preparers and Distributors of English Proprietory Medicines', *Dental Historian*, 32 (1997): 47–55.
Burnby, J. 'The Early Years of the Pharmaceutical Industry', in *The Pharmaceutical Industry: A Guide to Historical Records*, edited by Lesley Richmond, Julie Stevenson, and Alison Turton (Aldershot: Ashgate, 2003), 1–13.
Burnett, John, *Liquid Pleasures: A Social History of Drinks in Modern Britain* (London: Routledge, 1999).
Cody, Lisa Forman, '"No Cure, No Money," or the Invisible Hand of Quackery: The Language of Commerce, Credit, and Cash in Eighteenth-Century British

Medical Advertisements', *Studies in Eighteenth-Century Culture*, 28 (1999): 103–130.

Cook, Harold J, 'Roy Porter and the Persons of History', in *Medicine, Madness and Social History: Essays in Honour of Roy Porter*, edited by Roberta Bivins and John V Pickstone (Basingstoke: Palgrave Macmillan, 2007), 14–21.

Corley, T. A. B., 'The British Pharmaceutical Industry since 1851', in *The Pharmaceutical Industry: A Guide to Historical Records*, edited by Lesley Richmond, Julie Stevenson, and Alison Turton (Aldershot: Ashgate, 2003), 14–32.

Cox, Nancy, *The Complete Tradesman: A Study of Retailing, 1550–1820* (Aldershot: Ashgate, 2000).

Day, Julie, 'Elite Woman's Household Management: Yorkshire 1680–1810', (PhD thesis, University of Leeds, 2007).

Diaper, Stephanie, 'J. S. Fry & Sons: Growth and Decline in the Chocolate Industry, 1753–1918', in *Studies in the Business History of Bristol*, edited by C. Harvey and J. Press (Bristol: Bristol Academic, 1988), 33–54.

Doherty, Francis, 'The Anodyne Necklace: A Quack Remedy and Its Promotion', *Medical History*, 34 (1990): 268–293.

Gardner, Victoria E. M., *The Business of News in England, 1760–1820* (Basingstoke: Palgrave Macmillan, 2016).

Helfand, William, *Quack, Quack, Quack: The Sellers of Nostrums in Prints, Posters, Ephemera and Books* (New York: Grolier Club, 2002).

Holloway, S. W. F., 'The Orthodox Fringe: The Origins of the Pharmaceutical Society of Great Britain', in *Medical Fringe and Medical Orthodoxy 1750–1850*, edited by W. F. Bynum and Roy Porter (London: Croom Helm, 1987), 129–157.

Holloway, S. W. F., *Royal Pharmaceutical Society of Great Britain 1841–1991: A Political and Social History* (London: Pharmaceutical Press, 1991).

Isaac, Peter, 'Pills and Print', in *Medicine, Mortality and the Book Trade*, edited by Robin Myers and Michael Harris (Folkestone: St. Paul's Bibliographies, 1998), 25–47.

Jenner, Mark S. R., and Patrick Wallis, 'The Medical Marketplace', in *Medicine and the Market in England and its Colonies, c.1450–c.1850*, edited by Mark Jenner and Patrick Wallis (Basingstoke: Palgrave Macmillan, 2007), 1–23.

Jones, David Lewis, *Debates and Proceedings of the British Parliaments: A Guide to Printed Sources* (London: H M Stationery Office, 1986).

Loudon, Irvine, *Medical Care and the General Practitioner 1750–1850* (Oxford: Clarendon Press, 1986).

Loudon, Irvine, '"The Vile Race of Quacks with Which This Country Is Infested"', in *Medical Fringe and Medical Orthodoxy 1750–1850*, edited by W. F. Bynum and Roy Porter (London: Croom Helm, 1987), 106–128.

Mackintosh, Alan 'Authority and Ownership: The Growth and Wilting of Medicine Patenting in Georgian England', *British Journal for the History of Science*, 49 (2016): 541–559.

McIntyre, Sylvia, 'The Mineral Water Trade in the Eighteenth Century', *Journal of Transport History*, 2 (1973): 1–19.

Mui, Hoh-Cheung and Lorna Mui, *Shops and Shopkeeping in Eighteenth-Century England* (London: Routledge, 1989).

Newbery, Arthur Le Blanc, *Records of the House of Newbery* (Derby: Bemrose, 1911).

Porter, Dorothy, and Roy Porter, *Patient's Progress: Doctors and Doctoring in Eighteenth-Century England* (Stanford: Stanford University Press, 1989).

Porter, Roy, *Health for Sale: Quackery in England 1660–1850* (Manchester: Manchester University Press, 1989).

Rance, Caroline, *The Quack Doctor: Historical Remedies for all Your Ills* (Stroud, Gloucestershire: History Press, 2013).

Roscoe, S., *John Newbery and His Successors: A Bibliography* (Wormley, Herts: Five Owls Press, 1973).

Rousseau, G. S., editor, *The Letters and Papers of Sir John Hill, 1714–1775* (New York: AMS Press, 1982).

Rousseau, G. S., '"Stung into Action…": Medicine, Professionalism, and the News', *Prose Studies*, 21 (1998): 176–205.

Strachan, John, *Advertising and Satirical Culture in the Romantic Period* (Cambridge: Cambridge University Press, 2007).

Strathern, Paul, *Quacks, Rogues and Charlatans* (London: Royal College of Physicians, 2015).

Tweedale, Geoffrey, *At the Sign of the Plough: 275 Years of Allen and Hanburys and the British Pharmaceutical Industry 1715–1990* (London: John Murray, 1990).

Tweedale, Geoffrey, 'Archives of the Pharmaceutical Industry: Their Scope and Use', in *The Pharmaceutical Industry: A Guide to Historical Records*, edited by Lesley Richmond, Julie Stevenson and Alison Turton (Aldershot: Ashgate, 2003), 33–48.

Wallis, Patrick, 'Consumption, Retailing, and Medicine in Early-Modern London', *Economic History Review*, 61 (2008): 26–53.

Welsh, Charles, *A Bookseller of the Last Century* (London: Griffith, Farran, Okeden & Welsh, 1885).

Wild, Wayne, *Medicine-by-Post* (Amsterdam: Rodopi, 2008).

Woodcroft, Bennet, *Abridgements of Specifications Relating to Medicine, Surgery and Dentistry, 1620–1866* (London: Commissioners of Patents for Inventions, 1872).

CHAPTER 2

The Status of Patent Medicines

Patent medicines might have been praised or criticised in Georgian England, but all agreed that they were purchased and taken in large quantities. The criticism could be vigorous, especially amongst some regular practitioners who saw them as a threat, but such shrill denunciations created a potential paradox: if, according to these denigrators, they were a fraudulent imposition on a gullible public, why were so many so popular for prolonged periods? The damnation of patent medicines has been approvingly repeated by some later writers;[1] but closer observation reveals that the views of both the Georgian public and the medical practitioners were more complex. Opinions on the status of patent medicines were varied, and they were adjustable to suit the needs of the moment: for example, owners were prepared to be fiercely critical of other patent medicines while supporting their own.

The status of patent medicines evolved over time. The patenting of medicines dates from the beginning of the Georgian period, and it was rare until the middle of the eighteenth century, but similar owned medicines were available earlier. After the Restoration, secret, branded, medicines were distributed widely with the help of advertisements in the annual almanacs. By the 1750s, a patent medicines industry had been developed, and patent medicines were purchased across the country for the treatment of most acute and chronic conditions, including those confined to children, women and the elderly. Most patent medicines

A. Mackintosh, *The Patent Medicines Industry in Georgian England*, Medicine and Biomedical Sciences in Modern History,
https://doi.org/10.1007/978-3-319-69778-9_2

were recommended for several conditions, but they were advertised for a narrower range of problems than has been assumed by historians in the past. In this period, a specific medicine for a single problem was largely regarded as suspect by orthodox physicians: so prescribed medicines were also recommended for numerous conditions, with the choice of a particular medicine dependent on the patient's circumstances.

The professional standing of both the owners and critics of patent medicines is fundamental to understanding the status of the medicines themselves, and the standing of medical practitioners was expressed as 'regular' or 'irregular'. Such a distinction was in theory straightforward: regulars had received the required training as physicians, surgeons or apothecaries and practised in an appropriate manner, while irregulars had not. In practice, the absence of a medical register and the lack of formal qualifications for most practitioners meant that the boundary between regular and irregular status was hazy and porous, and that regular status was often a matter of local opinion. The word 'quack' rarely appears in this book outside quotations as it was largely a term of non-specific abuse, conveying little sense and running the risk of returning the reader to an earlier era of general criticism for all aspects of alternative medicine. 'Quackery', on the other hand, does have some meaningful content as it referred to a particular style of practice which was difficult to define but was readily recognised by contemporaries.

Georgian England was not a place for holding back critical opinions. Members of the royal family, including the King, were ruthlessly satirised, and members of all ranks of society could be condemned forcibly and possibly unjustly. In this environment, denunciations of patent medicines in general, and specific medicines in particular, could be vigorous from both the general public and medical practitioners. Their main targets were not so much the composition of the medicines, but more the secrecy of their recipes and the claims of their promoters. But the criticisms were far from universal, and those coming from physicians could be qualified, rather than blanket, condemnations. On the other side of the debate, influential members of the public took or recommended patent medicines, and prominent members of the medical professions supported them, or even owned them.

The backing by many members of the community was augmented by the inadvertent validation by the state through the patent system and the excise stamp. The royal patent, as it was usually described, was promoted as an official endorsement of both the novelty and the efficacy of patented medicines, though in reality it was a legal, not an executive, device which paid little attention to both of these properties. However, the patent was expensive and cumbersome to obtain, and the excise stamp, introduced in 1783, largely replaced it as an official commendation. The excise stamp, with a large central crown, was applied to every bottle and box, and thus carried a strong implication that the medicine was approved by the state. With both public and official support, patent medicines were an established, if sometimes controversial, component of the medical market, and they were regarded by many as a reasonable alternative to orthodox medicine.

Early Development of Secret Owned Medicines

Purchasing pre-prepared medicines for self-medication has a very long history, but the sale of what later became known as patent medicines becomes visible in the seventeenth century. This type of medicine, kept secret by an owner and with a wide distribution helped by advertising, was being sold pre-packaged, branded and at a fixed price in the mid-seventeenth century, though the legal patenting of medicines came later in the early eighteenth century.[2] As we shall see in the next chapter, a substantial English patent medicines industry was then developed in the middle of the eighteenth century; but examples of the techniques which were widely adopted in the industry to brand, publicise and distribute patent medicines can also be found in the seventeenth century.[3]

Louise Curth has studied almanacs as a vehicle for the promotion of secret medicines across the country. Almanacs had been developed in the Middle Ages for the annual publication of astrological predictions, and by the seventeenth century they contained additional practical advice, including health preservation and medical therapy. By the 1660s over 400,000 were printed annually in England, one almanac for every 2.5 households, and they were an important method of advertising goods nationally before provincial newspapers appeared in the early eighteenth century.[4] Some advertisements for secret medicines were printed in almanacs during the mid-seventeenth century, but they became commoner in the 1680s and 1690s, accompanied by the development of clearer branding.[5] For example, 'Spirits of Scurvy Grass' were advertised from the 1650s, but only from the 1680s did this type of medicine have a specific name, such as 'Robert Bateman's Scurvy Grass' which was now competing with at least three other scurvy grasses with different named owners.[6]

According to the almanac advertisements, some of these branded medicines were only sold at the producer's premises in London, but others were available at a variety of fixed retail sites, which included alehouses, taverns, inns and coffee houses, and to a lesser extent, booksellers.[7] By contrast, Elizabeth Furdell concluded that booksellers were the dominant vendors for owned medicines in the late seventeenth century.[8] The medicines could also be purchased from temporary outlets, such as fairs, hawkers and pedlars, and they could be delivered by post. Antedating some features associated with the Georgian patent medicines industry, a few of the advertised medicines could be bought at fixed retailers across England, or at least Lincolnshire southwards, and some owners produced a range of medicines, though many were only associated with a single product.[9]

The preservation of some of Anthony Daffy's account books and legal documents from the 1670s and 1680s means that we have more precise information on his Elixir than most other patent medicines in both the Stuart and Georgian periods. Anthony Daffy was a shoemaker who apparently inherited the recipe from a clergyman cousin and turned it into a very profitable medicine with 132 agents in England outside London and 38 in Ireland, continental Europe and New England. The books, which do not include London sales, record the dispatch of an average of 9000 half-pint bottles a year in the period 1678–1683 at a price of 2s 6d to 3s, demonstrating a substantial national market for an owned medicine before the Georgian period.[10] Daffy's Elixir was still a popular medicine for gout and stone in the nineteenth century, though the recipe may have changed: Thackeray mentioned it in several of his novels, it was used as a colloquial term for gin in the 1820s, and the recipe was published in the 1870s.[11]

In common with several later Georgian patent medicines, some of the secret medicines of the late seventeenth century were the property of regular practitioners. Two licentiates of the College of Physicians in the Oracle group on Cheapside owned and advertised their secret medicines in 1687.[12] Also, William Cockburn devised a remedy for dysentery in 1696 which proved very successful in a trial in the Navy, leading to a substantial practice in London and the sale of the medicine to the Fleet for forty years.[13] On land, John Coldbatch's Vulnary Powder for bleeding proved effective when tried out by the Army in Flanders in 1695: the following year, he sold the recipe to his publisher and bookseller, Daniel Brown, though he continued to 'supervise' its production.[14] Cockburn was already a licentiate of the College of Physicians when he created his medicine, while Coldbatch was initially an apothecary who became a physician after the success of his powder and was knighted in 1716.

The introduction of medicine patenting at the beginning of the eighteenth century was the next development in the market, potentially giving these owned medicines some official recognition. The first medicine patent is usually thought to have been Nehemiah Grew's in 1698 for medicinal salts from the spa at Epsom; but, as Josiah Peter who acquired the rights to the patent in 1700 pointed out, that patent was to protect the well-known production process of a naturally available medicine, and not for the compounding of a medicine itself.[15] The first patent to seek to protect the formulation of a medicine was taken out by Timothy Byfield in 1711 for sal. oleosum volatile. In a promotional treatise for his medicine, Byfield did not explain the reasons for seeking a patent; but he commented that there must be a clear distinction between his medicine and other dangerous 'factitious compounds' made from inferior ingredients.[16] In taking out a patent, he was probably aiming both to establish the uniqueness of his product and to prevent reproduction of its name. Medicine patents were rare in the early Georgian period, with only five more being granted before 1742; but owners now had an official, if cumbersome and expensive, mechanism for promoting their medicines and perhaps protecting their medicine names.[17] The further development of a patent medicines industry in the Georgian period is discussed in Chapter 3.

Demand for Patent Medicines

Several historians have noted the expansion of medicine-taking, both regular and irregular, during the eighteenth century.[18] The estimated population of England more than doubled in this era, but the increase in medicine-taking was more than just the result of a larger number of consumers.[19] Roy Porter's assertion that 'Georgian England was becoming a medicated society, drunk on self-drugging' is somewhat melodramatic, but it does emphasise the importance of medicines to the people of that period.[20] Within this general popularity of all medicines, patent medicines were being purchased in in increasing amounts.

How often were patent medicines taken and by whom? Undoubtedly, they were widely consumed. Three examples from across the country are: Mary Dennett, a member of the Isle of Wight gentry, describing her friends taking a patent medicine as an alternative to consulting a medical practitioner; Richard Latham, a Lancashire yeoman farmer, recording the purchase of at least thirty-five bottles or boxes of medicines in his accounts covering the last ten years of his life; and James Boswell's attraction to patent medicines for his recurrent episodes of venereal disease in London.[21] Also, Fanny Burney's respect for some patent medicines runs through her copious journals and letters, where she recommended Dr. James's Fever Powder at least six times. Her most revealing entry is in a letter from Brussels to her husband, an Allied general, dated four to two days before the Battle of Waterloo. With the opposing armies manoeuvring and fighting only a few miles away, she had managed to buy three, urgently needed, patent medicines for her husband from a commercial traveller from Manchester.[22]

Several sources confirm the widespread sale of a large number of patent medicines. Indeed, the multiple medicine advertisements in nearly every issue of every provincial newspaper are testimony to their popularity. The *Salisbury and Winchester Journal* printed an average of 14 advertisements for medicines in each issue during 1781, and *Aris's Birmingham Gazette* printed an average of nine in the same year. Such advertisements may have been fewer in other newspapers, but nearly all provincial newspapers from the mid-eighteenth century contained multiple advertisements for patent medicines in the large majority of issues. Medicine advertisements will be explored in detail in later chapters: each advertisement normally cost at least four to six shillings in the 1770s, including excise duty, and rather more later as the duty progressively increased.[23] Even if a newspaper printer who was a medicine retailer was advertising medicines for sale in his own newspaper, each advertisement still cost at least the stamp duty of a minimum of two shillings: there was no such thing as a free advertisement. This torrent of advertisements would not have been affordable unless significant sales of these medicines were anticipated. Other publications corroborate their frequent use and their many varieties, such as an educational book on the different English trades, whose description of the typical chemist and druggist included 'he also sells numerous quack medicines'.[24]

Contemporary estimates also consistently reported that the number of patent medicines taken was considerable and growing. Edward Harrison, a Lincolnshire physician who was leading an attempt at medical reform in 1806, asked for reports on the state of medicine from across the country and he concluded that 'empirical medicines of very pernicious effects are sold to an incredible amount'.[25] One surgeon reported to Harrison that sales of 'quack medicines' in his unidentified Suffolk town raised over £500 per year in stamp duty: this return implies a sale of many thousands, probably tens of thousands, of bottles or boxes in this single town each year.[26] These reports from practitioners, consistently describing a rising use of patent medicines, could be considered as special pleading for reform in the interests of regulars. However, this uniform increase is in contrast with the fluctuating number of empirics they recorded in their areas.[27] It seems unlikely that the consistent reports on the popularity of patent medicines in England during this period were exaggerations; and the quantity sold across the country is illustrated by the delivery of a *single* consignment in July 1802 of 767 bottles and boxes of patent medicines from Dicey and Sutton, a leading London wholesaler, to John Ware, the printer of the *Cumberland Pacquet* 310 miles away in Whitehaven, Cumberland.[28]

These estimates of large sales reported by regular practitioners are confirmed by the fortunes made by some patent medicine owners and wholesalers. As we shall see in the next chapter, Francis Newbery, Cluer Dicey and Nathaniel Godbold, amongst others, were able to purchase substantial estates and to join the local gentry. Amongst the less respectable owners, Samuel Solomon's large house and garden on the edge of Liverpool was for many years a symbol of homecoming as a traveller returned to Liverpool from London.[29] Even Francis Spilsbury, an ex-goldsmith described in the next chapter who apparently owned and promoted only one medicine, set up a trust fund of £4,000 in his will.[30] This potential for earning a fortune from a secret patent medicine was reflected in the award of £30,000 by Parliament to Edward Jenner for making his technique of vaccination freely available to all.[31]

The clearest evidence of the extensive sale of patent medicines comes from taxation reports. As will be described later in this chapter, after 1783 all patent medicine containers were required to have an attached excise stamp, which started at 1½d for medicines priced at one shilling or less and then increased progressively with the medicine prices. We can use the revenue raised to estimate the total annual sales of the medicines. For example, in 1810 the gross total of £41,201 was collected in England and Wales:[32] so if the average duty paid on each bottle or box is estimated, we can assess the total number sold. As part of the exploration of owners and wholesalers, all the medicine advertisements in runs of newspapers in Leeds and Birmingham during the first half of 1807 were reviewed, and these advertisements reveal that 4.4d was the average duty payable on the smallest advertised quantity of each medicine. With this figure for the average duty, the equivalent of about 2.2 million of the smallest bottles or boxes of patent medicines would have been sold across the country in 1810. The term 'equivalent' is used because some medicines were sold in larger containers, which would reduce the number of bottles and boxes, but would not significantly diminish the total volume of medicine. Of course, this figure is only an estimate because some advertised medicines would sell better than others and many available medicines were not advertised in these newspapers, both of which would alter the figure for the average duty; and the duty collection was unlikely to be completely efficient. But cheaper products often sell better than more expensive ones, so this calculation is more likely an underestimate rather than an exaggeration. It seems probable that at least two million bottles and boxes of patent medicines, or their equivalent volume in larger containers, were being sold annually in England and Wales by 1810. Loeb's comment that the mass market for patent medicines started around 1860 was wide of the mark.[33]

So the popularity of patent medicines is clear, but who took them is less certain. Historians have provided a wide range of opinions, with Porter concluding that the high prices meant a market amongst the affluent, with Mary Fissell writing that they were taken across society, including the poor, and with Irvine Loudon having 'little doubt' that they were predominately taken in the homes of the poor.[34] Contemporary accounts also came to different conclusions. For example, the surgeon Thomas Prosser described the takers of patent medicines as 'being mostly the lower sort of people, who live by industry or labour', while another source painted the opposite picture: 'the consumers of quack medicines are largely the wealthy but ignorant, superstitious old women and profligate rakes'.[35] Joseph Townsend, a Wiltshire rector who had trained in physic at Edinburgh before taking holy orders, observed their widespread consumption across society when he wrote that they were found 'not merely among the lower classes of society, but in respectable families, and almost every house'.[36] An anonymous physician chose to focus on the great and wise, who not only tolerated patent medicines, but could be among their most active supporters:

> You should consider, also, that many of the richest and greatest, and those who should be wisest, men in the nation (judges, bishops and peers), not only believe in quack medicines, and take great quantities of them, but are

> professed patrons of quacks, and allow their names to be used every day, in the common newspapers, as vouching for the efficacy of various quack medicines which they had employed in their own families, or on their own persons.[37]

A more precise glimpse of medicine purchasers can be found in the day-books of John Ware, the printer of the weekly *Cumberland Pacquet* and the leading patent medicine retailer in Whitehaven, Cumberland.[38] These day-books are a little frustrating as they only record account holders who bought medicines, with those who paid cash not being named, or their purchases being omitted altogether. During the three years 1800–1802, 64 different account holders who were unlikely to be medicine retailers purchased a patent medicine, and 20 bought the same medicine at least twice. Six account holders bought medicines from Ware in each of the three years. One was a surgeon who may have been acquiring the medicines for professional purposes. Of the remaining five habitual purchasers, two are described as 'esq.' and they can be identified as substantial members of the middling sort: one was Henry Ellison who owned a 'large and convenient dwelling house' with stabling and gardens, and the other was Edward Stanley who was sufficiently important for the marriages of his two daughters to be described in the *Cumberland Pacquet*.[39] The other habitual purchasers were Josiah Lewthwaite, an attorney who applied unsuccessfully to be a coroner in 1800, a clergyman and a glazier.[40] Another six purchasers described as 'esq.', five army officers, four clergymen and two attorneys bought patent medicines at least once. Ware's day-books are biased towards the recording of medicine sales to the higher levels of society where account holders were more likely to be found, and the purchasers were not necessarily the consumers; but they do confirm that members of the middling sort in the Whitehaven area bought patent medicines, sometimes repeatedly.

These, and other, contemporary reports and opinions suggest that patent medicine sales were particularly strong in the middle and upper levels of society: medicine prices ranged from a usual minimum of a shilling to over a pound in a few cases, and price would sometimes have been a barrier to poorer purchasers.[41] However, the poorer sections of society did purchase patent medicines, and retailers could supply a small quantity from a single bottle if necessary.[42] All but the poorest had access to patent medicines and substantial sales across the country were the result.

Regular and Irregular Medicine

Before exploring the varied opinions on patent medicines a digression is required to explain the terms 'regular' and 'irregular' as applied to Georgian medical practitioners and to medical practice. This is needed for several reasons. One is that this terminology positions patent medicines whose owners were practitioners: for example, Dr. James's Fever Powder, devised by the regularly trained Robert James, was widely prescribed for this and other reasons, while Dr. Brodum's Nervous Cordial, introduced by the obscurely trained and notorious William Brodum, was not recommended by other practitioners. More importantly, it positions the critics of patent medicines, enhancing our assessment of their comments. Many regular practitioners saw patent medicines as a threat and could be vigorous in their censures, though, as we shall see, others were more sympathetic. Irregulars could also be fiercely critical, especially of rival practitioners; but their comments were more focussed on particular issues, not on irregular practice as a whole. In general, irregular practitioners were regarded both as economic competitors and as threats to the status of the regulars, who therefore sought to identify them and to squeeze them and their medicines out of the market.[43] Consumers of the medicines were probably less concerned about the involvement of irregular practitioners, but the division into regular or irregular practitioners ran through many discussions on medicine benefits and side effects. In addition, an insight into the differences between regular and irregular practice is essential for understanding one of the main conclusions of this book, namely that the patent medicines industry was separate, though overlapping, from both types of practice.

The boundary between regular and irregular was fluid, porous and often controversial.[44] In principle, a regular could be simply delineated as a medical practitioner who had completed a recognised form of training as a physician, surgeon or apothecary: an irregular would be anybody who had not completed such training. For regular physicians, a university education and some practical training would be followed by approval in the form of an MD; while surgeons and apothecaries needed an appropriate apprenticeship. But recognised by whom? The achievement of an MD, the confirmation of the status of a physician, was at the discretion of the granting body, and practices varied widely. For instance, many English physicians had received their MD from a Scottish university. At Edinburgh University, the requirements for this degree were relatively rigorous, including attendance at the university: in contrast, Marischal College, Aberdeen, could hand one out in exchange for two letters of recommendation and 13 guineas, without the candidate needing to leave London.[45]

Apprenticeships for surgeons were also very variable, ranging from minding the shop for a few years in a small town then marrying the surgeon's daughter, to a substantial period of organised training in London or Edinburgh.[46] The required level of previous education was uneven, with Loudon commenting that at the end of the eighteenth century some surgeons were grammar school educated with a knowledge of classics, while a few were largely illiterate.[47] Before 1800, a prospective apprentice to a member of the Company of Surgeons was required to understand Latin, and this was tested by a governor of the Company:[48] but most surgeons were not members of the Company. After the creation of the College of Surgeons in 1800, the more ambitious surgeons took an examination to become a Member of the College, aiming to confirm their regular status: but examiners were accused of bribery and other abuses.[49] Until the Apothecaries Act of 1815, no organisation had any responsibility for regulating medical training across England, and even afterwards the powers under the Act to assess newly qualified apothecaries required time to develop and could be evaded.[50] To add to the uncertainties about what constituted regular training, practitioners who styled themselves as 'surgeon-apothecaries' were emerging as the forerunners of general practitioners: the necessary training for this dual competency was even less clear.[51]

Faced with these uncertainties, the full recognition of a practitioner as a regular required some local subjective assessment. Groups of regular practitioners would decide whether others could join them in cases of doubt. For instance, the notorious Dr. Brodum was rejected as a regular by the surgeons at Westminster Hospital in 1811, but the hospital's physicians felt that they had to accept his Aberdeen MD.[52] A more day-to-day illustration is an anonymous letter to a medical journal in 1806, probably written by William Hey, the well-known Leeds surgeon. It reported on a meeting called to discuss medical reform, which was attended by 21 invited physicians and surgeons who were known to have had some regular medical education 'and who are received by each other as regular practitioners'.[53] A specific example of the need for local assessment is Edward Harrison, the reforming Lincolnshire physician. In spite of his Edinburgh MD, opponents of reform suggested that he should not be regarded as a regular physician because he had gone into a partnership to run an advertised mental institution and he had also practised midwifery.[54] Harrison's supporters refuted these accusations that Harrison was not a regular physician when a meeting of the Faculty at Horncastle and a separate letter from the secretary of the Lincolnshire Medical Benevolent Society formally confirmed that Harrison was well respected, had not practised surgery or midwifery, and was definitely a regular physician.[55]

The linkage of regular training and an approved style of practice ensured Harrison's status as a regular. Contemporary reports from practitioners indicate that this combination was not always achieved and that some regulars were practising irregularly, often referred to as practising as quacks. One such report, which summarised the local practitioners with brutal clarity, was sent to Harrison following a meeting of 'local medical men' at an unspecified location. The 32 local practitioners and four druggists were tabulated in six groups:

> 14 regularly educated, and practise regularly
> 6 irregularly educated, and practise regularly
> 4 regularly educated, but incompetent, and practise regularly
> 3 regularly educated, and practise as quacks
> 5 irregularly educated, and practise as quacks
> 4 druggists, all interfering, more or less, with the practice of physic[56]

Thus, six irregulars were regarded as practising regularly, while three regulars were practising as quacks. The letter did not attempt to explain what style of practice earned the designation of quack.

Then as now, the word 'quack' brings a punch to a text, and it has been frequently employed in descriptions of alternative medicine and irregular practice. Yet historians have found difficulty in defining the term, concluding that it was malleable and subjective.[57] The word is best avoided as it is imprecise, and it harks back to the earlier era of universal condemnation for all forms of unorthodox medicine. However, the term 'quackery' does have a useful function today. It refers to a style of medical care which was different from the practice of most regular practitioners and recognisable as such by both regulars and lay people.

Quackery was subjective and had no single defining feature, but it was marked by combinations of itinerancy, showmanship, dishonesty, panaceas, magic charms and self-acknowledged separation from regular practice.[58] Whether quackery was coterminous with practising irregularly or a subset of it awaits further investigation.

Why Take a Patent Medicine?

Why did consumers purchase patent medicines in large and increasing quantities at a time when, if anything, England was becoming healthier: major epidemics and food shortages had become rare and commentators such as Malthus were becoming concerned about excess population growth. Questioning why consumers took patent medicines, rather than orthodox preparations, can be regarded as anachronistic as it imposes twenty-first century assumptions of the minimal value of these medicines onto the dissimilar Georgian daily life. Espousing the attitudes of an unregulated medical market, the reply to the question might be: Why not? Nevertheless, an awareness of the reasons why patent medicines were sometimes preferred to regular therapy increases our understanding of their status amongst the lay public, and it also clarifies some of the methods of promotion which will be explored in later chapters. The various concerns and inducements which swayed consumers' decisions on patent medicines will be discussed at several points in this book. This section will propose that the growth in irregular medicines was specifically encouraged by a wariness of both the theory and practice of official medicine, and that self-medication, that is therapy chosen by the consumers without the advice of medical practitioners, had practical benefits. Some of these irregular medicines were prepared domestically, but, in line with the growing consumption of finished goods in Georgian society, many were purchased pre-prepared as patent medicines.

Members of all ranks of society displayed a wariness of both medical practitioners and their prescriptions.[59] One example is Mary Dennett, who quickly fell out with her physician, Dr. Moysey, while staying at Bath in 1778. When he suggested a further consultation, she wrote to her fiancé 'I will see him hanged first', and finished 'Nonsense—Nonsense! So much for Dr Moysey'.[60] Dr. William Buchan, a strong advocate of lay medical information, found similar problems further down the social scale when he commented on self-dosing:

> Instances of this are daily met with amongst the ignorant peasants, who, while they absolutely refuse to take a medicine which has been prescribed by a physician, will swallow, with greediness, anything that is recommended to them by their credulous neighbours.[61]

Orthodox practitioners were intermittently distrusted for several reasons. Many laypeople and some medical practitioners felt that internal medicine was unable to deal with many problems and it had been held back from realising its full potential by the deficiencies of regular physicians. John Gregory, Professor of Physic in Edinburgh, was uncomfortable about the current state of medical knowledge in 1770:

> The science of physic has been sometimes advancing, sometimes declining; it has been subjected to the fate of the different systems of philosophy that have prevailed, besides being sometimes disgraced by peculiar follies of its own; its only genuine source, observation and experiment, has been corrupted by fraud, credulity, and a heated imagination, while men of genius and learning, because they were not physicians, have kept at a distance, as if it had been a matter in which they were not interested.[62]

Even in a long, critical, article on empiricism, Duncan Forbes, an Edinburgh physician, expressed similar thoughts when he wrote that the 'science of medicine' was behind other branches of human knowledge in 'progressing towards maturity'.[63]

Others, especially John Wesley, felt that corruption ran deeper and that practitioners were inclined to maximise the number of medicines prescribed and to prolong their administration as much as possible for financial gain.[64] For example, Boswell thought that his surgeon, Andrew Douglas, wanted to prolong treatment in this manner even though he was also a friend.[65] Francis Spilsbury, a patent medicine proprietor, similarly asserted that the faculty used a large quantity of useless medicines for financial benefit, subjecting the patient to an unnecessarily unpleasant and extended illness.[66] As a medicine owner, Spilsbury was not a disinterested observer, but in this fully argued pamphlet he was seeking support for patent medicines and would not have made this claim unless he felt that it would be believed by at least some of his readers. In addition, the regulars' therapy was often thought to be unnecessarily rigorous. As we shall see later, patent medicine advertisements sometimes emphasised the product's gentleness and lack of interference with daily life in contrast with regular therapy. This concern for the hardships of orthodoxy was also described by Spilsbury, who unpleasantly wrote that the regulars felt that they must open all the 'doors' to let the 'enemy', that is the illness, out:

> To this effect, the poor patient is served with a medicine to vomit, and sometimes (oh! dreadful to relate) to operate all-fours at once; that is to say, they purge, they vomit, they sweat and they urine all together.[67]

In addition to this distrust of the motives and methods of regular practitioners, some consumers were also concerned that the regular medications were not made up correctly. An anonymous 1830 book asserted that nine-tenths of drugs prescribed by regular practitioners were adulterated by 'unprincipled druggists' to increase their profits.[68] In 1811, 'Medicus', who was proposing Parliamentary legislation to control medicines, described in some detail how medicines were made carelessly by untrained 'boys' or deliberately adulterated: 'Scarcely a single article which is sold either to the public or the apothecaries, is prepared according to the directions of the Pharmacopeia'.[69] He also made a sweeping unproven estimate of the consequences of incorrect medicine compounding, asserting that it may have killed or disabled more people than the current war in Spain.[70]

Whether they were well founded or not, doubts about the quality of locally produced medicines favoured the purchase of the patent variety, normally made up by a single apparent expert to a uniform standard. The perceptions of dishonesty and incompetence amongst both medical practitioners and medicine suppliers would encourage the purchase of patent medicines and other forms of self-help, regardless of whether the accusations were correct. Forbes summed up a common assessment when he wrote that the backward state of medical knowledge and the actions of regular physicians ensured that the healing art was 'too frequently entrusted to the interested pretensions of nefarious quacks, and to the far less dangerous prescriptions of superannuated females'.[71]

Doubts about regular medicine promoted self-help. Medicines for self-help could either be made within the household or purchased from druggists or patent medicine vendors. The best-selling proponent of domestic self-help was John Wesley's *Primitive Physic* which was in press in multiple editions well into the nineteenth century: it provided lists of many self-prepared remedies which could be tried for all the common conditions. Wesley emphasised the safety and acceptability of domestic medicines in his preface: 'So that every man of common sense (unless in some rare cases) may prescribe either to himself or his neighbour: and may be very secure from doing harm, even where he can do no good'.[72] Domestic medicines also had practical and financial advantages, and both these types of benefit were well described in the preface to *The Country Housewife's Family Companion* which summarised some of the book's contents:

> [...] many receits of plain, cheap, experienced medicines and the cures they have made in country families, with many other most serviceable matters, by which poor families, and those that live some distance from a town, may become their own physician and surgeon, and probably many lives thereby saved, as well as chargeable bills prevented.[73]

For many, however, self-help took the form of buying pre-prepared medicines from booksellers, druggists and other retailers. Some of these purchased medicines had well-known recipes and could also be prepared at home, but many were patent medicines with a secret composition. The patent medicines came pre-packaged with clear indications for use and printed instructions, and they were designed to be taken without any local medical advice. The vendor could emphasise the care of its preparation at a central source in contrast to the dangerous efforts of local druggists. One reason amongst many for buying patent medicines was that they fulfilled a role as the commercial equivalent of domestic medicines. With the rapid urbanisation in the later Georgian period, many people were separated from the stable, often rural, domestic life, which was the normal setting for these home-prepared recipes. If city dwellers or travellers, especially those without family support, wanted to treat their ailments themselves, a patent medicine could provide the equivalent of a home-made medicine.[74] This desired resemblance is demonstrated by the names of some patent medicines which imply a similar composition to a domestic medicine made from simple vegetable ingredients, for example Essence of Coltsfoot and Balsam of Liquorice. For some, the patent medicines provided an available commercial equivalent of the familiar medicines prepared at home.

The environment of the eighteenth-century consumer society ensured that the upsurge in patent medicine use was more than just a straight replacement for prescribed or domestic remedies. A key feature of the consumer society was the increasing number of people who had surplus income to be spent on finished goods. Consumers were prepared to search out medicines with the help of publicity and to pay for them, and by this period many consumers had enough money to buy expensive medicines, with a large bottle of a high-priced medicine possibly costing over a pound.[75] The London physician Samuel Fothergill reflected the power of this growing consumption of finished products when he wrote that regular physicians needed to promote their orthodox medicines more effectively to compete with patent medicines and other types of therapy: 'If we do not promise more for the effect of our medicines than experience authorises, it is very likely they will not be taken.'[76] The opponents of patent medicines claimed that the public was being persuaded to buy medicines for diseases they did not have. As an anonymous pamphlet reported:

> Many credulous and foolish people in this island, especially in the metropolis, are very opulent, and often imagine themselves indisposed when only labouring under the torpor of indolence. Such beings will purchase any nostrum, however ridiculous.[77]

Fashion was an integral feature of this growing consumption of patent medicines, just as much as it was important for a wide range of activities such as the sale of clothes, tea drinking or horse racing.[78] Fashion was led by the superior members of society and, as we shall see in Chapter 8, some advertisements reported that the medicines were used by the aristocracy, the gentry and other members of the higher ranks. Other advertisements were utilising fashion, amongst other promotional tools, when they stressed the recommendations by friends and neighbours. In addition, the permitted conspicuous consumption of the time could also reinforce patent medicine sales by allowing proprietors who were also irregular practitioners to flaunt their wealth, encouraging a belief that the large income must be derived from a very effective product. One such proprietor was the well-known Samuel Solomon, a colourful medicine proprietor of humble origin and also without regular medical training in spite of his MD. A contemporary account described how he purchased an estate near Liverpool and rode around in a four-wheeled carriage for all to see.[79]

To summarise, in the Georgian medical market, choosing to take a patent medicine instead of, or as well as, a visit to a medical practitioner, whether regular or irregular, was a rational decision. It was encouraged by the practical convenience of self-help, and also by a distrust of regular medicine in general and more specifically the local compounding of medicines. The growing consumption of finished products of many kinds in the eighteenth century directed patients towards new, publicised and more expensive medicines, instead of just a withdrawal into domestic remedies or folk cures. The public were taking patent medicines in growing numbers: we now need to explore the response of the regular practitioners to this threatening development.

Regular Practitioners and Patent Medicines

The antipathy of the regular practitioners to patent medicines was not as deep and widespread as the leaders of the medical profession assumed it to be later in the nineteenth century, and the boundaries between regular therapy and patent medicines with similar ingredients were ill defined.[80] The use of patent, or secret, remedies was, at times, acceptable to regular practitioners and a few of these remedies became orthodox treatments. Criticisms of patent medicines could certainly be trenchant, especially from some surgeons and apothecaries who were in direct commercial competition, but the leading physicians of the day were often sympathetic to their use in certain circumstances. Further, many regularly trained practitioners developed secret remedies for their own use or for commercial exploitation, without suffering undue censure.[81]

Some patent medicines became recognised as part of orthodox therapy and were freely prescribed by regular practitioners. The best-known example was Dr. James's Fever Powder, which featured in the prologue to Chapter 1. Created in 1743, this powder was formulated and part-owned by Robert James, a London physician who was also well known for his three-volume *Medicinal Dictionary*.[82] With claimed sales of over 80,000 doses a year in the 1760s, the powder quickly became part of orthodox therapy.[83] Indeed, William Buchan, in his *Domestic Medicine*, regarded it as the accepted therapeutic standard for some fevers, and the apothecary William White referred to it as one of two orthodox antimonial preparations.[84] Attempts at copying were only partly successful and it remained a secret remedy from a single wholesaler: it was still being advertised as a patent medicine in newspapers in 1822 with a confident endorsement as 'the greatest discovery in medicine during the last century', and 'Pulvis Jacobi Vera' was being ordered for the Army and

Navy in the 1870s.[85] Other patent medicines which were commonly prescribed by regulars were Anderson's Scots Pills, recommended by William Cullen in a postal consultation in 1770, Godfrey's Cordial and Dover's Powder.[86] Thomas Fowler, a York physician, described Dover's Powder as 'a very efficacious remedy in the treatment of both the acute and chronic rheumatism'.[87] John Hunter expressed the feelings of some regular practitioners when he felt that the important consideration was whether a treatment worked, not whether it was derived from regular or 'quack' medicines.[88] This ill-defined boundary between regular and irregular therapy was well demonstrated by the difficulties of demarcating patent medicines for the purposes of the 1783 Medicines Act, as described later in this chapter: the replacement 1785 Act attempted to resolve the problem by listing 85 of them in an accompanying schedule.

Most patent medicines nevertheless remained outside orthodox therapy and could be described by regulars in very unflattering terms. Words used by a few of Harrison's correspondents included 'trash', 'abominable impositions', 'composed of the most pernicious materials'[89]; though more of these correspondents used relatively neutral descriptions such as 'quack' or 'empirical' medicines. Samuel Fothergill, a physician at the Westminster General Dispensary, considered patent medicine warehouses to be as dangerous as gin shops and lottery offices.[90] However, in general, the criticisms of patent medicines by the regulars were gentler up to about 1820 than later in the nineteenth century. Thus, even a 1777 pamphlet written by Thomas Prosser, a Wrexham surgeon, in order to condemn patent medicines, recognised that the risk of harm by these medicines was low, and that 'many excellent medicines are advertised, as they are imitations of the compositions of the common dispensatories'.[91] An anonymous 'eminent physician', writing to the *Medical and Chirurgical Review*, went a stage further by recognising that patent medicines were useful in resistant conditions and 'imaginary disorders', when regular therapy had little to offer:

> What can an honest physician do with an hysterical fine lady, or a fanciful hypochondriac who has got all the diseases in all systems of nosology, and ten times more; or with a gouty lord, or a guzzling alderman, or a greasy bishop?[92]

This type of low-key censure of patent medicines, accompanied by some understanding, is found in the writings of John Gregory and Thomas

Percival. As pioneers in medical ethics, they were concerned about the correct behaviour of medical practitioners and they were also physicians with impeccable reputations.[93] They criticised the medicines, but were also sympathetic to their use in certain circumstances. John Gregory issued a qualified condemnation in 1770:

> It is further alledged, that some of the best remedies were originally introduced as secrets, though discredited by the regular physicians. But allowing this to be true, yet I am persuaded, that these nostrums, on the whole, do much more hurt than good to mankind[94];

For Gregory, the problem with patent medicines was the lack of a trained practitioner to guide the patient, not the medicines themselves. As a result, he regarded them as 'one of the greatest public nuisances under which we labour in Great Britain'.[95] Two years later, in a revised version of the original work, he recognised that patent medicines and other forms of self-help did have a role:

> Cases are continually occurring of people labouring under diseases, who can have no access to the assistance of one of the faculty. It would be barbarous to hinder those from using such remedies as appeared to them most likely to afford them relief, or to prohibit a friend or a bystander from giving their assistance in such a situation.[96]

A generation later, Thomas Percival, a leading Manchester physician, wrote in *Medical Ethics*, a work which had received the 'approbation or assistance' of Erasmus Darwin, William Withering, Archdeacon Paley, and William Heberden amongst many others, that quack medicines 'should be discouraged by the faculty, as disgraceful to the profession, injurious to health, and often destructive even of life'.[97] He recognised however that some patients, especially those with 'lingering disorders', had confidence in them, and he observed that 'in these cases, some indulgence seems to be required to a credulity that is insurmountable. And the patient should neither incur the displeasure of the physician, nor be entirely deserted by him'.[98]

Percival also seemed to be making a distinction between secret, but potentially effective, medicines, and those based solely on bluff and salesmanship. In regard to the former, he observed that 'no physician or surgeon should dispense a secret *nostrum*, whether it be by his invention, or exclusive property'.[99] This implied that apothecaries and chemists *could* devise a patent medicine: one example might be the Calcined Magnesia created by his close friend Thomas Henry, a Manchester apothecary, though the secrecy of this medicine was debatable. Percival's observation also seems to suggest that physicians and surgeons could use a secret remedy if they did not own it, or had not been involved in its creation. In contrast, on the same page Percival roundly condemned ineffective 'quack' medicines, observing: 'And if mystery alone give it value and importance, such craft implies either disgraceful ignorance, or fraudulent avarice'.[100] Thus, two leading physicians were critical of patent medicines, but they regarded the prescription and consumption of patent medicines as being tolerable under certain circumstances.

Surgeons and apothecaries did develop secret medicines in this period, providing a link between orthodoxy and patent medicines, and physicians also did so occasionally. Physicians who devised patent medicines included: Robert James, who patented the Fever Powder (1747) and his Analeptic Pills (1774); Theophilus Lobb, a London physician and non-conformist minister, who patented a 'Family Tincture' (1762); and Robert Priestley, a Leeds physician, who advertised a secret anti-bilious powder.[101] However, most regular physicians were very cautious about creating secret remedies. One example of this reluctance was Glass's Magnesia, which had largely been invented by Thomas Glass, a prominent Exeter physician; but Thomas passed it to his brother Samuel Glass, an Oxford surgeon, to own and sell.[102] The cause of this reluctance was probably the need for a successful eighteenth-century physician to attain the qualities of a gentleman and remain at a distance from commercial activity.[103] The London physician William Fordyce supplied an example of the importance of such a reputation to a physician. As a surgeon, he had patented a stomach pill in 1763, but ten years later he had become a physician and, though aware of the commercial potential of his fever powder, he felt unable to sell it as a patent medicine:

> Had I been more ambitious of dying a rich man, than of living a useful member of Society, the powers of our Prophylactic Powder in preventing putrid fevers, or of nipping them in the bud, [...] would have remained a secret while I lived.[104]

The secret remedies devised by surgeons and apothecaries can be divided into two types. One type was a new formulation which could be named after the inventor/owner and then nationally publicised and distributed. Samuel Glass actively promoted his magnesia in the 1760s, and an advertisement claimed that it 'far exceeds every other' in purity and goodness.[105] Edmund Swinfen, a surgeon-apothecary and mayor of Leicester, provided other examples in the 1790s, with his Swinfen's Electuary for stone and gravel, Swinfen's Worm Cakes, and several other medicines bearing his name.[106] Edward Galliard, an Edinburgh apothecary, proposed that his antimonial febrifuge, the Edinburgh Powder, should be distributed from London, with the recipe kept secret.[107] In 1783, Edward Jenner, at the time an ambitious Gloucestershire surgeon-apothecary, intended to sell his own secret Tartar Emetic, and he corresponded with his friend and teacher John Hunter on the best methods of doing so.[108] Hunter emphasised the importance of maintaining the secrecy of the Tartar Emetic's recipe: 'I would also desire you to burn your book, for you will have all the world making it'.[109] For unknown reasons, Jenner did not go ahead with his plans.

The other type of secret medicine created by regular practitioners was a composition used solely in their own practice, or within a small circle of regulars. These medicines do not fulfil the criteria for a patent medicine as they were not advertised, but they do confirm that some regulars were comfortable with owned secret remedies. Richard Greene, a surgeon-apothecary who was also an alderman and sheriff in Lichfield, used such a remedy made from rhubarb in his practice, and was reluctant to divulge the composition even to his brother.[110] Bradford Wilmer, a surgeon, described in his published case records the recipe of a secret remedy for pulmonary disease which had been passed from one Coventry master apothecary to his apprentice over many years.[111] These secret unpublicised remedies used by regulars may have been common, but the records of them are necessarily scanty.

By the end of the eighteenth century, more pressure was being applied by regulars to exclude medicine owners from their ranks. Opinions differed, but the profession as a whole was becoming less tolerant of secret medicines. Galliard recognised that keeping his febrifuge secret would cause distress to his fellow practitioners: 'The proposal hurts you; I see it does: but there is no alternative'.[112] As we have seen, Percival's views on patent medicines at the beginning of the nineteenth century were more precise and forcible than Gregory's a generation earlier. A practical illustration of the desire to exclude medicine owners from orthodoxy is shown by the rules of several medical societies which barred them from membership: the Medical Society of London, the Society for the Relief of Widows and Orphans of Medical Men, and the Kent Medical Benevolent Society had all written this into their rules before the end of the eighteenth century.[113]

So the regulars recognised a separation between orthodox and patent medicines, and they frequently criticised the latter, sometimes vigorously. Nonetheless, there were also links between the two, as the regulars prescribed patent medicines at times, and they often recognised that both types of medicines had similar ingredients and effects. Physicians tempered their criticisms with practical understanding, while the surgeons and apothecaries sometimes created their own secret remedies. Thus, patent medicines were not shunned by regular practitioners as they were to be, at least officially, later in the nineteenth century. These views of the regular practitioners are relevant to the positioning of patent medicines within the medical market, and perhaps to the attitudes of the consumers: but the fate of the medicines in that market was determined by the views of potential purchasers, not by the opinions of the practitioners, and these views were influenced by the state's official recognition.

Official Recognition by the State

In the competition for consumers, patent medicines had a big advantage over other forms of therapy: from the mid-eighteenth century, they were apparently endorsed by the government, which could be expressed as coming from the King himself. This endorsement came in two forms. First, the patent system was interpreted as an official device to confirm the efficacy of a medicine, and then the medicine excise stamp, introduced in 1783, seemed to demonstrate that the approval of the government had been obtained. Neither assumption was intended or true, but they raised the status of patent medicines at a time when other types of therapy had no official support or regulation.

The promotional advantages of the patent system, accompanied by the ability to stretch them to unpatented medicines, have been touched upon in the Introduction, and the advantages and disadvantages of obtaining medicine patents have been recently reviewed.[114] Briefly, the minority of owners who obtained a patent were able to use it as an apparent confirmation by the government of the efficacy of their medicine, and also as a form of copyright of the medicine's name. They did not use it to protect the composition of the medicine, relying on secrecy to achieve this. Patenting medicines was rare before the 1740s, rose to a peak in the 1750s, and became infrequent from the start of the next century. Patenting had disadvantages as well as these advantages, with the former including the expensive and cumbersome process of obtaining a patent, the risk that the patent specification might reveal too much detail of the recipe, and a threat to the reputation of the owner.

When the excise duty stamp for patent medicines was introduced in 1783, it was intended solely for the purpose of raising revenue. However, the excise stamp was applied to every single bottle and box of a patent medicine sold in England, and it had the inadvertent effect of bestowing an apparent, visible, government approval on both the medicines and, to a lesser extent, their vendors. This not only enhanced the standing of patent medicines, but did so more conveniently than the patent. The stamp also provided a form of copyright which was simpler, cheaper and possibly more effective than a patent. In addition, the accompanying regulations for the excise duty required vendors to take out an official licence which augmented the status of some, and these regulations influenced the sale of patent medicines by discouraging druggists from the publicised vending of patent medicines for a period of twenty years.

Although the six Medicines Acts related to the taxation of medicines between 1783 and 1812 created a duty on many medicines lasting a century-and-a-half, and had several unintended consequences, no coherent account of them has been written. Even authoritative histories of pharmacy largely ignore them; for example, Holloway's chronicle on the Royal Pharmaceutical Society contains only a brief mention of them.[115] Yet they raise a series of important questions which impinge on both the status of patent medicines and the methods of the patent medicines industry. Why were the initial Medicines Acts ineffective, for instance? How much revenue was indeed raised and were they successful in the longer term? What were the effects of the apparent official recognition of patent medicines on selling and using them? What were the consequences of the initial discouragement of selling patent medicines by regulars? To answer these questions, the effects of these Medicines Acts will first be summarised, followed by an evaluation of them in providing official authority and a means of copyrighting, and finishing with an assessment of their influence on the involvement of druggists in the selling of patent medicines.[116]

One problem in answering these questions is that the Medicines Acts were regarded in Parliament as purely tax-raising devices and seem to have generated minimal interest, apart from the amount of revenue obtained, with the result that contemporary descriptions and comments are sparse. No Parliamentary commission or committee seems to have specifically looked into medicines between 1780 and 1840. As regards the outcome of the legislation, sources outside Parliament have proved useful, particularly *Kearsley's Tax Tables*, a recurring publication which explained current, especially new, taxation arrangements for a general commercial readership; several editions have survived for the period 1786–1808. Taking the evidence together, the end-product of the legislation is fairly clear, but it is much harder to see at which stage of each Parliamentary process the previous law was changed and why. The practical interpretation of the acts can also be uncertain as, in the absence of considered official judgments, the surviving descriptions inevitably have an element of special pleading by aggrieved witnesses.

Lord John Cavendish introduced the excise tax on patent medicines amongst a string of additional revenue raising measures in his 1783 Budget. An additional £560,000 per year was needed to finance the extra interest on the now enormous national debt after the American War. Cavendish expected that this new tax would generate £15,000 per year, quickly revised to an estimate of £30,000, once the practical issues had been sorted out. He was reported as saying 'Quack medicines he thought very proper objects of taxation; and he believed the House would be surprised at the sum that he had good grounds that a tax on them would produce'.[117]

He was initially proved wrong in both assumptions: the 1783 Medicines Act produced strong opposition and very little revenue. Overall, it required five more acts of Parliament before Cavendish's aims were fully achieved nearly thirty years later. The 1783 Medicines Act was poorly drafted in that it sought to tax both *some* medicine vendors and *some* medicines, without being able to define either satisfactorily. Vendors 'who were not regularly bred to the profession of doctors, exc.' had to pay for an annual licence, which generated additional revenue and allowed their sales to be monitored:[118] regular vendors, defined as those who had been apprenticed to a surgeon, apothecary or druggist, or had been a military surgeon, were not included. Also excluded were other vendors who had dealt solely in medicines, without selling anything else, for the previous three years; this additional exemption inevitably led to disputes as the boundary between medicines and other articles was difficult to delineate.

The result of the 1783 Act was confusion. Three of the largest London medicine owners and wholesalers (Francis Newbery, Thomas Dicey and Hilton Wray) initially took out licences: then, acting in concert, they refused to renew them after a year and were taken to Court by the Stamp Office.[119] Newbery and Dicey were acquitted; but Wray was found guilty because he had sold two small toothbrushes and tooth powder which were categorised as perfumery, and he was consequently no longer exempt as a long-standing specialised medicine seller. No significant revenue seems to have been obtained under this Act, as later Parliamentary revenue figures regard the stamp duty income from medicines as starting with the 1785 Act, with 1784 not appearing in tables of annual totals.[120]

The main thrust of the 1785 Act was to tax the medicines, rather than to tax both the medicines and the medicine vendors.[121] The principles stated in the Act were that all medicines with a patent at any time were to be included, accompanied by unpatented medicines which remained secret in composition, had a claimed ownership and were advertised. In practice, deciding which medicines should be taxed remained difficult, and a schedule of 85 medicines was produced with the Act, with others to be included if they fulfilled these criteria. Vendors were required to take out an annual licence, costing 20 shillings in London and five shillings elsewhere, with exemptions being confined to 'regular bred' surgeons, apothecaries, chemists and druggists. Depending on the medicine price, the stamp duty varied from 'one penny halfpenny' to one shilling for a medicine priced at 5 shillings or more. The 1785 Act did produce revenue from medicines but not as much as was anticipated or required, with the income received remaining less than half the intended amount for the rest of the century.[122]

The rush of Parliamentary activity on medicine taxation from 1802 to 1804, with Medicines Acts in the first two years and a more general Stamp Act in 1804, aimed to increase the medicine revenue by ensuring that more medicines were taxed, by increasing the duties, and by promoting better compliance. These three Acts were a small component of a widespread increase in excise duties to help finance the war with France. In the absence of any detailed Parliamentary reports, it is hard to discover which Act introduced a particular change, and the Acts can best be regarded as a single piece of legislation, introduced over three years. As a result of these, all vendors, proprietors and manufacturers dealing with patent medicines had to be licensed; regular practitioners were no longer excluded.[123] The number of specified medicines to be taxed rose from 85 to 440,[124] and others were included under general descriptions such as all tooth powders and tinctures, and all lozenges.[125] It was made clearer that the first person to handle the medicine was responsible for fixing the stamp, and rewards for informers were specified. The new duties with more high-level bands are shown in Table 2.1.

These Acts more than doubled the revenue from medicines, and the earlier revenue target of £30,000 per year was exceeded.[126] The regular apothecaries and druggists were unhappy about their need to take out a licence and several petitions were received by both Houses of Parliament, for example from the druggists of Plymouth, Tavistock and Sheffield.[127] However, complaints were not received from other patent medicine vendors, mostly booksellers, stationers, and printers, who were perhaps gratified that the regulars were now being taxed in a similar fashion to themselves. The 1812 Medicines Act produced so little initial impact that it was largely ignored by the Parliamentary records. Its main effect was to widen the scope of medicines to include mineral waters and other products, and the number of taxed medicines in the official schedule rose to over 1300 in 1830.[128]

The medicine stamps were an efficient method of tax collection once the initial problems had been overcome, and they also had an important influence on patent medicine selling. Designs varied over time, but all the stamps had an engraving of the crown which was positioned over the cork and the stamps were then stuck to the side of the bottle by two wings (Fig. 2.1). The same stamps were also stuck to boxes of pills or other containers. Some of the stamps up to 1819 had four wings in a

Table 2.1 Medicine stamp duties from 1802 (Kearsley, *Tax Tables 1808*)

Medicine price	*Stamp Duty*
1s or less	1½d
>1s to 2s 6d	3d
>2s 6d to 4s	6d
>4s to 10s	1s
>10s to 20s	2s
>20s to 30s	3s
>30s to 50s	10s
>50s	20s

cruciform pattern.[129] For each value of excise duty, the stamps had a different design and colour, and the names of the larger owners and wholesalers were engraved on them. The stamps were bought from the Stamp Office in London or one of its nationwide agents, with a discount for large orders. The method was in itself simple and effective, with the government's revenue being obtained in advance of any sale; but universal implementation was crucial. As always in this period, forgery was taken very seriously, and Thomas Collicott, a London medicine vendor, was sentenced to death, later commuted to transportation, at the Old Bailey in 1812 for what seems to have been a fairly crude forgery of the sixpenny stamp.[130] Maximising revenue required strict enforcement: it is hard to know whether this was successfully accomplished, but it was certainly attempted.

So the Medicines Acts eventually achieved their single purpose of raising the required amount of government income. What, however, were their other consequences? The most significant was that they enhanced the standing of *all* patent medicines by bestowing a degree of official

Fig. 2.1 1½d medicine excise stamps printed with the names of Francis Newbery and Dicey & Co. (courtesy of the Thackray Medical Museum, Leeds, first published in Mackintosh, 'Authority' by Cambridge University Press)

recognition and authority which was denied to orthodox therapy. The physical presence of the official stamp, with a crown at its centre, on every bottle or box gave each medicine an unintended respectability, which was visible with every new purchase. Furthermore, allowing the larger vendors to have their own names engraved on the stamp linked them with the apparent official endorsement; this promoted both the creation and the recognition of a brand, essential components for maintaining a premium price. The stamp also carried a strong implication that the medicine was effective. The newspaper advertisers were not slow to promote this authority as we shall see in Chapter 8.

A particular advantage of the excise stamp in comparison to the expensive cumbersome patent was that there was no additional cost or inconvenience in obtaining this form of official endorsement for all patent medicines; the duty had to be paid anyway. Like the patent, the excise stamp could be a form of copyrighting for the name of the medicine, but it was backed by much clearer laws than the patenting system, was not limited to 14 years, and posed no threat to the secrecy of the recipe. The printing of the names, or the signatures, of the owners or wholesalers on the excise stamp made it more difficult for a counterfeiter to sell his own medicine under the original name. Copying the stamp would be a clear forgery. Furthermore, in comparison to the uncertain case law supporting a patent, the excise stamp had been created by recent statutory law with clear mechanisms and penalties, particularly after the revisions of 1802–1804. Advertisements could point out that imitating the stamp was a felony, with the possibility of the death penalty.[131] Some owners did continue to obtain patents after 1783, but they now had a new, cheaper universal, system available to them and the excise stamp contributed to the declining importance of the medicine patent in the late Georgian period.

In addition, the licensed medicine retailers, as well as the wholesalers, also obtained a form of official recognition before any general licensing of regular surgeons, apothecaries and druggists. For example, Francis Spilsbury claimed in 1785 that medicine vendors were starting to add M. L. (Medicinae Licentiatus) after their names, though the practice does not seem to have been widespread, and in 1829 a Court was told that the licence made the production of a patent medicine honest and legal.[132] By 1830, opposition to this official recognition of both the medicines and their vendors could generate intemperate views, as shown in a Commons petition from a member of the Inner Temple who had 'long regretted that the vile and destructive trade of tampering with the lives and health of the community should have received a legal and authoritative sanction and protection by virtue of the Stamp Duty on the villainous trash'.[133] Thus, whatever the grumbles from patent medicine vendors in the 1780s about the new Medicines Acts, a vociferous opponent recognised the considerable commercial benefits conferred by the Acts upon the patent medicines industry. Only from 1885 did the stamps carry the caveat 'This stamp implies no government guarantee'.[134]

An additional consequence of the 1783 and 1785 Acts on the retailing of patent medicines was that these Acts probably discouraged the regular druggists from selling them. Under the Acts, regular chemists and druggists could avoid taking out a licence, and thereby elude the excise duty, if they did not sell medicines which were owned, secret and advertised. The day-to-day application of the Acts is unclear and may have been variable, but there was probably a risk that once a druggist had a licence, more of his stock would be subject to taxation. This would have discouraged druggists from publicising any sale of patent medicines, and may have deterred some from even stocking them. Certainly, as we shall see in Chapter 5, druggists were infrequently listed as agents for advertised medicines in Leeds, Birmingham and Salisbury newspapers in 1794. Druggists became more prominent in newspaper advertisements in 1807 and 1822, partly because they were treated identically to other medicine vendors once the exemptions for regular practitioners were removed in the 1802–1804 Acts. So for twenty years non-medical medicine vendors, who were commonly members of the print trades, were freer to publicise their sales of advertised medicines than the druggists.

The Medicine Acts eventually achieved their aim of raising a significant amount of revenue, with a minimum of controversy and a modest collection cost. Indeed, unlike most excise duties of the period, the stamp duty on medicines continued throughout the nineteenth century and nearly half of the twentieth, again making an extra contribution to raising funds for a war when it was doubled in 1915, before being finally abolished in 1941 when purchase tax took over.[135] Overall the Acts had the unintended effect of raising the status of patent medicines, and also of their vendors. The 1783 and 1785 Acts also seem to have discouraged the regular druggists from promoting these medicines, and so, for a time, these medicines would be predominately in the hands of non-medical entrepreneurs who perhaps would be more willing than the druggists to advertise vigorously and denigrate competitors.

Indications for Patent Medicine Use

Previous writers on patent medicines have rarely had much to say about their indications apart from a few specific examples. Porter did not attempt to define their therapeutic scope, but he did note that the medicines had become more targeted by the late eighteenth century.[136] Rawlings felt that advertisers 'aimed at patients suffering from painful, unpleasant, serious, but not immediately fatal' conditions; but his adjectives could cover most significant medical problems.[137] Some accounts of patent medicines assert that they were panaceas, but this is based on a few selected, and particularly colourful, examples.[138] To assess the status of patent medicines we need a much clearer idea of the range of conditions for which the owners recommended their products, but this is not straightforward. A comprehensive analysis of the medicines is impossible as several hundreds were produced, many with little or no surviving documentation. The specifications for the patents might provide some information for the minority of medicines which received one, but there was no obligation to promote the medicines for the same conditions. Lists of patent medicines subject to the medicine stamp duty were published by Parliament from time to time; but these lists were just names with no further information. The best available guide is the published indications for medicines in continuous runs of provincial newspapers. Taken as a whole, the indications printed in the advertisements were aiming at a broad range of both acute and chronic conditions at all ages, confirming that the medicines were a significant proportion of the medical market. Considered individually, most medicines were aimed at a relatively small assortment of related illnesses, though others were publicised as universal remedies for a wide range of indications.

The advertised indications were documented in the advertisements for medicines in the studied newspapers from Leeds, Birmingham and Salisbury, as described in Chapter 1. Five periods between 1769 and 1822 were investigated, with medicines being excluded if they only appeared in lists or as brief addenda of three lines or less. The first advertisements of the remaining *featured medicines* were explored, a total of 559 advertisements in all. Eight advertisements did not mention an indication, leaving 551 for analysis. These advertisements are a sample of the newspaper medicine advertisements of the time, so they are not a comprehensive survey of all the medicines advertised in newspapers, and obviously they do not include medicines which were only advertised locally by handbills or other means. Nevertheless, they are derived from three localities across the country and they are based on five periods spread over 54 years; so they should provide a good impression of the range of indications of Georgian patent medicines.

The indications were assessed using a categorisation of diseases derived from the chapter headings in Part 2 of William Buchan's *Domestic Medicine*.[139] Numerous editions were published of this enduring work, but the chapter headings do not seem to have varied significantly from edition to edition. The most popular medical book from this period was John Wesley, *Primitive Physic*, but this work was unsuitable as Wesley just listed the diseases in no clear order without attempting to organise them into chapters. First published in 1769, *Domestic Medicine* was one of the best sellers of all books, not just medical ones, in Georgian England, and would have been read not only by the consumers of patent medicines, but also almost certainly by the producers who made the decisions on their indications.[140] Beside its enormous popularity, the book also fits well with this analysis as it followed the principles of orthodox medicine, unlike *Primitive Physic* which encouraged consumers to avoid orthodox practitioners as far as possible. Patent medicines were also positioned close to orthodox medicine in this period, and so they shared underlying theories with Buchan's book.

The categorisation of diseases based on Buchan's chapter headings is listed in Table 2.2. Buchan devoted seven chapters to fevers and agues, but they were uncommon indications for patent medicines and so fevers and agues have been grouped together for clarity. Patent medicines were normally not recommended for acute casualties such as dislocations, drowning and suffocation, and these problems have been excluded. The treatment of corns, which was the sole indication for a few medicines, did not feature in *Domestic Medicine*: it has been added as an extra category. Apart from being a useful tool to classify patent medicines, Buchan's arrangement of his chapters is interesting as it seemed to reflect his understanding of the readers' priorities. For example, inflammation of the eyes, quinsy and throat inflammation, and worms benefitted from a chapter each, while nervous diseases, which may have been less important in day-to-day life, were all included in a single chapter with an impressive range of problems.

First, these categories can be used to discover whether the medicines were aiming to be panaceas or were targeted at a limited range of conditions. Table 2.3 shows the number of featured medicines whose

Table 2.2 Categories of diseases based on condensed chapter headings in *Domestic Medicine* (*DM*)

Category	*Diseases*	*DM chapter numbers*
1	Fevers, agues	13–15, 19–22
2	Pleurisy, inflammation of the lungs	16, 17
3	Consumptions	18
4	Smallpox	23
5	Measles, scarlet fever, bilious fever	24
6	St. Anthony's Fire	25
7	Inflammation of the brain	26
8	Inflammation of the eyes	27
9	Quinsy, inflammation of the throat	28
10	Colds and coughs, hooping cough	29
11	Inflammation of the intestines, kidneys, bladder, liver; colic	30
12	Cholera & excessive bowel discharges	31
13	Disorders of the kidney & bladder, diabetes, incontinence	32
14	Discharges of blood including in spit, vomit & urine	33
15	Headache, toothache, earache	34
16	Worms	35
17	Jaundice	36
18	Dropsy including ascites and hydrocephalus	37
19	Gout, rheumatism (acute and chronic)	38
20	Scurvy, leprosy, scrophula, evil, itch	39
21	Asthma	40
22	Apoplexy	41
23	Costiveness, loss of appetite, indigestion, heartburn	42
24	Nervous diseases	43
25	Disorders of the senses (sight, hearing, taste, smell, touch)	44
26	Scirrhus and cancer	45
27	Poisons including bite of a mad dog	46
28	Venereal disease	47
29	Specific diseases of women	48
30	Specific diseases of children including croup, teething, rickets, convulsions, water on the head	49
31	Surgery including wounds, fractures, burns, bruises, leg ulcers, sprains, strains	50, part of 52
32	Corns	Not in *DM*

Table 2.3 Numbers of featured medicines recommended for one, two, three, or more than three categories of diseases (% of all featured medicines)

	1769		*1781*		*1794*		*1807*		*1822*		*Total*	
No. of medicines	**48**	%	**128**	%	**93**	%	**114**	%	**168**	%	**551**	%
1 category	**18**	*38*	**47**	*37*	**43**	*46*	**39**	*34*	**62**	*37*	**209**	*38*
2 categories	**3**	*6*	**24**	*19*	**15**	*16*	**22**	*19*	**42**	*25*	**106**	*19*
3 categories	**13**	*27*	**21**	*16*	**15**	*16*	**35**	*31*	**35**	*21*	**119**	*22*
>3 categories	**14**	*29*	**36**	*28*	**20**	*22*	**18**	*16*	**29**	*17*	**117**	*21*

indications were confined to one, two, three, or more than three categories of diseases. We should remember that these categories could include several conditions, as can be seen in Table 2.2, but Buchan felt that there was some commonality in either the conditions or their management when he grouped them in the same chapter. Around a fifth of the medicines were indicated for more than three categories, a wide range of problems, with the percentage diminishing a little over time. Some medicines were indeed proudly proclaimed in the advertisements as cures for a very wide range of conditions, though few went as far as the one for Turlington's Balsam of Life which described indications in nine of these categories and then finished by summarising the indications as 'in short, almost every disorder incident to the human frame'.[141] In contrast, over a third of the medicines were recommended for a single category of diseases and this proportion seems to have been roughly constant over the years. A number of these medicines were only indicated for a single disease, particularly for the itch, worms, deafness or corns. Long lists of indications, which seemed irrational and absurd to later medical practitioners, can provide vivid examples for historians, but this detailed analysis shows that many patent medicines in this period were promoted for a relatively small number of problems. Both Porter and Helfand came to a similar general conclusion without providing evidence.[142] This focussed approach corresponded to developments in regular therapy where prescribed medicines were becoming more specific for particular conditions.[143]

Second, these categories can provide strong guidance on which conditions the medicines were seeking to improve. The four most popular categories were scorbutic conditions (category 20 in Table 2.2), joint problems (19), nervous disease (24) and inflammation of abdominal organs (11), with the first being indicated for nearly a third of the medicines. Many, but by no means all, of the conditions in these four categories were unlikely to be fatal and could have taken a long time to treat: some may have been resistant to regular therapy and could be recurrent. Thus, the market encouraged the treatment of diseases which might result in the sale of a substantial quantity of a patent medicine. Amongst other categories, venereal diseases were an indication for several medicines, but this category was not as common as isolated examples of medicine advertising or an exploration of medical books for the general public might suggest.[144] When the categories were placed in rank order, venereal diseases appeared as twelfth, below both asthmas and consumptions. The two categories at the bottom of the rank order were inflammation of the brain (7) and cholera and excessive bowel discharges (12), with no medicines recommended for them. At least one patent medicine was sold for the remaining thirty of the thirty-two categories. Also, contrary to some claims by historians, the medicine indications were not confined to conditions in the middle years of life.[145] Some advertisements mentioned that the treatment could also be used in infants, children, nursing mothers or in old age, and other medicines were indicated for conditions confined to women or children (categories 29 and 30). The promotional content of the advertisements is discussed in detail in Chapter 8.

Although the indications were biased towards chronic conditions, acute diseases were not neglected, with the category of colds, coughs and hooping cough (10) appearing fifth in the rank order. These acute respiratory illnesses were not regarded as trivial in this period when consumers were aware that the complications of a cold could prove fatal and that many infants died from hooping cough. A small number of medicines were specifically promoted for acute conditions, especially Dr. James's Fever Powder. Strikingly, one of these, Dr. Sibly's Reanimating Solar Tincture, was indicated for sudden death due to blows, fits, falls, suffocation, drowning or other problems: it is difficult to suggest a more acute indication than this.[146] When we compare individual years, the frequencies of the categories of diseases were broadly similar, with no category appearing or disappearing during the period.

Overall, the patent medicines industry could provide something for nearly all diseases, amongst all sections of the population. Nothing suggests that this wide range was planned; rather it shows an industry responding to the demands of the market. The range was biased towards chronic, troublesome, complaints, such as scorbutic conditions, which might be particularly profitable; but the numerous consumers sought a wide array of patent medicines for their many problems. Within this broad range, these consumers were often supplied with a medicine focussed on a limited group of conditions, rather than one aimed at curing everything.

Conclusion

Patent medicines were extensively used by all but the poorest, and they could be purchased for almost all diseases: but the status of patent medicines in this period was contested both by the public and by the practitioners, and it is hard to pin down. This is to some extent inevitable when many individuals made their own decisions on whether to take, recommend or prescribe a patent medicine, and these decisions were based on a number of factors, both current and in their past. Summarising the actions and opinions of millions of people and thousands of medical practitioners of all types is inevitably challenging. But the task is made more difficult by a lack of explanations from the actors which would otherwise have put the actions in a clearer context. Thus we know that patent medicines were widely taken and some reasons for this can be revealed, yet hardly any documentation of an individual's motives for taking a particular patent medicine has survived, or has been allowed to survive, outside promotional material: so we are largely ignorant about the relative importance of these reasons. Did consumers purchase the medicines because they had confidence in their superiority, or was it because buying a patent medicine was cheaper and more convenient than consulting a regular practitioner? How important were the concerns about regular therapies? How often did consumers take a patent medicine as a convenient commercial equivalent of domestic medicines? These comparative questions are unanswerable with the currently available evidence. The public's views on patent medicines remain indistinct.

The attitude of the public, practitioners and governments to patent medicines was not only indistinct, but also ambivalent. The public seemed to want them to be new and different from regular therapies, but as we shall see in Chapter 8, they also preferred them to be close to these regular therapies, and not too new or too different. The practitioners were often sharply critical of 'quack' medicines in general, while many amongst their ranks prescribed them and some were owners. Successive governments claimed to be neutral, or critical, on the desirability of taking patent medicines, yet they structured the medicine excise stamp in such a way that it implied government approval, and they did not seek to reverse this until the late nineteenth century. Critics accused governments of trying to maintain or increase patent medicine sales to maximise tax revenues, regardless of the consequences.[147] So the status of patent medicines was not only contested but also fluid: views and actions could change depending on the aims of an individual or organisation at a particular moment.

However, the most important conclusion of this chapter is not the varied and sometimes fluctuating standing of patent medicines; it is the finding that they had a specific status which was distinct from the activities of irregular practitioners. The public regarded them as a popular form of self-help which was separate from alternative therapies provided by the irregulars. Their legal and official status was unique within the medical market: no specific taxation was imposed by governments on other participants in the market. Regular practitioners were sometimes less sure in their general condemnations about a separation between 'quack medicine' (the practice of quacks) and 'quack medicines' (mostly patent medicines); but once they analysed irregulars and patent medicines in more detail, their recommendations show a clearer distinction between them. Arguments will persist on the precise status of patent medicines in this period, but the important observation is that they had a status of their own.

In conjunction with this status, an industry was created to supply these widely used consumer products. Providing medicines in large quantities across the country required the investment, the skills and the organisation of an industry, and many of its main operators were respectable tradesmen or regular practitioners rather than colourful irregulars. The next chapter reveals the range of owners and discusses why the production and sale of patent medicines should be considered as an industry.

Notes

1. For example—Carter, *Doctors*, 224–237; Cody, 'No Cure', 103; Strathern, *Quacks*, 53.
2. Cook, 'Good Advice', 5; Curth, 'Commercialisation'; Furdell, *Publishing*, 135–154; Mackintosh, 'Authority'.
3. Curth, *English Almanacs*; Curth, 'Commercialisation'; Furdell, *Publishing*; Hancock and Wallis, 'Quacking'.
4. Curth, *English Almanacs*, 81.

5. Curth, 'Commercialisation', 54–59.
6. Bateman's Scurvy Grass was advertised 151 times in Curth's analysis of almanac advertisements from the last two decades of the century (Curth, *English Almanacs*, 190).
7. Curth, *English Almanacs*, 193–196.
8. Furdell, *Publishing*, 131–133.
9. Curth, *Commercialisation*, 57, 62–63; Furdell, *Publishing*, 132–133, 152.
10. Hancock and Wallis, 'Quacking', 14.
11. Hancock and Wallis, 'Quacking', 34; Amateur, *Life*, vol. 1, 393; Curth, 'Commercialisation', 58.
12. DeLacy, *Germ*, 33.
13. Creighton, 'Cockburn, William'.
14. Cook, 'Colbatch, Sir John'; Furdell, *Publishing*, 132.
15. Peter, *Truth*, 58.
16. Byfield, *Rise*, 13.
17. Appendix B; Mackintosh, 'Authority'.
18. Wallis and Pirohakul, 'Medical Revolutions'; Holloway, *Pharmaceutical Society*, 37; Porter and Porter, 'Rise', 280; Kett, 'Medical Practice', 20; Fissell, *Patients*, 56.
19. Wrigley and Schofield, *Population History*, 208.
20. Porter, *Health*, 43.
21. Pedley, *Marriage*, 113, 228; Weatherill, *Richard Latham*; Porter and Porter, *Patient's Progress*, 104.
22. Hughes, *Fanny Burney*, 209.
23. Spilsbury, *Free Thoughts*, xxxiii; Nevett, *Advertising*, 18.
24. *Book of English Trades*, 82.
25. *MCR*, 13 (1806), cxlix.
26. *MCR*, 13 (1806), xxxviii.
27. See a reduction of empirics recorded in Essex and Middlesex, *MCR*, 13 (1806), lxxiv and lxxvii.
28. Cumbria Archives, Day Books of John Ware.
29. Picton, *Memorials*, vol. 2, 223.
30. National Archives, Will of Francis Spilsbury.
31. MacLeod, *Inventing*, 193.
32. *House of Commons Papers (Accounts and Papers)*, ix, Finance Accounts of Great Britain, 20–21.
33. Loeb, 'Doctors', 409.
34. Porter, *Health*, 52; Fissell, *Patients*, 73; Loudon, *Medical Care*, 213.
35. Prosser, *Oeconomy*, 13; *Deadly Adulteration*, 135.
36. Townsend, *Guide to Health*, vi.
37. *MCR*, 12 (1806), cliii.
38. Cumbria Archives, Day Books of John Ware.

39. *CP*, 23 July 1805, 30 October 1804 & 12 January 1808.
40. *CP*, 9 September 1800.
41. In the 1790s, the weekly wage of an unskilled labourer was usually 7s to 12s, and few wage earners received more than £50 per year (M. J. Daunton, *Progress and Poverty: An Economic and Social History of Britain 1700-1850* (Oxford: Oxford University Press, 1995), 421–428; Jerry White, *London in the Eighteenth Century: A Great and Monstrous Thing* (London: Vintage, 2013), 234).
42. Spilsbury, *Discursory Thoughts*, 14.
43. Loudon, 'Vile Race', 109–110. For a vigorous condemnation of several irregular practitioners by a physician, see Adair, *Essays*.
44. Porter, *Health*, 4–8; Bynum, 'Wages of Sin', 6; Corfield, *Power*, 143.
45. Ietros, 'Quacks', 70.
46. Puzzle-Pate, 'Medical Science', 261.
47. Loudon, *Medical Care*, 31, 35.
48. Kearsley, *Trades*, 52.
49. Puzzle-Pate, 'Medical Science', 267.
50. Loudon, 'Vile Race', 119; Lawrence, 'Private Enterprise', 66. In theory, the Colleges of Physicians and Surgeons in London had some powers to regulate training, but they were rarely applied, even within the capital.
51. Loudon, *Medical Care*, 24–26.
52. Medicus, 'Brodum'.
53. *MCR*, 13 (1806), lxvi.
54. *MPJ*, 16 (1806), 352. Physicians had to avoid advertising their skills and traditionally they did not practise surgery or midwifery, though many of them did.
55. *MCR*, 14 (1807), xxvii and lvi.
56. *MCR*, 13 (1806), xvi.
57. Helfand, *Quack*, 11–14; Porter, 'Before the Fringe', 2–3.
58. Barry, *Publicity*, 31; Porter and Porter, *Patient's Progress*, 23; *MCR*, 1806, *13*, xliii & clxxxiv.
59. Porter and Porter, *Patient's Progress*, 101; Wild, *Medicine-by-Post*, 19.
60. Pedley, *Marriage*, 126.
61. Buchan, *Domestic Medicine*, 6th edn, xx.
62. Gregory, *Observations*, 172.
63. Forbes, 'Empiricism', 438.
64. Wesley, *Primitive Physic*, vii.
65. Wild, *Medicine-by-Post*, 23.
66. Spilsbury, *Free Thoughts*, xi.
67. Spilsbury, *Free Thoughts*, 77.
68. *Deadly Adulteration*, 127.
69. Medicus, *Adulteration*, 31.

70. Medicus, *Adulteration*, 31.
71. Forbes, 'Empiricism', 438.
72. Wesley, *Primitive Physic*, viii.
73. Ellis, *Country Housewife*, ii.
74. Fissell, *Patients*, 38.
75. For an extreme example, a large bottle of Solomon's Cordial Balm of Gilead was priced at 33 shillings (*ABG*, 14 January 1822).
76. Fothergill, 'Practice', 285.
77. *Quack Doctors*, 25.
78. McKendrick, 'Commercialisation', 40–45; Berg, *Luxury*, 247–250; Plumb, 'Commercialisation', 273–274.
79. Denizen 'Solomon', 297.
80. For comments on the similarities of ingredients, see Porter, *Health*, 24; Helfand, *Quack*, 15; *Lancet*, 1 (1823): 30, 62, 89 and 138; Prosser, *Oeconomy*, 2.
81. As we have seen in Chapter 1, a secret recipe was the key feature of a patent medicine.
82. Welsh, 21.
83. Porter, *Health*, 45.
84. Buchan, *Domestic Medicine*, 6th edn, 726; White, *Observations*, 6.
85. *LI*, 4 February 1822; 'Messrs. Newbery', 115.
86. Wild, *Medicine-by-Post*, 201; Porter and Porter, *Patient's Progress*, 107.
87. *MCR*, 2 (1795), 176.
88. Moore, 'John Hunter', 395.
89. *MCR*, 13 (1806), xcviii and cvi.
90. Fothergill, 'Practice', 286.
91. Prosser, *Oeconomy*, 2.
92. *MCR*, 12 (1806), clvi.
93. For summaries of their contributions to medical ethics: Robert Baker and Laurence McCullough, 'The Discourses of Philosophical Medical Ethics', in *The Cambridge World History of Medical Ethics*, edited by Robert Baker and Laurence McCullough (Cambridge: Cambridge University Press, 2009), 281–309; Laurence McCullough, 'The Discourses of Practitioners in Eighteenth-Century Britain', in *The Cambridge World History of Medical Ethics*, edited by Robert Baker and Laurence McCullough (Cambridge: Cambridge University Press, 2009), 403–413.
94. Gregory, *Observations*, 56.
95. Gregory, *Observations*, 57.
96. Gregory, *Lectures*, 230.
97. Percival, *Ethics*, 44.
98. Percival, *Ethics*, 45.
99. Percival, *Ethics*, 45.

100. Percival, *Ethics*, 45.
101. Woodcroft, *Titles;* Priestley, *Interesting Remarks.*
102. Goodwin, 'Glass, Thomas'.
103. For a discussion of the need for physicians to display gentlemanly conduct see Wild, *Medicine-by-Post*, 10–21.
104. From the final paragraph of Fordyce, *Enquiry*, 228.
105. *ABG*, 1 January 1781.
106. *ABG*, 21 April 1794; Holloway, *Pharmaceutical Society*, 50.
107. Galliard, *Antimonial Medicines*, 38.
108. Paget, *John Hunter*, 164–167.
109. Paget, *John Hunter*, 164.
110. *Correspondence of Rev. Greene*, 133.
111. Lane, 'Medical Practice', 375.
112. Galliard, *Antimonial Medicines*, 38.
113. *Regulations*, 12; *Medical Diary*, 28; *Laws and Regulations*, 11.
114. Mackintosh, 'Authority'.
115. Holloway, *Pharmaceutical Society*, 36.
116. The Acts for medicine taxation were passed in two main batches, one in 1783 and 1785 followed by another in 1802, 1803 and 1804, together with a tightening-up Act in 1812.
117. Cobbett, vol. 23, 934.
118. Cobbett, vol. 23, 935.
119. Spilsbury, *Discursory Thoughts*, 48–52.
120. For example, *House of Commons Sessional Papers of the Eighteenth Century*, vol. 50, 361.
121. Kearsley, *Tax Tables 1786*, 88–94.
122. *House of Commons Sessional Papers for the Eighteenth Century*, vol. 50, 409; *Parliamentary Register 1780–1796*, vol. 28, Appendix C; *House of Commons Papers (Accounts and Papers)*, iii, Public Income of Great Britain, 28–29.
123. Kearsley, *Tax Tables 1808*, 119.
124. *House of Commons Papers (Accounts and Papers)*, viii, 636.
125. Kearsley, *Tax Tables 1808*, 155.
126. For example, gross revenue of £41,201 in 1810 (*House of Commons Papers (Accounts and Papers)*, ix, Finance Accounts of Great Britain, 20–21).
127. *House of Commons Journal*, 58 (1803), 35, 60 and 79.
128. *House of Commons Papers (Accounts and Papers)*, xxv, 518.
129. Booth, *Catalogue*, vol. 2, A138.
130. 'Trial of Thomas Collicott'.
131. For example, Dicey and Co., whose stamp is shown in Fig. 2.1, reminded readers in an advertisement for Dr. Bateman's Pectoral Drops (*LI*, 10 March 1794) that imitation of the stamp 'is, of course, a capital offence'.

132. Spilsbury, *Discursory Thoughts*, 23; 'May v Jordan'.
133. *House of Commons Journal*, 85 (1830), 623.
134. Booth, *Catalogue*, vol. 2, A147.
135. Booth, *Catalogue*, vol. 2, A138.
136. Porter, *Health*, 119.
137. Rawlings, 'Medicines', 6.
138. Strathern, *Quacks*, 80; Furdell, *Publishing*, 138.
139. Buchan, *Domestic Medicine*, 14th edn.
140. Sher, *Enlightenment*, 219.
141. *ABG*, 1 January 1781.
142. Porter, *Health*, 119; Helfand, *Quack*, 32.
143. Cook and Walker, 'Circulation', 341; Maehle, *Drugs*, 2.
144. Loudon, 'Vile Race', 114; Fissell, 'Marketplace', 125; Lane, *Social History*, 151.
145. Loudon, 'Vile Race', 113; Rawlings, 'Medicines', 6.
146. *SWJ*, 15 June 1807.
147. *MCR*, 12 (1806), clii; *MCR*, 13 (1806), clxxi; *MPJ*, 15 (1806), 259.

Bibliography

Archives

Cumbria Archives, Whitehaven, Day Books of John Ware, DA 276A.
National Archives, Kew, Will of Francis Spilsbury, Prob.11/1236/50.

Newspapers and Journals

Aris's Birmingham Gazette (ABG).
Chemist and Druggist.
Cumberland Pacquet (CP).
Lancet.
Leeds Intelligencer (LI).
Medical and Chirurgical Review (MCR).
Medical and Physical Journal (MPJ).
Salisbury and Winchester Journal (SWJ).

Printed Primary Sources

Adair, James, *Essays on Fashionable Diseases* (London: T. P. Bateman, 1790?).
An Amateur, *Real Life in London*, 2 vols (London: Jones & Co., 1821).
Book of English Trades (London: Rivington, 1821).
Buchan, William, *Domestic Medicine*, 6th edn (Edinburgh, 1779).

Buchan, William, *Domestic Medicine*, 14th edn (London, 1794).
Byfield, Timothy, *A Short but Full Account of the Rise, Nature and Management of Smallpox and Other Putrid Fevers* (London, 1711).
The Correspondence of the Reverend Joseph Greene: Parson, Schoolmaster and Antiquary 1712–1790 (London: H. M. Stationery Office, 1965).
Deadly Adulteration and Slow Poisoning (London: Sherwood, Gilbert and Piper, 1830).
Denizen of Liverpool, 'Samuel Solomon MD', *The Scourge*, 2 (1811), 287–303.
Ellis, W, *The Country Housewife's Family Companion* (London, 1750).
Forbes, Duncan, 'On the Origins and Progress of Empiricism', *MPJ*, 14 (1805), 437–448.
Fordyce, William, *A New Enquiry into the Causes, Symptoms and Cure, of Putrid and Inflammatory Fevers* (London: T. Cadell, 1773).
Fothergill, Samuel, 'An Account of the Practice of one of the Physicians of the Westminster General Dispensary', *MPJ*, 18 (1807), 284–286.
Galliard, Edward, *The Use and Abuse of Antimonial Medicines* (London: John Murray, 1773).
Gregory, John, *Observations on the Duties and Offices of a Physician* (London: Strahan & Cadell, 1770).
Gregory, John, *Lectures on the Duties and Qualifications of a Physician*, new edn (London, 1772).
House of Commons Journal.
House of Commons Papers (Accounts and Papers).
House of Commons Sessional Papers for the Eighteenth Century.
Ietros, 'Of Quacks and Empiricism', *MPJ*, 13 (1805), 66–75.
Kearsley, George, *Kearsley's Tax Tables 1786* (London, 1786).
Kearsley, George, *Kearsley's Table of Trades* (London, 1787).
Kearsley, G, *Kearsley's Tax Tables 1808*, (London, 1808).
Laws and Regulations of the Benevolent Medical Society ... in the county of Kent (Canterbury, 1799).
Medical Diary for the year 1799 (London, 1798).
Medicus, 'Dr Brodum's Intrigues with the College of Physicians', *The Scourge*, 2 (1811), 491–493.
'Messrs. May v. Doctors J & C Jordan', *Monthly Gazette of Health*, 14 (1829), 522–523.
Parliamentary Register 1780–1796, vol. 28, Appendix C.
Percival, Thomas, *Medical Ethics* (Manchester: Russell, 1803).
Peter, Josiah, *Truth in Opposition to Ignorant and Malicious Falsehood* (London, 1701).
Priestley, Robert, *A Few Interesting Remarks on Bilious Disorders and Yellow Fever* (Leeds: Thomas Gill, 1798).
Prosser, Thomas, *The Oeconomy of Quackery Considered* (London, 1777).

Puzzle-Pate, Peter, 'Medical Science Exemplified', *The Scourge*, 2 (1811), 260–268.

Quack Doctors Dissected, (Gloucester, 1805?).

Regulations of the Medical Society of London (London, 1775?).

Spilsbury, Francis, *Free Thoughts on Quacks and Their Medicines* (London, 1776).

Spilsbury, Francis, *Discursory Thoughts*, 2nd edn (London, 1785).

Townsend, Joseph, *A Guide to Health* (London, 1795).

'Trial of Thomas Collicott', 1812, *Old Bailey Proceedings Online*, www.oldbaileyonline.org (accessed 2 June 2017).

Wesley, John, *Primitive Physic*, 23rd edn (London, 1791).

White, William, *Observations on the Use of Dr. James's Powder, Tartar Emetic and Other Antimonial Preparations in Fevers* (London: T. Cadell, 1774).

Secondary Sources

Barry, Jonathan, 'Publicity and the Public Good: Presenting Medicine in Eighteenth-Century Bristol', in *Medical Fringe and Medical Orthodoxy 1750–1850*, edited by W. F. Bynum and Roy Porter (London: Croom Helm, 1987), 29–39.

Berg, Maxine, *Luxury and Pleasure in Eighteenth-Century Britain* (Oxford: Oxford University Press, 2005).

Booth, R. G, *A Catalogue of the Revenue Stamps of the U K, Isle of Man, Channel Islands and Eire*, 2 vols (Hexham, Northumberland: Tim Clutterbuck, 1982).

Bynum, W. F, 'Treating the Wages of Sin: Venereal Disease and Specialism in Eighteenth-Century Britain', in *Medical Fringe and Medical Orthodoxy 1750–1850*, edited by W. F Bynum and Roy Porter (London: Croom Helm, 1987), 5–28.

Carter, Robert Brudenell, *Doctors and Their Work, or Medicine, Quackery and Disease* (London: Smith, Elder, & Co., 1903).

Cobbett, William, *Cobbett's Parliamentary History of England*, 36 vols (London: R Bagshaw, 1806–1820).

Cody, Lisa Forman, '"No Cure, No Money," or the Invisible Hand of Quackery: The Language of Commerce, Credit, and Cash in Eighteenth-Century British Medical Advertisements', *Studies in Eighteenth-Century Culture*, 28 (1999): 103–130.

Cook, Harold J, 'Good Advice and Little Medicine: The Professional Authority of Early Modern English Physicians', *Journal of British Studies*, 33 (1994): 1–31.

Cook, Harold J, 'Colbatch, Sir John', *ODNB* (accessed 11 November 2016).

Cook, Harold J, and Timothy D. Walker, 'Circulation of Medicine in the Early Modern Atlantic World', *Social History of Medicine*, 26 (2013): 337–351.

Corfield, Penelope, *Power and the Professions in Britain 1700–1850* (London: Routledge, 1995).

Creighton, Charles, 'Cockburn, William', rev. Anita Guerrini, *ODNB* (accessed 11 November 2016).

Curth, Louise, 'The Commercialisation of Medicine in the Popular Press: English Almanacs 1640–1700', *Seventeenth Century*, 17 (2002): 48–69.

Curth, Louise, *English Almanacs, Astrology and Popular Medicine: 1550–1700* (Manchester: Manchester University Press, 2007).

DeLacy, Margaret, *The Germ of an Idea: Contagionism and Society in Britain, 1660–1730* (Basingstoke, Hampshire: Palgrave Macmillan, 2016).

Fissell, Mary, *Patients, Power, and the Poor in Eighteenth-Century Bristol* (Cambridge: Cambridge University Press, 1991).

Fissell, Mary, 'The Marketplace of Print', in *Medicine and the Market in England and its Colonies, c.1450–c.1850*, edited by Mark Jenner and Patrick Wallis (Basingstoke: Palgrave Macmillan, 2007), 108–132.

Furdell, Elizabeth Lane, *Publishing and Medicine in Early Modern England* (Rochester, NY: University of Rochester Press, 2002).

Goodwin, Gordon, 'Glass, Thomas', rev. Alick Cameron, *ODNB* (accessed 21 Dec 2016).

Hancock, David, and Patrick Wallis, 'Quacking and Commerce in Seventeenth-Century London: The Proprietary Medicine Business of Anthony Daffy', *Medical History,* Supplement 25 (2005).

Helfand, William, *Quack, Quack, Quack: The Sellers of Nostrums in Prints, Posters, Ephemera and Books* (New York: Grolier Club, 2002).

Holloway, S. W. F, *Royal Pharmaceutical Society of Great Britain 1841–1991: A Political and Social History* (London: The Pharmaceutical Press, 1991).

Hughes, Peter, ed., *The Journals and Letters of Fanny Burney (Madame D'Arblay): Vol.VIII* (Oxford: Clarendon Press, 1982).

Kett, J. F, 'Provincial Medical Practice in England 1730–1815', *Journal of the History of Medicine and Allied Sciences*, 19 (1964): 17–29.

Lane, Joan, 'Eighteenth-Century Medical Practice: A Case Study of Bradford Wilmer, Surgeon of Coventry, 1737–1813', *Social History of Medicine*, 3 (1990): 369–386.

Lane, Joan, *A Social History of Medicine*, (London: Routledge, 2001).

Lawrence, Susan C, 'Private Enterprise and Public Interests: Medical Education and the Apothecaries Act, 1780–1825', in *British Medicine in an Age of Reform*, edited by Roger French and Andrew Wear (London: Routledge, 1991), 45–73.

Loeb, Lori, 'Doctors and Patent Medicines in Modern Britain: Professionalism and Consumerism', *Albion*, 33 (2001): 404–425.

Loudon, Irvine, *Medical Care and the General Practitioner 1750–1850* (Oxford: Clarendon Press, 1986).

Loudon, Irvine, '"The Vile Race of Quacks with which this Country is Infested"', in *Medical Fringe and Medical Orthodoxy 1750–1850*, edited by W F Bynum and Roy Porter (London: Croom Helm, 1987), 106–128.

Mackintosh, Alan, 'Authority and Ownership: The Growth and Wilting of Medicine Patenting in Georgian England', *British Journal for the History of Science*, 49 (2016): 541–559.

MacLeod, Christine, *Inventing the Industrial Revolution: The English Patent System, 1660–1800* (Cambridge: Cambridge University Press, 1988).

Maehle, Andrea-Holger, *Drugs on Trial: Experimental Pharmacology and Therapeutic Innovation in the Eighteenth Century* (Amsterdam: Rodopi, 1999).

McKendrick, Neil, 'Commercialisation and the Economy', in *The Birth of a Consumer Society: The Commercialization of Eighteenth-Century England*, edited by Neil McKendrick, John Brewer and J. H Plumb (London: Europa, 1982), 7–194.

Moore, Wendy, 'John Hunter: Learning from Natural Experiments, "Placebos", and the State of Mind of a Patient in the 18th Century', *Journal of the Royal Society of Medicine*, *102* (2009): 394–396.

Nevett, T. R, *Advertising in Britain: A History* (London: Heinemann, 1982).

Paget, Stephen, *John Hunter: Man of Science and Surgeon (1728–1793)* (London: Fisher Unwin, 1897).

Pedley, Avril, ed., *A Georgian Marriage: The Family Papers of Sir Nash and Lady Grose 1761–1814* (Wimborne Minster: Dovecote Press, 2007).

Picton, J. A, *Memorials of Liverpool*, 2nd edn, 2 vols (London: Longmans, Green & Co, 1875).

Plumb, J. H, 'Commercialisation and Society', in *The Birth of a Consumer Society: The Commercialization of Eighteenth-Century England*, edited by Neil McKendrick, John Brewer and J. H Plumb (London: Europa, 1982), 263–334.

Porter, Dorothy, and Roy Porter, *Patient's Progress: Doctors and Doctoring in Eighteenth-Century England* (Stanford: Stanford University Press, 1989).

Porter, Roy, 'Before the Fringe: 'Quackery' and the Eighteenth Century Medical Market', in *Studies in the History of Alternative Medicine*, edited by Roger Cooter (Basingstoke, Hampshire: Macmillan Press, 1988), 1–27.

Porter, Roy, *Health for Sale: Quackery in England 1660–1850* (Manchester: Manchester University Press, 1989).

Porter, Roy, and Dorothy Porter, 'The Rise of the English Drugs Industry: The Role of Thomas Corbyn', *Medical History*, 33 (1989): 277–295.

Rawlings, F. H, 'Old Proprietary Medicines', *Pharmaceutical Historian*, 26 (1996): 4–8.

Sher, Richard, *The Enlightenment and the Book: Scottish Authors and Their Publishers in Eighteenth-Century Britain, Ireland and America* (London: University of Chicago Press, 2006).

Strathern, Paul, *Quacks, Rogues and Charlatans* (London: Royal College of Physicians, 2015).

'The Messrs. Newbery', *Chemist and Druggist*, 15 (1874): 112–116.

Wallis, Patrick, and Teerapa Pirohakul, 'Medical Revolutions? The Growth of Medicine in England, 1660–1800', *Journal of Social History*, 49 (2016): 510–531.

Weatherill, Lorna, ed., *The Account Book of Richard Latham 1724–1767* (Oxford: Oxford University Press, 1990).

Welsh, Charles, *A Bookseller of the Last Century* (London: Griffith, Farran, Okeden & Welsh, 1885).

Wild, Wayne, *Medicine-by-Post* (Amsterdam: Rodopi, 2008).

Woodcroft, Bennet, ed., *Titles of Patents of Invention, Chronologically Arranged* (London: Patent Office, 1854).

Wrigley, E. A, and R. S Schofield, *The Population History of England 1541–1871: A Reconstruction* (Cambridge: Cambridge University Press, 1989).

CHAPTER 3

Constructing the Industry

By the mid-eighteenth century, patent medicines were a distinct and growing constituent of the medical market, widely consumed by all sections of society and capable of providing therapy for most conditions. As a result, many men, and some women, derived much of their income from devising, producing, wholesaling and vending patent medicines which were often sold all over the country. This chapter is a systematic and across-the-board study of the ownership and production of patent medicines in Georgian England, and it puts forward two main arguments. One is that these activities were part of an established industry in that they were mostly organised, respectable, stable and profitable, and with their own business practices. The other is that, contrary to the views of many historians, the ownership and the production were mostly distinct from irregular medicine and quackery. The wholesaling and vending of patent medicines will then be discussed in the following two chapters. As well as exploring the medicine proprietors and their medicines, this chapter also acts as a *Dramatis Personae* for some of the participants in the industry who have already been mentioned in this book or will appear in later chapters.

The first section of this chapter investigates the development of the patent medicines industry in the early years of the Georgian era. The emphasis switched from patent medicines as an extension of an individual's practice to owners distributing medicines, which they had often not devised, across the country. The chapter then describes the range

A. Mackintosh, *The Patent Medicines Industry in Georgian England*, Medicine and Biomedical Sciences in Modern History,
https://doi.org/10.1007/978-3-319-69778-9_3

of medicine owners, placing them in the six groups of market leaders, tradesmen, medical professionals, elite, irregular practitioners and local owners, with a section allocated to each group. Examples of owners within each group show that much of the production of patent medicines was organised and it could be very profitable. Most of the owners were regarded as being respectable in the sense that they were considered to be honest according to the business practices of the time and they could maintain a position in society. Many of their businesses were stable, persisting for several decades and sometimes being inherited by widows, sons or daughters. The industry employed its own practices, for instance maintaining the secrecy of the recipe and advertising heavily, and in later chapters we shall see that the industry used distinctive advertising techniques, nationwide distribution from a single source, and specialised retailers. After these six sections on the groups of owners, the following section provides numerical confirmation that the majority of the owners of the widely distributed medicines did not indulge in irregular medical practice. The last section before the Conclusion explores the possible origins of the medicines. The patent medicines had similarities to regular therapy, but most were not copies of prescribed medicines and they were often created for the industry.

This chapter contends that patent medicine ownership was largely separate from quackery, both in the techniques used and in the aims of the participants. Most medicine owners were seeking to make money by selling their medicines as widely as possible to consumers they had not met: they were not giving medical advice or providing therapies other than their medicines. A minority were irregular medical practitioners, and some of these were colourful itinerants who were described as quacks by contemporaries. In practice, the mountebank selling medicines in a public space was becoming rare by the late eighteenth century, and most itinerant owners sought to remain as close as possible to regular medicine by mimicking the actions of orthodox physicians, rather than those of quackery.

Development of the Patent Medicines Industry

The formation of the industry was a prolonged process which can be traced back to at least the mid-seventeenth century. As discussed in the last chapter, many of its features, such as branding, fixed prices, national advertising, distribution across the country and ownership of a range of medicines, can be observed in Stuart England, and some medicines, especially Daffy's Elixir, were sold in large quantities. Nevertheless, these features are much less conspicuous than they were to become later, and, as far as we can tell, most medicines were only sold locally in modest quantities, perhaps as an extension of the owner's medical practice.

The pace quickened in the early Georgian period when several developments encouraged the creation of a substantial patent medicines industry. Advertising in the provincial newspapers provided a more flexible, more locally responsive and potentially more repetitive means of national promotion compared to the service delivered by annual almanacs. The first provincial newspaper was published in 1702, and by 1720 about 25 provincial newspapers were available each week in England.[1] Advertisements of all types were sparse in these early newspapers, but several popular medicines, such as Daffy's Elixir, Stoughton's Cordial and Dr. Bateman's Pectoral Drops were promoted in newspapers across England in the early 1720s. In addition, from 1711, the patenting of a medicine's recipe provided both an additional form of promotion and a means of copyrighting a medicine's name, though applications were uncommon in the early years with only six medicines patented up to 1741.[2] Then the turnpike boom in the 1750s produced a coherent road network which further enhanced the industry by delivering simpler, faster and less weather-dependent national distribution, accompanied by improved postal communication.[3]

Two owners, Joshua Ward (1685–1761) and John Hill (1714–1775), were amongst the first to develop and maintain a range of profitable well-publicised medicines for national promotion and distribution. For nearly thirty years, Ward maintained both a lucrative medical practice and a range of branded medicines for distribution across the country, and he can be seen as a bridge between patent medicines as an extension of a practice and the commercial medicines divorced from a practitioner which were common in later years. Ward's origins were obscure and he spent 16 years in exile in France from about 1717, probably due to Jacobite sympathies.[4] Following a pardon, he returned to London in 1734 with his Pill and his Drop, which rapidly became popular medicines, producing an estimated income of £12,500 per year by 1736 according to his critic, the apothecary Joseph Clutton.[5] Ward's success was assisted by royal patronage after he had successfully treated George II's dislocated thumb which had previously been diagnosed as gout, but he also exploited puffing, advertising and other forms of promotion to keep himself and his growing number of medicines in public view. The medicines were prepared by two chemists, White and D'Osterman, not by Ward himself.

Joshua Ward became notorious, attracting the accolade of being depicted alongside two famous irregular practitioners in William Hogarth's 'The Company of Undertakers'.[6] The composition of Ward's medicines varied over the years, but the Pills usually contained antimony, accompanied by arsenic and cobalt according to Clutton, and they generated considerable controversy.[7] They were described as dangerous by critics and Ward attempted to prosecute the publisher of the *Grub-Street Journal* for libel in 1735.[8] Others, such as Horace Walpole and Henry Fielding, supported Ward and his medicines, and Edward Gibbon was treated by Ward for multiple symptoms in childhood.[9] At Ward's death in 1761, nine medicines bearing his name were on sale, and their recipes were made public with the help of funds from the government.[10] The aim was to continue production with the profits going to charitable causes, but, unlike many later patent medicines, Ward's medicines did not outlive their creator for long. They were probably too closely linked to Ward to survive his demise.

The polymath John Hill was a somewhat less controversial figure who introduced, from the 1740s, a range of herbal medicines which contrasted with the chemical, and supposedly more dangerous, medicines of Joshua Ward.[11] In addition to being a regular apothecary, Hill was a well-known naturalist, playwright, actor and prolific author who married an aristocratic wife. His medical role was limited, so his medicines did not depend on links to his practice and they were more enduring than Ward's medicines. At least eight medicines branded with his name, and from 1774 his Swedish knighthood, were sold across England, all with reassuring herbal names such as Pectoral Balsam of Honey and Essence of Waterdock. In common with many medicines in the fully established patent medicines industry, Hill's medicines remained popular after their creator's death in 1775: Hill's widow and later his daughter maintained both the secret of the recipe and profitable sales until Benjamin Shaw bought the recipes in 1802.[12] Shaw, and his later partner Edwards, continued advertising Hill's medicines for at least the next 20 years.

By the mid-eighteenth century, several patent medicines were being developed as consumer products subject to the normal commercial practices of the day. The owner directly linked to a single patent medicine, which he or she had invented, persisted throughout the Georgian period, but several owners and wholesalers were now involved with a range of medicines which they had not created. They were acquiring the medicines by one means or another, including purchase and inheritance. We have already seen in the prologue to Chapter 1 that John Newbery was a part-owner of Dr. James's Fever Powder. In addition, he was in a four-man group that bought the right to sell Hooper's Pills in 1743; he agreed to sell three medicines owned by James Grosett in 1757; and he owned a share in a version of Dr. Bateman's Pectoral Pills from 1761.[13] A major interest in five patent medicines was recorded in his will, though he created none of them himself.[14] At the same time, Cluer Dicey (1715–1775) was building a range of medicines for national distribution, advertising 22 medicines together in 1769.[15] Joseph Collett, who styled himself as a 'dealer in medicines', took out patents for four different medicines between 1744 and 1758, and he later patented a urinary catheter.[16] Patent medicines had become consumer products to be handled by the commercial practices of the time, not just medical therapy at a distance. Both regular and irregular practitioner owners endured, sometimes with a national distribution; but, as will be discussed below, many owners no longer provided any form of medical therapy apart from the sale of their medicines.

Types of Medicine Owners

Who owned the patent medicines and, in a few cases, made a fortune out of them? With hundreds of medicines, the variety of proprietors was considerable, ranging from large London businesses which owned and distributed a range of medicines across the country to rural owners who just sold a small quantity of a single medicine in their locality. To simplify this variety, the owners have been divided into six groups (Table 3.1), which have been created to explain the wide assortment of owners and are not

Table 3.1 Groups of owners

Market leaders
Tradesmen and tradeswomen
Medical professionals
Elite owners
Irregular practitioners
Local owners

based on any primary or secondary source. Each group contains some similarities in the owners' backgrounds and actions, while exhibiting distinguishing features from other groups. For each group, except the last one, a case study describes an owner's participation in the patent medicines industry, followed by an exploration of the methods of some other proprietors within the same group. Many of these owners and medicines reappear later in the book.

These case studies and the discussion of other owners fulfil three principal functions. The first is the demonstration that producing patent medicines was usually organised, stable and profitable: it was part of an established industry. The second is an assessment of the range of owners placed in their social background, and it shows that most were respectable in the context of their period and distant from quackery: only a minority were irregular medical practitioners. The third function is the introduction of some of the methods employed by the industry to ensure that the medicines would be commercially successful over a prolonged period: these methods will be explored in more detail in later chapters. Overall, proprietors in the first four groups ran normal commercial enterprises, with an additional element of philanthropy amongst the elite owners; only those in the last two groups can be connected to quackery.

Each of these groups contained a range of proprietors and variable practices, but each had some common features which reflect the different modes of medicine ownership. The *market leaders* owned, or part-owned, several medicines, might produce some of them, and also acted as wholesalers for these and other medicines. London-based and few in number, they nevertheless controlled a large section of the medicines industry judged by their volume of sales. The more numerous owners who were *tradesmen and tradeswomen* could be found in London and across the country: they relied on a smaller number of medicines, perhaps just a single one, and producing medicines might only be part of their livelihood. The *medical professionals* were the regular physicians, surgeons and apothecaries who owned patent medicines. If they wished to maintain their regular status, they had to justify their ownership and perhaps present their medicine as being somehow different from the mass of secret patent medicines. The *elite owners* were members of the upper classes or held high public offices. They were rare, but nevertheless significant, as they show that owning a patent medicine could have an element of philanthropy on top of profit, and that ownership was not a bar to an elevated status.

The last two of the six groups were more diffuse and had rather less in common. The *irregular practitioners* combined the production and sale of their medicines with irregular medical practice. Some were itinerant for at least part of the time and a few of these became nationally known and controversial figures. The *local owners* are often largely unknown to us, some only appearing as a brief reference in an advertisement. They produced one or more medicines outside London for local distribution, and they were probably commoner than the newspaper advertisements suggest as they may have employed other forms of publicity, such as locally distributed bills. Many owners in these last two groups would be regarded as quacks by contemporaries, but they still might follow the practices of the established patent medicines industry.

Categorising the wide range of medicine owners is not straightforward, especially when the required amount of information is not available, and it can remain difficult even when we know rather more. One example of this uncertainty is the continuum which existed between the successful tradesmen owners of several medicines and the market leaders; but this is of little practical importance as their methods were similar. A more challenging problem is whether a tradesman proprietor also practised medicine and should be regarded as an irregular practitioner. For this classification, a tradesman was entitled to give verbal or written advice on his own medicine, such as a published treatise, without being considered an irregular: but once he took up medical practice, perhaps by claiming to be a surgeon, he moved into the irregular group. The greatest difficulty is in deciding whether an acknowledged practitioner was regular or irregular. As discussed in Chapter 2, regular practitioners were in theory identified by having completed the required training or experience, and by the nature of their practice; but the criteria were flexible and the final decision often depended on the opinions of local regular practitioners. Fortunately, some contemporary, or nearly contemporary, opinions, particularly those of that forceful critic of his fellow practitioners James Adair, are available to help decisions on which side of the dividing line some well-known owners fell.[17]

Market Leaders

The market leaders ran successful businesses which often persisted for decades, sometimes for over a century. They were both owners and wholesalers for several medicines; so the market leaders feature both in this chapter and in the next, which deals with wholesaling. The concept of an established patent medicines industry from the mid-eighteenth century is reinforced by these substantial, respectable and long-standing, businesses, accompanied by national distribution networks. The three most prominent market leaders were the Newbery family, the Dicey family and their successors, and Thomas Jackson along with his partner and successor James Barclay. The fortunes of the Newbery family were created by ownership, especially Dr. James's Fever Powder, and so Francis Newbery will be presented here, whereas those of the Diceys and the Jackson/Barclay business were based on wholesaling and will be discussed in the next chapter.

Case Study: Francis Newbery

Francis Newbery (1746–1818), usually referred to as 'junior' up to 1781 to distinguish him from his older first cousin of the same name, was the best-known market leader. We have already met his father, John, at several points. Inheriting a thriving medicines business from his father in 1767, along with part of his bookselling/publishing interests, Francis concentrated on the more lucrative medicines business. As the proprietor of several popular medicines and the wholesaler for others, he was able to develop a business utilising national advertising, contracted local agents, and fixed price brands. But his commercial success with patent medicines did not alienate him from contemporary polite society.

At the time of his father's death in 1767, Francis had spent five years at Oxford and had studied anatomy in London, with the intention of becoming a physician.[18] Francis, John's only surviving son, was the sole inheritor of the medicines business, but received only a minority interest in his father's printing and bookselling business.[19] With the help of advice from Dr. Robert James and Samuel Johnson, Francis gave up training to be a physician, and his interest in the publishing business lasted only a few years.[20] Francis's inheritance, and the advice he received concerning it, show that medicine vending was an acceptable occupation which could be pursued in place of a career as a physician or a publishing bookseller, something that would be regarded as inconceivable by later generations. Following a dispute in 1779 with his step-brother, Thomas Carnan, he moved along the road from 65 to 45 St. Paul's Churchyard, and took no further part in the book trade.[21] The business was still owned by his descendants in the twentieth century.[22]

Francis's close involvement with patent medicines did not diminish his position in society. At the housewarming party in Francis's impressive, double-fronted, premises at 45 St. Paul's Churchyard, Samuel Johnson commented that Francis was enabled 'by your father's industry and your own, to lead the life of a gentleman'.[23] Medicine selling became Francis's only paid occupation, but it was not his only activity. He had many cultural accomplishments suitable for somebody of his status in polite society: he played the violin, his translations of Horace were published and some of his poems were set to music.[24] His portrait was painted by Gainsborough;[25] and in 1791 he joined the gentry by purchasing the Sussex estate of Heathfield Park. Now a sporting country gentleman, he was appointed sheriff of East Sussex in 1795. Francis was reputed to be almost a millionaire when he died at Heathfield in 1818.[26]

How did he run his medicines business? As discussed later in this and subsequent chapters, his success seems to have been due to maintaining a good stock of sought-after products, and to keeping a tight grip on his business across the country by directly controlling his frequent advertising and by licensing his retailers. The key medicine in his stock was Dr. James's Fever Powder, which he had inherited from his father. This preparation, containing mercury and antimony, was enormously successful, and was frequently prescribed by regular practitioners: it was still being sold by Francis's descendants in the 1870s.[27] He sought to keep it in the public eye by painting the words 'Warehouse for Dr James's Powder' on the front of his premises and printing his address in advertisements as 'The only warehouse for Dr James's Powder, 45 St Paul's Churchyard'.[28] Francis was responsible for producing the Powder, with the ingredients of antimony and cream of tartar being provided by a well-known chemist, William Jones, who received a supply of the finished product as part payment.[29] Francis distributed and partly owned several other popular medicines, including Dr. James's Analeptic Pills and Dr. Steers's Opodeldoc, as well as wholesaling and retailing a number of other medicines which he did not own.

The market leaders' businesses showed substantial longevity, and they could also act together in their mutual interest, a characteristic feature of an established industry. One who did so was Hilton Wray, who in the 1780s was in partnership with his aunt Martha Wray, the niece of Robert Turlington, the creator of a very successful balsam (patented in 1744).[30] Hilton Wray was portrayed by the tradesman owner Francis Spilsbury (see below) as 'a regular wholesale and retale chemist and druggist, in an extensive medical line of many years standing'.[31] As described in Chapter 2, in 1785 he combined with Francis Newbery and Thomas Dicey in refusing to renew his licence to sell patent medicines. They were prosecuted, and Newbery and Dicey were found not guilty by the jury, but Wray was convicted on a technicality.

The market leaders can be regarded as the central stable core of the patent medicines industry. However, they were few in number and not typical of the mass of medicine owners. A larger group of proprietors comprised the tradesmen who owned medicines, and their engagement in the industry will be explored next.

Tradesmen and Tradeswomen Owners

Georgian London teemed with tradesmen and tradeswomen who would turn their hand to anything which might make money.[32] Other towns were not far behind, and as the market for medicines increased, many tradesmen and tradeswomen in London and across the country started to produce their own remedies, either in addition to, or instead of, their normal occupations. The majority were men, and for simplicity they will all be grouped as 'tradesmen'. Little is known about many of them, but some of the more successful ones can be seen in a clearer light. The best documented was Francis Spilsbury.

Case Study: Francis Spilsbury

Francis Spilsbury (1733–1793) and Francis Newbery were both medicine proprietors and they had similar names, but otherwise they had very little else in common. Newbery was a cultured man who developed an already successful business and joined the gentry; while Spilsbury created his business from nothing, and any participation in polite society, or the commissioning of fashionable portrait painters, has remained unrecorded. Spilsbury advertised heavily in the provincial newspapers, publicised himself as a worthy Englishman, and was quick to provide his views on controversial subjects, including some unrelated to medicines. We can regard him as an aggressive patent medicine entrepreneur, who conformed to the normal commercial standards of the time.

Like his father, Francis Spilsbury was initially a substantial London tradesman and a member of the Goldsmiths' Company who concentrated on silverware. Then, for uncertain reasons, he started medicine production around 1770 and ran a successful medicines business for about 25 years.[33] At least initially, he seems to have continued as a goldsmith, giving evidence to a Parliamentary Committee in 1773 on the unauthorised leakage of silver plate designs from the Assay Office.[34] However, his own publications make no mention of persevering as a goldsmith amongst their comments on a wide range of topics, and it is probable that medicine production and vending became his only occupation.

Spilsbury's medicines business was unlike Francis Newbery's. It consisted largely, perhaps entirely, of manufacturing and distributing a single product, Spilsbury's Antiscorbutic Drops, in contrast to Newbery's ownership of a selection of medicines and the wholesaling of others. With this single medicine, Spilsbury was able to create a lucrative business, as shown by his will, which set up a trust fund of £4000 for his wife Dorothy and their children.[35] Dorothy continued the business herself until at least 1807, and the Drops were still available in 1823, one of many examples of a medicines business continuing after the death of the original proprietor.[36]

So how did Spilsbury make, publicise and sell his Antiscorbutic Drops? Like many medicine proprietors, he wanted not only to keep the recipe hidden, but also to demonstrate that only he, and perhaps one or two others, knew the secret, so that any counterfeit version would be ineffective and perhaps dangerous. His will stated, for the benefit of the public, that only his wife knew the recipe and that she should continue with the business.[37] His advertisements confirm that the Drops were made at his premises ('dispensary'), and that he was responsible for distributing the Drops throughout England.[38] The records of an Old Bailey trial for the theft of a consignment of 39 bottles of Spilsbury's Drops from a coach at Charing Cross, support this by showing that, a year after Spilsbury's death, his wife was sending this consignment to a bookseller in Newbury in Berkshire without any other wholesaler being involved.[39]

Spilsbury relentlessly used newspaper advertisements and other forms of publicity to promote his Drops in their unique bottle, which was embossed with his name and the words 'King's Patent'. As we shall see in Chapters 6 and 8, his advertisements are notable for their frequency, their variety and their determination to get the reader's attention. However, Spilsbury's publicity was not confined to advertisements: indeed, he searched for opportunities to keep his medicine in the public eye, and to create the impression that he was honest and well intentioned. Such publicity could be the promotion of his views on the treatment of scurvy in general or on the use of his Drops in particular, or it could be comments on more general issues relating to the sale of patent medicines, such as his initial opposition to the medicine excise duty, or unrelated topics, such as the new horse tax.[40] He explained the virtues of patent medicines as a group, pointing out that their owners were sure that they worked, unlike a regular physician who was 'the real trier, the true experimenter'.[41] Like many proprietors, he wrote a separate treatise extolling the virtues of his treatment methods, and he also gave free medicines to the poor.[42] In addition to this self-publicity and puffery, Spilsbury was capable of astute observations: unlike many regular physicians of the time, he recognised the difference between sea scurvy, which responded to vegetables, and land scurvy, which did not.[43] Spilsbury remained within the normal commercial standards of the era, using an image of honesty and benevolence as a business asset, and even his vigorous critic Thomas Prosser praised him for his 'candour'.[44]

Other Tradesmen Medicine Owners

Thus, heavy advertising and the grasping of any opportunity for self-promotion enabled Francis Spilsbury to build up a successful long-term medicines business based on a single agent. Another successful tradesman owner was Nathaniel Godbold (1730–1799), originally a baker/confectioner who also speculated in property.[45] Godbold was less prone to self-publicity that Spilsbury, but he also frequently advertised his single medicine. He was thought to earn £10,000 a year from his Vegetable Balsam, which he patented in 1785; and in 1790, he bought Westbrook Place, a house with a hundred-acre park near Godalming for £30,000, while continuing to produce his balsam in Bloomsbury Square, London.[46] After his death, a memorial plaque was erected in Godalming Parish Church, describing him as 'inventor and proprietor of that excellent medicine, the Vegetable Balsam'.[47] An obituary in the *Gentleman's Magazine* described him as 'proprietor and inventor of the much-celebrated vegetable balsam' and commented 'in him, the world has lost a valuable member of society'.[48] His sons, Nathaniel and Samuel, continued to distribute the medicine from the same address until at least 1822.[49] The Godbold family are an example of a lucrative patent medicine business, occupying the same premises over several decades. Nathaniel senior also provides evidence that owning a successful secret medicine did not diminish social acceptance: indeed, his obituary suggests that it might even have enhanced it.

A rather different type of tradesman medicine owner was Thomas Wilson, who owned and distributed several medicines in Birmingham and surrounding towns, without apparently seeking a national market. No biographical details are available for him; but an advertisement in 1794 revealed that he was making and selling ('wholesale and retail') several medicines under his own name in Edgbaston Road, Birmingham, including antiscorbutic drops and worm cakes.[50] By 1807, he was running what seems to have been a larger business in Worcester Street, Birmingham, with eleven of his own retail agents in towns across the West Midlands.[51] The same growth of his business can be seen in the commercial directories of the time. He was not listed in the *Universal British Directory* of the early 1790s, but *Chapman's Directory* (1801) has an impressive three-line entry under his name as 'proprietor of the improved antiscorbutic drops, worm cakes, Scott's and Hooper's pills, nervous pills, British Oil, exc. exc.', with no other occupation being mentioned.[52] Thomas Wilson is an example of a tradesman who devoted himself to a seemingly successful regional medicine business without achieving a national distribution.

Spilsbury, Godbold and Wilson were all examples of tradesmen who ran established, respectable, businesses with a national or regional market. The tradesmen also included established chemists and druggists such as William Jones (died 1789). Jones supplied regular drugs to London hospitals and physicians, to hospitals and apothecaries in the Midlands, Chester and the West Country, and to agents abroad; and as we have seen, he supplied Francis Newbery with antimony and cream of tartar.[53] He acted as a banker and fire insurance agent, and in addition, he sold his own secret Tincture of Peruvian Bark at 3s 6d per bottle, arranging for Benjamin Collins to advertise it in the *Salisbury Journal*.[54]

These four tradesmen produced widely available and sought-after products from fixed publicised premises over many years. Owners in this large group could run significant enterprises for the national market, making considerable profits from the appropriate capital. They did not normally practise medicine in the sense of providing a range of treatments, although they were knowledgeable on the available treatments for particular conditions and would sell their own medicine direct to the public. So they were not irregular practitioners who sold medicines: rather, we should regard them as artisans who acquired the necessary skills to provide a product for the medical market. As such, they resembled the many other artisans who provided an increasing range of goods for the expanding number of Georgian consumers.

Medical Professionals and Their Medicines

In this period, several proprietors of advertised patent medicines claimed to be regular physicians, surgeons or apothecaries, and used their apparent training and experience to enhance the branding of their products. Some of the claims were contested, but accepted regular practitioners did own patent medicines. In the mid-eighteenth century, a regular physician such as Robert James could own two patent medicines without an undue threat to his professional status, but later in the century, as we saw in the last chapter, ownership of a medicine by a regular practitioner was becoming less acceptable. In particular, the secrecy of the composition was coming under increasing attack from some orthodox practitioners by the end of the century. So regulars with their own medicines had to steer a middle course between their commercial interests and the need to maintain their medical respectability. As we shall see, some found it difficult to do so. However, the subject of our case study, Thomas Henry, maintained this balance with conspicuous success, not only creating a family business which provided a very good income for himself and the next two generations, but also achieving a high medical and intellectual reputation in Manchester.

Case Study: Thomas Henry

Thomas Henry FRS (1734–1816) was a well-trained apothecary and part of the established order in Manchester.[55] For half a century, he was involved in medicine, natural philosophy and education in the town. He was an apothecary to Manchester Royal Infirmary, an experimenter on the production and use of 'fixed air' (later called carbon dioxide), and one of the founders of the Manchester Academy, or 'New College'. A close friend of Thomas Percival, he was part of the core of Manchester intellectual life, helping to form the Manchester Literary and Philosophical Society in 1781, and becoming its first joint secretary and later its president. His international reputation is strikingly illustrated by Alessandro Volta's planned appointments on a visit to Britain in 1782. The Italian physicist and future inventor of the voltaic pile saw Henry,

Percival and the industrial dyer Charles Taylor in Manchester, while he met Priestley, Boulton and Watt in Birmingham as well as the ironmasters at Coalbrookdale.[56]

A significant proportion of Henry's income came from his version of magnesia, 'Calcined Magnesia'. Magnesia was a popular laxative, and its manufacture as Epsom Salts had been granted the first patent for a medicine in 1698. During the eighteenth century, a number of purer versions of magnesia had been devised, and from the 1760s the most prominent was produced by Samuel Glass, an Oxford surgeon, with the help of his brother Thomas, an Exeter physician.[57] Henry developed a new method of preparing magnesia, and Richard Warren's reading of Henry's paper on the topic at the Royal College of Physicians in 1771 provides one of the few examples of accurate documentation of the origins of a patent medicine.[58] The London physician and chemist William Saunders (1743–1817) then suggested to Henry's close friend Thomas Percival that a process of calcination would further improve the magnesia, and Henry added this process for his own version, thereby reducing the major side effects of abdominal distension and discomfort.[59] The national distribution of Henry's magnesia was placed in the hands of Joseph Johnson in London, the radical bookselling publisher. Magnesia was a very lucrative medicine: Samuel Glass sold the recipe of his version to the Weymouth bookseller Peter Delamotte in 1772 for the enormous sum of £1500, while Henry's magnesia was the foundation of the family's fortunes for three generations.[60]

How did Henry manage to maintain his reputation as a respectable apothecary and pioneering chemist while making a good income from an extensively advertised patent medicine? He achieved this in a subtler way than the usual blanket secrecy to prevent the recipe being imitated. Henry boldly proclaimed his opposition to concealment and published the details of his manufacturing process so that it was available to all.[61] Yet other apothecaries and chemists did not rush to reproduce Henry's Calcined Magnesia, and its advertisement for over fifty years at a good price shows that any competition was limited.[62] Thus, Henry seemed to achieve the same commercial outcome as secrecy, while proclaiming that he was very happy for any apothecary to make his magnesia, and also criticising rivals for keeping their methods of production secret.[63] How he managed to attain this desirable combination is not clear, but it is probable that his published method was not as easy to imitate as he asserted. The method had over twenty stages, each involving a physical process such as dissolving, boiling, heating, drying or decanting, and Henry emphasised the purity of the water and the need to avoid dust or uncleanliness. Other apothecaries and chemists may have found it difficult or uneconomic to copy the long process. Whatever the reasons, the very public commotion between Henry and the supporters of Glass's magnesia in 1773–1774 confirms that Henry had every intention of preventing imitation.

Perhaps anticipating that all publicity is good publicity, Henry initiated this dispute with a venomous, eight-page published attack on the purity of Samuel Glass's magnesia and the character of its new owner, Peter Delamotte.[64] Henry had bought samples of Glass's magnesia in Manchester, Preston, Chester and London, and he had a duty to inform the public that it had gone up in price and was 'now impure and calcerous', though he did not intend 'to enter into any controversy on the subject'.[65] Not surprisingly, Delamotte and Thomas Glass, Samuel's brother, did not lie down so easily, and they replied with vigorous attacks on both Henry's reputation and the scientific basis of his allegations. Delamotte described nine experiments, performed 'under the inspection' of four experts with appointments at Oxford University, which demonstrated that Glass's magnesia was purer and lighter than Henry's version.[66] Delamotte summed up his argument by attacking both Henry's character and his Calcined Magnesia: 'his *Veracity* had been found to be as *light*, as his *Magnesia* was *heavy*'.[67] Thomas Glass added his authority as a well-known Exeter physician, rejecting all of Henry's assertions with the help of experiments on 21 samples of Glass's magnesia, half the samples coming directly from Delamotte and half being bought in Oxford, London, Bath and Exeter.[68] Moreover, Glass asserted that Henry could not be trusted and mockingly referred to him as 'that *ingenuous* apothecary in Manchester'.[69]

Henry's published reply to Delamotte's and Glass's publications also dealt with both character and science, and it was uninhibited by any professional courtesy.[70] Henry commented that Delamotte was only intermittently in Oxford to supervise the preparation of Glass's magnesia, describing him ironically as 'a stationer in Weymouth, and, no doubt, a very *ingenious chemist*'.[71] Delamotte had paid so much for the medicine that he could not afford to throw it away when it had not been prepared correctly, and Henry described several experiments which showed that Glass's magnesia was impure and calcerous, including some of Thomas Glass's experiments but with different results. A supporting letter from Thomas Percival confirmed the experiments and was 'bearing testimony to your character as a man of ingenuity, honour, and the strictest integrity'.[72] Henry added another supporting letter from his friend John Aikin, a leading dissenter and a Warrington surgeon at that time, and finished by reprinting his original eight-page critical appendix, just to make sure that there were no misunderstandings of his position. No further replies from Delamotte or Thomas Glass are available.

The participants in this well-publicised quarrel sought to justify their positions both by the examination of their characters and by methodical experiments which reputable physicians and academics were prepared to support. This investigation of a patent medicine is a long way from the recent belief that these medicines were essentially a confidence trick on the gullible public. Under the pressures of the market, owners could spend considerable efforts in perfecting their products: few regular medicines would have been explored as thoroughly as these two versions of magnesia. Claims that a patent medicine was superior to more conventional therapy might have been true.

Overall, this account of Thomas Henry and his magnesia shows that a respected medical practitioner could devise and produce a patent medicine in Georgian England without damaging his professional standing. However, this type of ownership became unacceptable in the nineteenth century, and its existence was often suppressed, as illustrated by a biography of Henry's son William, a physician and distinguished chemist who managed the family magnesia business in the first two decades of the nineteenth century. William was the author of Henry's Law on the solubility of gasses, and he was still sufficiently highly regarded in the 1860s to appear in the famous Royal Institution painting *Distinguished Men of Science of Great Britain Living in the years 1807-8*; but his accompanying memoir, written by his son William Charles (Thomas Henry's grandson), eliminated the patent medicine trade from the record, only briefly mentioning that his father had had 'superintendence of a chemical business previously established by his father'.[73] As we see elsewhere in this book, contemporaries were usually more tolerant of producing or selling patent medicines than were their descendants.

Other Professional Medicine Owners

Several medical practitioners who were medicine owners have already been mentioned in Chapter 2. How did proprietors who were regular practitioners combine the commercial necessity of keeping the recipe of their medicine secret with the professional requirements to be seen to be open and honest in dealing with the public? Some tried to remain within regular practice by justifying the secrecy of the recipe, while others decided not to worry unduly about their reputation and to give priority to selling their medicine.

Thomas Henry's example of publishing his recipe without attracting competition was difficult to follow, and some medical practitioners sought to provide a good reason why the composition of their medicine should remain secret. Robert Priestley, who was listed in the 1783 *Medical Register* as a surgeon-apothecary in Leeds, was one example.[74] By 1798 he was a physician with an MD, and he was promoting his own Anti-bilious Powder in advertisements and by a treatise.[75] In the treatise he claimed that he originally intended to publish the composition of the medicine; then he realised that the ingredients were difficult to acquire and handle as 'they are only to be obtained at suitable seasons, and by a judicious selection, and they require a tedious and careful preparation'.[76] If the recipe was generally known, others might use the wrong ingredients to prepare a medicine which would be both ineffective and have side effects. It was therefore in the interests of the public that the recipe should remain secret.

Edward Galliard, an Edinburgh apothecary, took a similar approach to Priestley but expressed it more dramatically. In a 1772 paper before an ad hoc group of local physicians, Galliard introduced his Edinburgh Febrifuge Powder as an antimony-based competitor to Dr. James's Fever Powder. After explaining the characteristics of the ideal antimony febrifuge, he stated that the process of preparing the ideal powder 'now lies on the table'.[77] But the printed page then had a blank space. Using a similar argument to Priestley, he went on to explain that excellent chemistry, genuine ingredients and expensive utensils were needed; so the interests of the public would be safeguarded by confining the knowledge of the composition to just a few practitioners and employing only one wholesaler in Edinburgh and one in London.[78] Galliard recognised that his suggestions would be controversial amongst regular physicians: 'The proposal hurts you; I see it does: but there is no alternative'.[79] So he sought to mollify his audience by suggesting that the medicine could be sold at cost price, or that some of the profits could be used to support a medical library.

Other practitioners were less concerned about preserving their professional reputations, particularly in the early Georgian period. As James

Adair, a trenchant critic of both regular and irregular medicine, put it, 'some of these nostrum-mongers have been appendages to the profession'.[80] John Hill, a regular apothecary, kept the recipes of his eight advertised medicines concealed, and his will emphasised that they should remain secret during his wife's lifetime so that his children could benefit as well.[81] John Norton, a regular London surgeon, earned a fortune from his Maredant's Drops, which he advertised frequently across the country in the 1760s and 1770s.[82] These practitioners preferred to earn as much from their medicines as possible, if necessary at the expense of their professional standing. For a late eighteenth-century practitioner, possessing a successful medicine was more complicated than ownership by a non-medical tradesman. The medical professionals had to work out a method of maintaining their reputations while benefiting from their secret discoveries, or they needed to earn enough from their medicines to replace the diminution of their practice.

Elite Owners

The elite owners were in the upper classes or held high offices. They were rare, but they show particularly clearly that possessing a medicine was not a bar to social acceptance. Also, they illustrate that philanthropy could be an additional reason for owning a medicine. A striking example, initially described by Amanda Vickery, was Elizabeth Shackleton, a member of the Lancashire gentry, who inherited her late husband's secret recipe for a medicine for the bite of a mad dog, and then used the resulting philanthropic reputation from its sale to expand her social contacts across northern England.[83] Elizabeth did not patent her medicine and she does not seem to have paid for advertisements, though the *Leeds Intelligencer* did describe it as 'the never-failing remedy'.[84] Otherwise, it had the characteristics of a patent medicine and it is reasonable to regard it as one.

Case Study: Elizabeth Shackleton

The three case studies of medicine proprietors which I have discussed so far (Francis Newbery, Francis Spilsbury and Thomas Henry) have one feature in common: they all aggressively defended and commercially exploited their medicines as much as possible. However, Elizabeth Shackleton (1726–1781), was different not only in being a woman and living in the northern countryside, but also in showing restraint in developing and promoting her medicine for the bite of a mad dog. Elizabeth was a member of the minor gentry at Colne in Lancashire, making and distributing her medicine as an act of philanthropy, but also, in my opinion, with an element of money making.

The core of this account has been provided by Amanda Vickery who has used correspondence and diaries to build up an enthralling description of the day-to-day life of both Elizabeth and the section of provincial society which included her relatives, friends and acquaintances. In addition, I have inspected copies of Elizabeth's diaries in the Lancashire Records Office. Elizabeth's first husband, Robert Parker, had owned a medicine for the bite of a mad dog, whose efficacy was praised by a correspondent to the *Gentleman's Magazine* in 1753.[85] On her husband's death in 1758, Elizabeth took over the preparation and distribution of this 'Colne Medicine', and she continued making it until her death in 1781, selling it all over the North of England. Elizabeth managed her lower gentry household which prepared a wide range of produce, such as wine, butter, conserves and perfumes, together with brewing and distilling.[86] Some additional funds were generated by this domestic production: for example, in 1776, butter was sold to neighbours, producing the equivalent of two maidservants' annual wages.[87]

Fatal madness accompanied by hydrophobia secondary to the bite of a mad dog was not a common problem, but its sudden and sporadic appearance produced considerable anxiety and many treatments were recommended for prevention. Lancashire was recognised as an area with special expertise in prevention. Specifically, the 'Ormskirk Medicine' from that county was highly regarded as an effective treatment for the bite of a mad dog.[88] John Berkenhout, in his essay on the bite of a mad dog, devoted six pages to criticising it, but still recognised its fame: 'The Ormskirk medicine, which in many parts of this kingdom, particularly in the North, is deemed infallible'.[89] The origin of the Ormskirk Medicine was disputed by the competing owners of rival versions, and it need not concern us here. By the end of the century, Francis Newbery had acquired the sole London agency and national distribution for one version, which he still called the Ormskirk Medicine.[90] Colne is about forty miles from Ormskirk, and Elizabeth's medicine had been originally sold by her first husband in conjunction with Mr. Hill, one of the claimed Ormskirk owners. As a result, some of the successful aura of the Ormskirk product spread to Colne, and so an origin in rural Lancashire was not a handicap, but rather a source which promoted confidence.

Why did Elizabeth continue to produce the medicine for twenty two years after the death of her first husband? Many widows did so to preserve the business, and the secrecy of the recipe, until a son was of an age to take over, but she continued to make the medicine even after she had passed the secret recipe to her second son John in 1776.[91] One reason to continue was philanthropy. The rural gentlewoman providing medicines for her neighbourhood, for example Lady Grace Mildmay in Northamptonshire 150 years earlier, has been recognised by historians.[92] Closer to Colne, the wives of the eighteenth-century Yorkshire gentry devised and passed on medicinal recipes for the use of their families and staff[93]; and Elizabeth's actions can be seen as a similar form of genteel philanthropy. As Vickery has pointed out, the wide distribution of her medicine would ensure that the philanthropy was recognised not only by her immediate neighbours, but also by the socially superior titled gentry and nobility across the north.[94]

However, Elizabeth seems to have been concerned about making money as well as making a reputation. Although her first husband Robert Parker was a gentleman and county office holder, the rents from his estate were only £290 a year.[95] The minimum for the life of a country squire was about £300 a year, so additional income would have been welcome during her first marriage.[96] Vickery is unconvinced that profit was a motive, but the evidence suggests that Elizabeth's medicine was intended to produce an income, in addition to the philanthropy, during her second marriage to the wealthier John. Her husband looked after the financial side of the medicine business, determining the price and arranging payments from distant customers, and, as we have seen, she did make money by producing extra domestic products, such as butter, for sale.[97] When she passed the details of the recipe on to her son in 1776, she exhorted him to keep it secret, which was essential for a continuing profit.[98] This moneymaking does not negate the value of a humanitarian reputation to Elizabeth: she was preparing and selling the medicine for both purposes.

The Catholic Medicine-Makers

Although Elizabeth Shackleton aimed to make some money, her social position and background kept her a long way from the aggressive commercialism of the previous three case studies. Other elite owners are difficult to find: any proprietor seeking to maintain an elevated position in society was unlikely to indulge in self-justifying treatises or other publications as written by Francis Spilsbury, and the small scale of their enterprises would not have warranted extensive advertising. Furthermore, the universal condemnation of patent medicines by Victorian writers did not encourage the long-term retention of any relevant collections of letters or accounts. So documented elite owners are rare, but surprising examples are provided by three leading eighteenth-century Catholics. In this era, Catholics rarely participated in the upper reaches of civil society, but this trio were certainly influential in Catholic life and sought to be amongst its leaders. Like Elizabeth Shackleton, their motives in creating their own medicines combined philanthropy with money making.

The information on these Catholic medicine-makers is derived from the extant correspondence between James Coghlan, the leading Catholic publishing bookseller in London, and some of the principal Roman Catholics of late eighteenth-century Britain. It reveals both the existence, and some of the detail, of medicine-making activity by Coghlan and by two other Catholics, namely Bishop George Hay, the Vicar-Apostolic for the Scottish Lowlands, and Father Henry Francis Xavier Chappel, a Dominican priest in Leicester. A collection of 152 letters sent to Coghlan by various Catholic friends and contacts between 1770 and 1800 has been preserved.[99] In addition, many outgoing letters from Coghlan, including some to Hay, have been preserved in the Scottish Catholic Archives in Edinburgh and elsewhere, and they have been published alongside the Preston records.[100] In-between the specifics of religious organisation and practice, the letters reveal details of medicine production, ownership and distribution by these three men. Studying these letters poses two related questions. Why did this respectable trio with existing sources of income, and busy lives, become involved in making medicines for sale, and how did this affect their position in Catholic, and wider, society? The trio probably would not have referred to their unpatented products as patent medicines, but their remedies had an owner, were intended to be advertised, and had a secret recipe; and thereby they fulfil my criteria for designation as patent medicines.

James Coghlan (1731?–1800) was the leading Catholic printing and publishing bookseller in London.[101] From 1764 he published annually *The Laity's Directory*, which was the closest publication to a Catholic periodical at the time. An essential conduit for many types of information, not just printed works, he was well known to both the Catholic priesthood and laity in Britain, with many of them calling frequently at his shop or exchanging letters.[102] As a publishing bookseller, he could be placed in the tradesman category of medicine owners, but his role as one of the leading Catholic laymen in England and his added philanthropic motives for medicine production bring him into the elite group.

Coghlan was making, advertising and distributing up to five of his own medicines from 1779 until his death in 1800. At least one of his medicines was advertised at the end of most editions of the *Laity's Directory*, and also in some of his other publications, continuing after his death in 1800.[103] Coghlan does not seem to have had any medical training or notable experience, and he asserted that the recipes originated from the Jesuits' Library or other Catholic sources.[104] The scale of Coghlan's medicine production is unclear, but he claimed to sell the medicines abroad and he was described as having a 'large medical warehouse'.[105] The production must have been lucrative as it continued for over twenty years after his death, with the profits being assigned to a Catholic charity.[106]

Why did Coghlan produce and sell medicines? The obvious answer was to make money. But the reasons seem to be more complex than just profit. Much, perhaps all, of Coghlan's advertising was in his own publications for Catholics, such as *The Laity's Directory* and his own catalogues. This would have reduced the costs of advertising, but Catholics were only 1% of the English population at this time, and a desire to maximise profits above all else would have required advertising to a wider readership.[107] In addition, three of his five medicines had 'Jesuits' in their title and a fourth, a medicated snuff, was 'prepared from the original receipt found in the Jesuits Library'. He clearly wanted to give his medicines Catholic origins, and associating four of his five medicines with the Jesuits would have appealed much more strongly to the small Catholic community than to the Protestant majority. Thus, Coghlan seemed to be devising and selling medicines for consumption by Catholics, aiming to help his co-religionists as well as making money. Blom and colleagues have noticed a mingling of philanthropy with profit making in Coghlan's publishing business: 'his practical charity, [...], was a natural extension of his business interest to the Church as a whole'.[108] We can see a similar linkage with his medicines: for Coghlan, making money and helping fellow Catholics were inseparable.

Coghlan maintained a correspondence with George Hay (1729–1811) for thirty years. From 1778, Hay was the Vicar-Apostolic for the Lowland district and joint head of the Catholic Church in Scotland, resigning on health grounds in 1805. A stern figure, he was widely respected both within and outside the Church in Scotland at a time when it was moving from persecution to partial acceptance.[109] He set up the first Catholic seminary in Scotland and was a recognised scholar with four major religious works, as well as numerous pamphlets. He also supervised a new translation of the Bible- and devised his own medicine.

Hay had originally trained as a surgeon, and he produced his own Antiscorbutic Tincture in Scotland for charitable purposes.[110] In 1798, Hay sent Coghlan fifty copies of his new translation of the Bible for sale in London and four bottles of the Antiscorbutic Tincture to try as a patent medicine. Hay provided Coghlan with precise instructions on the use of the medicine and suggested some background reading, but unfortunately the tincture was not a success.[111] Hay promised that he would make 'more trials of its virtues' and provide an improved version in due course, but no further attempts to sell the Antiscorbutic Tincture in London were recorded.[112]

Why did Hay, a very active religious leader and scholar, devise his own medicine? In Scotland, he seems to have produced and dispensed it as charity, but the attempted sale in London was probably to make money. Hay's letters show that his Church was suffering financial hardship in 1797/1798 due to the combination of the unexpected extra costs of building the new seminary and the suspension of funds from Rome caused by a French invasion of central Italy.[113] These financial problems are a plausible explanation of Hay's desire to introduce his medicine to London at the beginning of 1798, though such a link was not specifically mentioned in his surviving letters. Hay showed a similar mingling of income with philanthropy in his books: he told Coghlan that his main aim in translating the Bible was to spread the Word among the people, but that the 2d profit on each copy was also important.[114]

Our knowledge of the third member of the trio is limited as one letter represents the only documentation of Father Henry Chappel's medicine making. Henry Chappel (1749–1825) was attached to the secretive Dominican mission at Leicester in the late eighteenth century.[115] Chappel wrote to Coghlan in 1799 with an apparently unsolicited request for Coghlan to sell his 'specific for the cure of stone and gravel', which he had used successfully in over a hundred cases.[116] Chappel offered a business deal with a partial refund to the consumer if the medicine was unsuccessful, but no reply from Coghlan is available. With this limited evidence, little can be said about Chappel's motives, but they could represent once again a combination of charity and profit: the Leicester mission may have been suffering from the same financial pressures at the end of the century as the Scottish Church.

The desire of these leading Catholics to be involved in the medicine trade again demonstrates that making and selling patent medicines was often an acceptable form of business without any associated social stigma. This trio, professing a faith which was still frequently condemned, did not show any need to conceal or minimise their attempts to sell medicines. They were not worried that their reputations would be diminished by their activities, or that their leadership in the Catholic Church would be compromised.

The finding that elite medicine owners could combine profit and philanthropy should produce a small, but distinct, tilt in our understanding of patent medicine owners. Such a combination has rarely been considered as a reason for owning medicines: claims of charitable intentions in advertisements have not been taken seriously, though several tradesmen owners were reported to be giving their medicine without charge to the poor, or at a reduced price, including Spilsbury and Godbold.[117] In the past, historians have regarded these claims as being largely misleading verbiage, solely designed to present the owner in a favourable light. But if we accept the possibility of a combination of profit and philanthropy as a motivation, then some of these claims may be at least partly true. Declarations in an advertisement that an owner had charitable intentions should be carefully assessed, rather than being quickly rejected: it could be that an owner did indeed want to benefit other sufferers, as long as a profit was made as well.

Irregular Practitioners

In the early Georgian period, colourful quacks and mountebanks selling medicines in a public space were common. The diary of Thomas Turner, a Sussex shopkeeper, described in 1760 the weekly visit of a mountebank who, standing on a permanent stage, was 'selling packets which are to cure people of more distempers than they ever had in their lives for one shilling each'.[118] One shilling was a common price for a patent medicine, but it is not clear whether these mountebanks were selling branded and advertised patent medicines, or merely their own anonymous products which were not available from other sources. One irregular who did sell patent medicines in the open air was Richard Rock, memorably portrayed selling medicines in Covent Garden in Hogarth's 1738 painting *Morning*.[119] Hogarth painted Rock's advertising board with the royal arms signifying a patent, though Rock did not possess a medicine patent until he acquired one in 1751 for his venereal disease treatment. Rock, who allegedly started as a porter, styled himself as a 'licentiate in medicines' on his patent application.[120]

By the end of the eighteenth century, the travelling mountebank selling medicines in public spaces was becoming rarer in England. In the 1790s, Adair started a paragraph on the former occupations of quacks with 'Whilst itinerant mountebanks were in fashion: though the breed is almost extinct in this country [...]'.[121] A correspondent from Essex reported in 1806 to Edward Harrison, the medical reformer, that itinerants were still visiting but in fewer numbers, while another from Middlesex reported that his area contained no quacks and indeed 'the learned and celebrated Dr Brodum' (see below) no longer visited.[122] As the patent medicine industry developed in the later Georgian period, mountebanks and other irregulars selling medicines in public spaces had no significant role in the ownership and production of widely advertised patent medicines, though the remaining itinerants could have devised their own products for immediate sale. The organised industry had superseded the small-scale operator who lacked the skills and capital to participate in an increasingly national market.

At this time, several popular and widely advertised patent medicines were owned by irregular practitioners, who attempted to imitate regular practice on their travels rather than act as mountebanks. Self-publicity was essential for these owners as the practitioner and his medicine were linked together: the medicine represented the practical result of the owner's superior knowledge. So, itinerant irregulars who were medicine owners advertised extensively and justified their actions in treatises, producing a considerable amount of inevitably biased information about their actions and their medicines. One consequence was ridicule in the monthly journals, resulting in a rich brew of claim and counterclaim. So we have a lot of information about some of them: the difficulty is deciding how much is accurate.

Case Study: William Brodum

The most infamous medicine proprietor of the 1790s and 1800s was William Brodum (died 1824), a well-known irregular practitioner with medicine production as part of his practice. Seeking to be regarded as an established London physician in spite of a lack of recognised training, he bought an MD and aggressively publicised both himself and his patent medicines. He also travelled widely across England, offering consultations and promoting his medicines; and for a time he was very successful, but he attracted widespread criticism and satire.

Brodum's origins and early life were murky, with a number of conflicting accounts. The accounts agree that he was Jewish and foreign-born, and that he had gained some experience with irregular medicine vendors in England before promoting himself as a physician from around 1790 and obtaining an MD degree from Marischal College, Aberdeen in 1791.[123] He claimed to have been trained as both a naval and army surgeon in Europe.[124] According to his 1790 handbill, he was able to treat a very wide range of disorders, especially 'a certain disorder' (venereal disease), with the minimum of inconvenience, by consultations, letters, or the examination of the urine.[125] Later newspaper advertisements offered an elegant apartment with attendants and a carriage which could be rented by a gentleman or lady wishing to consult Dr. Brodum.[126] For a time he was very successful, allegedly earning £5000 a year from selling his medicines.[127]

From the early 1790s, Dr. Brodum's Nervous and Restorative Cordial for nervous conditions, consumptions and deafness and Dr. Brodum's Botanical Syrup, used in a wide range of complaints, were generally available, and he patented both medicines in 1799.[128] Brodum's medicines conformed to his practice's image of gentility with the lowest price for the smallest bottle or packet of any of his medicines being 5s 5d, with larger ones costing up to 2 guineas.[129] At a time when a labourer's daily wage was often little more than one shilling, these prices indicate that he was targeting the more prosperous members of the community. Brodum's *Guide to Old Age*, first published in 1795, ran to 344 pages in two volumes and went through multiple editions over several years.[130] In line with his pursuit of gentility, the *Guide* was dedicated to the King and mentioned that the royal family took his medicines. This type of substantial publication, superficially concerned with furthering medical self-help, was a common method of promotion by medicine-owning irregular practitioners.

The combination of imitating regular physicians, blatant self-publicity, frequent newspaper advertising and unabashed money making made Brodum the subject of much contemporary comment. He was accused of planning to bribe his way to becoming President of the Royal College of Physicians.[131] He was frequently satirised and his name used to represent irregular practitioners in general.[132] Perhaps the ultimate indication of Brodum's celebrity was an elaborate masquerade at Foley House in 1802, attended by the Prince of Wales and two of his brothers. Artificial village shops were created in the great hall and manned by the local aristocracy and gentry. One shop was 'Doctor Brodum's shop', and the whole scene 'produced all the comic effect that may be imagined to arise from the characters that composed it'.[133]

Other Irregular Medicine Owners

Two other prominent irregulars who owned medicines were Samuel Solomon and John Lignum. Based in Liverpool and Manchester respectively, they also travelled to promote themselves and their medicines. Samuel Solomon (1768/1769–1819) was Jewish, like Brodum, and had also obtained an MD from Marischal College, in his case in 1796;[134] as a result they were often satirically linked together. After a period as a spectacle salesman, Solomon started selling his famous Cordial Balm of Gilead, which was recommended for a wide range of nervous and debilitating disorders.[135] He also owned and produced an Anti-Impetigines to purify the blood for scorbutic and other complaints, and an Abstergent Lotion to be applied on scorbutic eruptions. In common with other irregular medicine owners, Solomon was an aggressive publicist for himself and his medicines. In his own book, *A Guide to Health*, he asserted that his practice was as successful as that of any physician in Europe, and he claimed to spend at least £5000 per year on advertising.[136] Continuing to travel to promote himself and his medicines, he built the impressive Gilead House on the eastern edge of Liverpool in 1804 and he became 'one of the institutions' of that town.[137]

John Lignum (probably died in 1826) was less itinerant than Brodum and Solomon, and he claimed a less elevated medical status. Leaving Edinburgh as an apothecary called John Wood, he became a surgeon called John Lignum in Manchester.[138] His travelling was apparently confined to the North of England and, probably as a result, he escaped the attention of the critics and satirists. His Antiscorbutic Drops and a Lotion, together with pills for venereal disease, were produced for at least thirty years at his premises in Thomas Street and then Bridge Street in Manchester, and they were extensively advertised in the newspapers.

These three irregulars, who each owned several patent medicines, were regarded by many contemporaries as quacks. But even so, they do not fit with our traditional image of an irregular selling medicines in the open air with the help of a vigorous sales technique. They aspired to behave as regulars, and they tried to stick as closely as possible to orthodox practice at home and on their travels. This was shown when Brodum and Lignum, by coincidence, both visited Leeds in July 1793. Rather than addressing public meetings, they placed advertisements in the Leeds newspapers, listing the premises where they could be consulted and the hours they would be available, so mimicking conventional medical practice.[139] Indeed, Brodum made some effort to be recognised as a regular physician, attending Westminster Hospital as a 45-year-old medical student, stopping his travelling, and discontinuing selling the medicines himself.[140]

Unlike the owners in the first four groups, the prominent irregular medicine owners described in this section had clear links with quackery. Yet they also adopted many of the practices of the patent medicines industry in that they kept their medicines secret, they operated from a fixed base where their medicines were produced in large quantities over several decades, and they arranged their own distribution across the country.[141] They followed the business methods of many market leaders and tradesmen owners with widespread advertising in newspapers and handbills, obtaining supporting testimonials, publishing self-justifying treatises and seeking a national market. So they imitated many of the practices of other owners within the patent medicines industry, though they practised medicine as well. The prominent irregular owners straddle the boundary between patent medicines as a distinct industry and quackery.

Local Owners

The last category of proprietor to be discussed, and the one we know least about, comprised the small-scale owners who usually only sold one or two medicines in their immediate locality. Most remain obscure and so this section will be short without a case study. Their names were revealed in newspaper advertisements with some brief details, sometimes claiming to be a medical practitioner, but little else can be discovered about most of them. Some thirty four medicines with a named provincial owner and apparently only a local distribution can be found in the list of medicines in Appendix A.2. No further details were available on thirteen of these local owners, while eleven claimed to be a chemist or druggist, nine a surgeon or surgeon-apothecary, and one just an unspecified doctor. The absence of any career details on nearly all the claimed medical practitioners suggests that many were irregulars; regulars often provided evidence to justify their status.

A few of these local owners emerge as clearer individuals with the help of the advertisements or other contemporary evidence. For example, an event such as a family dispute might allow more to be discerned about an owner, such as Amelia Ings, who claimed to have been selling Foot's Cathartic Mixture in Wiltshire, for nineteen years, having been instructed in its preparation by her grandfather Henry Foot.[142] A week later, Mary Foot asserted that the only correct version of this medicine had come to her from another member of the family.[143] Mr. West, a salt proprietor at Lymington, is another example of a small local owner. His new crystallisation method had improved the safety of Epsom salts and his Lymington Marine Epsom Salts were now available from his premises, with agents to be appointed in the near future.[144] By the following week, he had engaged a glass merchant in Salisbury and an unspecified tradesman in Romsey as agents.[145]

Some medicines developed by local owners eventually achieved a wider distribution, with the name of the original locality providing a degree of branding. By 1831, the Lymington Epsom Salts were being made by T. K. Welsh in Lymington, but they were also being sold by John Sanger in Oxford Street, London, and distributed across the country by Barclay and Sons.[146] The Trowbridge Pills for stone and gravel, and for female disorders, provide another example. They had been invented by Jane Hanney of Trowbridge in the early eighteenth century, and were then produced and sold in Wiltshire by Jane, her daughter and grand-daughter for the rest of the century.[147] By the beginning of the next century, they were being distributed across the country by Howard and Evans in London.[148] In 1826, her granddaughter Jane Wynne sold the Pills out of the family to Joshua Vardy, a Warminster chemist and printer, and they were still available in the 1850s.[149]

It is tempting to regard these local owners as the commercial successors to the local herbalists of earlier years. Certainly, the passing of the recipe through two or more generations of the same family, as we have seen, was a common claim. Other illustrations of local owners inheriting medicines would be Joseph Wright, a miller at Wortley Windmill near

Leeds, whose Medicine for the Bite of a Mad Dog had been used successfully by his parents at Colne for fifty years, and Mrs. Walter's Recipe for Pulmonary Complaints which had been prepared by Mrs. Walters for over fifty years and was now being made by her niece, Miss Hall.[150] However, in the absence of almost any details of the medicines' recipes, a direct line back to the herbalists of earlier generations cannot be traced.

The local owners included a wide variety of both men and women who wished to make money by selling their medicines in their area without committing to the expense of a distribution network or widespread advertising. For many, it was probably only a supplement to their normal employment, which might include medical practice. Their diversity and the shortage of records mean that the group is difficult to summarise. Although their production of medicines may be broadly similar to the other categories of owners, the potentially sporadic nature of their work and the lack of national distribution mean that they do not fit easily into the concept of an industry. Some of the local owners would have been regarded as established tradesmen or regular practitioners, but many were probably irregulars. The true place of these local owners in the medical market remains unclear.

Frequency of Ownership Groups

The full picture of medicine ownership as a component of a patent medicines industry requires some idea of the numerical importance of these groups. If the first three groups, which were suggestive of an organised, respectable trade, were more prominent than the last two, which carried connotations of quackery, itinerancy, and short-term gain, then the concept of an established industry gains substance. Numerical analysis of all the medicines owners is impossible, but we can study the proprietors of the widely distributed 'national' medicines (Appendix A.1), and also the medicine patentees. The finding is that the market leaders, tradesmen and medical professionals combined were commoner in these two sources than the irregular practitioners, and so much of the ownership of patent medicines was distinct from irregular medicine and quackery.

The 'national' medicines, that is the medicines that were clearly promoted and distributed for a wide market in the studied newspapers, are defined and listed in Appendix A.1. The names of the owners were available for 115 of them, and some occupational information could often be found about these proprietors in the advertisements themselves, the

Oxford Dictionary of National Biography, the online *World Biographical Index System*, and other sources including James Adair. Assignment to an ownership group was often based on a strong probability rather than certainty. When the ownership changed over the years, the first owner with any biographical information was selected. Some owners could not be assigned, either due to a lack of information on whether a practitioner was regular or irregular, or just due to a shortage of any information, and they have been placed in separate categories (Table 3.2). When the owners could be assigned to one of my six groups, 83% were in the first three groups. Even if we make the unlikely assumption that *all* the owners who could not be assigned were irregulars, the first three groups would still be a clear majority with 63% of the total. So most of the national medicines were owned by one of the large medicines businesses, tradesmen or medical professionals.

The medicine patentees did not necessarily remain the owner of their medicines, and they differed from the general run of medicine owners by including fewer regular practitioners. Nevertheless, when we look at medicine patentees between 1740 and 1830 (Table 3.3), the results are broadly similar to those from the owners of the advertised national medicines. All but two of the patentees included their occupation in the application, but the occupation could be vague, such as 'gent', with insufficient information to be assigned to a group. For others, it is again unclear whether they were regulars or irregulars. The main finding is that 77% of the assignable patentees were in the first three groups. Once

Table 3.2 Types of owners of 115 national medicines (% of assignable owners)

Owner group	*Number*	*%*	*Comment*
Market leaders	15	*17*	Market leaders all owned several medicines
Tradesmen	31	*36*	
Professionals	26	*30*	18 apothecaries
Elite	0		Did not advertise in newspapers
Irregulars	15	*17*	
Local	0		By definition, local only
Unassignable			
Professional or irregular?	11		
Insufficient information	17		

Table 3.3 Classification of the medicine patentees, 1740–1830 (Appendix B), (% of assignable patentees)

Owner group	*Number*	*%*	*Comment*
Market leaders	2	*3*	
Tradesmen	37	*53*	
Professionals	15	*21*	
Elite	0		Unlikely to seek a patent
Irregulars	16	*23*	
Local	0		Not aiming for widespread sale
Unassignable			
Professional or irregular?	20		
Insufficient information	19		

again, if we make the improbable assumption that *all* the unassignable patentees were irregulars, the first three groups would still contain 50% of the patentees. This lower percentage in comparison to the national medicines seems to be due to the market leaders rarely patenting medicines: for example, neither Francis Newbery nor his father patented a medicine themselves. Allowing for this, the numbers of owners in each group was comparable amongst the different series of proprietors/inventors. These analyses confirm that the market leaders, tradesmen or medical professionals owned and produced a clear majority of the most popular medicines.

Origins of Patent Medicines

The origins of most patent medicines were as secret as their recipes. For promotional purposes, the owners might provide some vague information, but the truth of any such assertions, made to promote branding, is uncertain. Nevertheless, we can use these reports to get an impression of the *possible* sources of the medicines; at least they present the type of origins which would have been plausible to contemporaries. The examples of ownership already mentioned in this chapter, together with the recorded origins of other medicines, indicate that patent medicines came from a wide variety of conceivable sources. Both contemporary writers and recent historians have suggested that patent medicines were essentially copies of regular medicines.[151] This may be true for some, but it is an oversimplification; they were certainly based on the same medical principles and had similar constituents, but the immediate origins of patent medicines were probably outside the regular pharmacopoeia.

Some owners, particularly those in the professionals group, devised their own medicines or introduced medicines from elsewhere. As we have seen in this chapter, the physician Robert Priestley and the apothecaries Thomas Henry, Edward Galliard and John Hill created their therapies, and in the case of Henry and Galliard provided some detail on how this was done.[152] Some tradesmen owners also created their own medicines, such as Nathaniel Godbold with his Vegetable Balsam, and William Jones who apparently used his skills as a chemist to create his Tincture of Peruvian Bark. Other owners claimed to have acquired existing inventions, such as Francis Spilsbury who wrote that his medicine had come from an unnamed chemist via a third party, and James Coghlan who apparently used extant recipes from the Jesuits' Library and other Catholic sources.[153]

The recipes for medicines could be bought, or dishonestly obtained, from existing owners. As already described, Peter Delamotte paid the enormous sum of £1500 to Samuel Glass for his recipe for magnesia. William Brodum was accused of copying medicines belonging to other irregular practitioners.[154] Amongst Edward Harrison's correspondents, an anonymous Suffolk physician described two examples of a good recipe being passed on. The legitimate transmission was:

> A poor woman some years ago sold a bookseller here a receipt for a pill for £5. He compounded it and advertised it with great success for several years, then sold it to a druggist of the same place for a high price, who now vends it with increased reputation and sale.[155]

A less honest transfer was carried out by a 'lad of all works' for an apothecary who helped in the compounding of a secret medicine and then opened up his own druggist shop across the street, apparently making £400–500 per year from producing the same medicine for a different indication.[156]

These owners were selling medicines which had been recently created, either by themselves or by others; but other medicines had been around for a long time as herbal or patent medicines, and their current ownership could be disputed. Elizabeth Shackleton's medicine for the bite of a mad dog was one of several for this condition which had been derived from a long-standing local treatment in Ormskirk, Lancashire, and, as we have seen, the Wiltshire origins of Foot's Cathartic Mixture were controversial. In the second half of the eighteenth century, the Dicey family, one of the market leaders whom we will discuss in more detail in the next chapter, distributed a number of old medicines whose ownership was unclear. These medicines, which included versions of Daffy's Elixir, Anderson's Scots Pills, and Bateman's Pectoral Drops, had been sold as patent medicines for many decades with their ownership becoming hazy. Anderson's Scots Pills had been available since the early seventeenth century and Daffy's Elixir had been on the market since the 1670s.[157] Bateman's Pectoral Drops were patented in 1726, and a version was part-owned by John Newbery in 1761.[158] Rather than creating new medicines, the Diceys were using their powerful marketing and distribution skills to capitalise on old ones.

We can deduce that patent medicines could be new or old, original or purchased, skilfully designed or just traditional: but, whatever their origins, they were not identical to regular medicines. They did contain similar ingredients to regular medicines as corroborated by the *Lancet*, which published the 'compositions of quack medicines', 24 of them, in its first four issues.[159] However, many of these medicines were complex mixtures created specifically for the patent medicines industry. For example, Spilsbury's Antiscorbutic Drops allegedly contained 'corrosive sublimate, gentian root, dried orange-peel, of each two drachms; crude antimony, red saunders, of each one drachm; rectified spirits of wine, water, of each eight ounces'.[160] The industry needed distinctive products to capture its share of the medical market.

Conclusion

The overall verdict is clear: most of the ownership and production of patent medicines was part of a substantial and respectable industry which was largely distinct from quackery and irregular practice, and which followed its own methods. Some exceptions to this finding are evident: the production of patent medicines by some irregular practitioners and by local owners produced an overlap with irregular medical practice, and regular practitioners could also claim, not always convincingly, that their medicines were solely an adjunct to their orthodox practice and not a separate business. Nevertheless, most of the ownership and production was a component of a separate industry, and it is worth summarising the arguments in this chapter, and the previous one, which lead to this novel conclusion.

First, the medicines were derived and utilised specifically for the industry. Up to the 1820s, they were based on similar theories, and contained similar ingredients, to the regular medicines of the time; but they were not copies. Their sources were varied, and they had been created to be supplied as patent medicines for the medical market. In addition, by the standards of the time, the majority were each targeted at a relatively small range of conditions as defined in Buchan's arrangement of diseases. We have to be careful about anachronism with this last finding as today's belief in the virtues of specific therapies for specific conditions was often not exhibited by eighteenth-century physicians, and we must not assume that the relative specificity of many patent medicines was some sort of anticipation of future therapeutic developments. Orthodox physic then saw every patient's condition as unique; therefore specific therapy for a disease was impossible and the physician had to select the best treatment for that patient from a broad range.[161] Many regular practitioners of the time would not have regarded therapy targeted at a limited range of conditions as achievable or necessary: the knowledge and experience of the practitioner was the key to successful therapy. Aiming patent medicines at a limited range of conditions was more a practical issue. Unlike regular therapy, patent medicines were designed to be taken without advice from practitioners, perhaps without any advice at all, and they had to have a manageable number of indications which were understandable by the consumers. So the relative specificity of patent medicines was a response to the demands of the market, not an attempt to imitate or lead prevailing medical orthodoxy. Nevertheless, regular therapy was also becoming more specific for particular conditions in this period, and this raises an interesting question. Could this developing selectivity of regular therapy be partly a response to the commercial pressures in the medical market from the more focussed patent medicines? The common assumption that regular medicines became more targeted secondary to new medical understanding may be an oversimplification for this period: the success of patent medicines could have been changing regular therapy ahead of advances in pharmaceutical knowledge.

Second, most of the owners were respectable and followed similar practices. A considerable amount of evidence has been presented in this chapter to demonstrate that most owners conformed to the normal commercial standards of the time and maintained, or even enhanced, their position in society. The proprietors in the groups of market leaders, tradesmen, professionals and elite owners, which included the majority of owners, suffered no apparent embarrassment from their activities and were accepted into their social class, though the medical professionals had to be careful on how they conducted their business. In addition, members of these groups followed similar practices such as preserving their property by secrecy, following the conventions of regular medicine where possible, and promoting their medicines by branding and advertising. Owners in the irregular and local groups were less likely to be considered as respectable, but they still mostly followed the commercial practices of the owners in the other groups.

The third underpinning of the concept of the patent medicines industry was the stability of many of the owners and medicines. Of course, several owners and medicines came and went in a short time, and two medicines already mentioned provide examples of this. The Edinburgh Febrifuge Powder was very popular in the years 1772–1774 before going rapidly out of favour, and Priestley's Anti-Bilious Powder seems to have been widely available only in 1798–1799.[162] In contrast, this chapter has described several owners (such as Francis Newbery, Francis Spilsbury, Nathaniel Godbold, Thomas Henry, John Hill, Elizabeth Shackleton and Jane Hanney) who not only produced the medicines themselves for several decades, but also passed their business on to a widow, son or daughter. Further, some medicines, such as Anderson's Scots Pills, Daffy's Elixir, Bateman's Pectoral Drops and Dr. James's Fever Powder were popular for over a century. This combination of respectable owners operating from fixed publicised premises for long periods of time, and standard products being purchased for many years, confirms that the ownership and production of patent medicines was a genuine industry. We are already a long way from Porter's conception of patent medicines as the observable manifestation of quackery.

The owners required revenue from extensive sales across the country, and the next stage in the exploration of the industry is the investigation of medicine wholesaling and distribution. The necessity to send patent medicines from a single source to the whole of England created a pioneering national market for these branded goods.

Notes

1. Gardner, *Business of News*, 18: Wiles, *Freshest Advices*, 373.
2. Mackintosh, 'Authority'.
3. Daunton, *Progress*, 299–305; Gerhold, 'Stage Coaching'.
4. Corley, 'Ward, Joshua'; Nicolson, 'Ward's Pill'.
5. Ward, *Letter*.
6. 'Company of Undertakers'.
7. Ward, *Letter*; Clutton, *Relation*.
8. Clutton, *Relation*, 55.
9. Nicolson, 'Ward's Pill'.
10. Page, *Receipts*, 28. Seven of Ward's medicines are listed in Appendix A.2.
11. Rousseau, *Letters*; O'Connor, 'Hill, Sir John'.
12. Rousseau, *Letters*, xxxiii.
13. Welsh, *Bookseller*, 18, 36 & 64–65.
14. National Archives, Will of John Newbery.
15. *ABG*, 30 January 1769.
16. See Appendix B.
17. Adair, *Essays*, 180–190.
18. Welsh, *Bookseller*, 120–134.
19. National Archives, Will of John Newbery.
20. Welsh, *Bookseller*, 135.
21. Roscoe, *John Newbery*, 18.
22. For a 1910 photograph of the imposing 'Newbery Business House' in Charterhouse Square, London, see Newbery, *Records*, 80.
23. Newbery, *Records*, 47.
24. Maxted, 'Newbery, Francis'.
25. Public Catalogue Foundation, National Trust.
26. Roscoe, *John Newbery*, 17 (footnote).
27. Messrs. Newbery', 115.
28. For example, Dr. James's Analeptic Pills, *LI*, 5 June 1781.
29. Watson, 'Trading Accounts', 56.
30. Turlington's Balsam of Life, *ABG*, 1 January 1781.
31. Spilsbury, *Discursory Thoughts*, 52.
32. For an exposition of the diversity of trades, see White, *London*, 212–215.
33. Grimwade, *Goldsmiths*, 667; *GM*, 63 (1793), 773. 'Goldsmith' included tradesmen, like Spilsbury, who largely worked with silver.
34. Grimwade, *Goldsmiths*, 667.
35. National Archives, Will of Francis Spilsbury.
36. *LI*, 26 January 1807; *Lancet*, 1 (1823), 30.
37. National Archives, Will of Francis Spilsbury.
38. For example, 'Prepared at his Dispensary, Mount Row, Westminster Bridge, Surrey', *LM*, 27 March 1781.

39. Trial of Philip Gibson (1795).
40. Spilsbury, *Discursory Thoughts*, 23.
41. Spilsbury, *Free Thoughts on Quacks*, 3–4 (footnote).
42. Francis Spilsbury, *Free Thoughts on Scurvy*; Prosser, *Oeconomy*, 15.
43. Spilsbury, *Free Thoughts on Quacks*, 85.
44. Prosser, *Oeconomy*, 65 (footnote).
45. *GM*, 91 (1821), 490.
46. Rawlings, 'Proprietary Medicines', 7; *GM*, 91 (1821), 598.
47. *GM*, 91 (1821), 598. The large plaque still occupies a prominent position at the east end of the church.
48. *GM*, 70 (1800), 84.
49. *SWJ*, 4 February 1822.
50. *ABG*, 24 March 1794.
51. *ABG*, 12 January 1807.
52. Chapman, *Birmingham Directory*, 93. Most entries in this directory were only online.
53. Watson, 'Trading Accounts', 48, 53 & 68.
54. Watson, 'Trading Accounts', 58 & 75.
55. Greenaway, 'Henry, Thomas'; Kirkby, 'Thomas Henry'.
56. Jones, *Enlightenment*, 17.
57. Handbill for Glass's Magnesia Alba, c. 1764, John Johnson Collection, Patent Medicines 3 (34); Goodwin, 'Glass, Thomas'.
58. *Medical Transactions*, 2 (1772), 226–234.
59. Henry, *Account*, 8.
60. Delamotte, *Refutation*, 12.
61. Henry, *Preparation*, 7.
62. Calcined Magnesia was still being advertised in the newspapers in the 1820s, for example *ABG*, 4 February 1822.
63. Henry, *Preparation*, Appendix, 8
64. Henry, *Preparation*, Appendix, 3.
65. Henry, *Preparation*, Appendix, 3–8.
66. Delamotte, *Refutation*, 15–22.
67. Delamotte, *Refutation*, 8.
68. Glass, *Examination*.
69. Glass, *Examination*, 26. Original italics. Note that 'ingenuous' means honourable and straightforward, not ingenious.
70. Henry, *Letter*.
71. Henry, *Letter*, 16. Original italics. 'Ingenious' probably conveys the eighteenth-century meaning of possessing a high intellect.
72. Henry, *Letter*, 23.
73. Walker, *Memoirs*, 87.
74. *Medical Register for the year 1783*.

75. *LI*, 5 November 1798; Priestley, *Remarks.*
76. Priestley, *Remarks*, vi.
77. Galliard, *Antimonial Medicines*, 36.
78. The London wholesaler was the first John Murray.
79. Galliard, *Antimonial Medicines*, 38.
80. Adair, *Essays*, 187.
81. Rousseau, *Letters*, 186.
82. Adair, *Essays*, 188.
83. Vickery, *Daughter*, 153–55.
84. *LI*, 13 August 1765.
85. *GM*, 23 (1753), 368.
86. Vickery, *Daughter*, 147–153.
87. Vickery, *Daughter*, 152.
88. Hunter, 'William Hill'.
89. Berkenhout, *Bite*, 51.
90. *Swinney's Birmingham Chronicle*, 21 February 1793.
91. Vickery, *Daughter*, 154.
92. Pollock, *Faith.*
93. Day, 'Household Management', 210–212.
94. Vickery, *Daughter*, 155.
95. Vickery, *Daughter*, 20.
96. Porter, *English Society*, 66.
97. Lancashire Records Office (LRO), Diary of Elizabeth Shackleton, DDB/81/11.
98. LRO, DDB/81/27.
99. LRO, Correspondence of James Coghlan, RCBu/14.
100. Blom et al., *Correspondence.*
101. Steenbrink, 'Coghlan, James Peter'.
102. Blom et al., *Correspondence*, xiv–xvi.
103. All five medicines were in the 1788 edition of the *Laity's Directory.*
104. Coghlan, *Publications*, 12.
105. LRO, RCBu/14/97.
106. Blom et al., *Correspondence*, xxxii.
107. O'Gorman, *Eighteenth Century*, 312.
108. Blom et al., *Correspondence*, xxxiii.
109. Halloran, 'Hay, George'.
110. LRO, RCBu/14/92.
111. LRO, RCBu/14/84.
112. LRO, RCBu/14/94.
113. LRO, RCBu/14/70.
114. LRO, RCBu/14/84.
115. Rogers, Nicholas, *Catholics*, 59.

116. LRO, RCBu/14/135.
117. Spilsbury, *Free Thoughts on Quacks*, 126; *GM*, 70 (1800), 84.
118. Vaisey, *Thomas Turner*, 208.
119. Wagner, 'Satire', 204.
120. Adair, *Essays*, 185; Appendix B.
121. Adair, *Essays*, 183.
122. *MCR*, 13 (1806), lxxvii & lxxiv. Brodum was itinerant, but not a mountebank.
123. *House of Commons Journal*, 41 (1798), 589; Edgar Samuel, 'Brodum, William'; Ietros, 'Quacks'; 'Anthony Daffy Swinton', 34.
124. *MPJ*, 13 (1805), 266.
125. 'Dr Brodum, Oculist'.
126. *LM*, 11 January 1794.
127. Ietros, 'Quacks', 75.
128. Appendices A & B.
129. *LM*, 8 February 1794.
130. Brodum, *Guide*.
131. Medicus, 'Brodum's Intrigues', 492.
132. 'Lecture 3', 22; Denizen, 'Solomon' 289; 'Historical and Practical Treatise'.
133. 'Sketch of the Foley-House Masquerade'.
134. Corley, 'Solomon, Samuel'.
135. Denizen, 'Solomon', 289.
136. Solomon, *Guide*, ix and xii.
137. Picton, *Memorials*, vol. 2, 223.
138. Slugg, *Reminiscences*, 57. Wood/Lignum probably thought that the Latin version of his name was more erudite.
139. *LI*, 1 July and 22 July 1793.
140. Medicus, 'Brodum's Intrigues', 491; 'Anthony Daffy Swinton', 36.
141. Brodum, Solomon and Lignum all sent medicines to John Ware, printer of the weekly *Cumberland Pacquet* in Whitehaven at the far end of England (Cumbria Archives, Day Books of John Ware).
142. *SWJ*, 3 June 1822.
143. *SWJ*, 10 June 1822.
144. *SWJ*, 20 May 1822.
145. *SWJ*, 27 May 1822.
146. *Hampshire Advertiser*, 25 June 1831.
147. Rogers, K. H., *Medical Matters*, 37.
148. *Hull Packet*, 17 April 1804.
149. Rogers, K. H., *Medical Matters*, 38.
150. *LM*, 4 May 1793; *SWJ*, 8 April 1822.
151. Adair, *Essays*, 193; Cody, 'No Cure', 106; Helfand, *Quack*, 32.

152. Galliard, *Antimonial Medicines*, 36; Henry, *Preparation*, 5–7.
153. Spilsbury, *Free Thoughts on Quacks*, 111; Coghlan, *Publications*, 12.
154. 'Anthony Daffy Swinton', 34.
155. *MCR*, 13 (1806), xxxvi.
156. *MCR*, 13 (1806), xxxvi.
157. Rawlings, 'Proprietary Medicines', 5; Hancock, and Wallis, 'Quacking', 3.
158. Welsh, *Bookseller*, 64.
159. *Lancet*, 1 (1823), 30, 62, 89, 138.
160. *Lancet*, 1 (1823), 30. 'Corrosive sublimate' was mercuric chloride and 'red saunders' was sandalwood.
161. For a contemporary attack on specific therapies, see Forbes, 'Origins'.
162. Zachs, *John Murray*, 48.

Bibliography

Archives

Cumbria Archives, Whitehaven, Day Books of John Ware, DA276A.
John Johnson Collection, Bodleian Library, Oxford.
Lancashire Records Office (LRO), Preston: Diary of Elizabeth Shackleton, DDB/81; Correspondence of James Coghlan, RCBu/14.
National Archives, Kew: Will of John Newbery, Prob. 11/935; Will of Francis Spilsbury, Prob. 11/1236.

Newspapers and Journals

Aris's Birmingham Gazette (ABG).
Gentleman's Magazine (GM).
Hampshire Advertiser.
House of Commons Journal.
Hull Packet.
Lancet.
Leeds Intelligencer (LI).
Leeds Mercury (LM).
Medical and Chirurgical Review (MCR).
Medical and Physical Journal (MPJ).
Medical Transactions.
Salisbury and Winchester Journal (SWJ).
Swinney's Birmingham Chronicle.

Printed or Painted Primary Sources

Adair, James, *Essays on Fashionable Diseases* (London: T. P. Bateman, 1790?).

'Anthony Daffy Swinton', *The Scourge*, 1 (1811), 27–46.

Barfoot, Peter, and John Wilkes, *Universal British Directory, 1793–1798* (London, 1793–1798).

Berkenhout, John, *An Essay on the Bite of a Mad Dog* (London: R. Baldwin, 1783).

Brodum, William, *Guide to Old Age, or a Cure for the Indiscretions of Youth* (London, 1795).

Chapman, Thomas, *Chapman's Birmingham Directory* (Birmingham, 1801).

Clutton, Joseph, *Relation of the Good and Bad Effects of Joshua Ward's Pill and Drops* (London, 1736).

Coghlan, J. P., *New Publications* (London: Coghlan, 1787).

'Company of Undertakers', British Museum, 1868,0822.1542.

Delamotte, Peter, *Refutation of Mr Henry's Strictures on Glass's Magnesia* (London, 1774).

Denizen of Liverpool, 'Samuel Solomon M. D.', *The Scourge*, 2 (1811), 287–303.

'Dr Brodum, Oculist, and Herbalist', 1790, British Library, Eighteenth Century Collections. http://0-find.galegroup.com (accessed 5 June 2016).

Forbes, Duncan, 'On the Origins and Progress of Empiricism', *MPJ*, 15 (1806), 362–370.

Galliard, Edward, *The Use and Abuse of Antimonial Medicines* (London: John Murray, 1773).

Glass, Thomas, *An Examination of Mr Henry's Strictures on Glass's Magnesia* (London, 1774).

Henry, Thomas, *The Preparation, Calcination, and Medicinal Uses of Magnesia Alba* (London: Joseph Johnson, 1773).

Henry, Thomas, *A Letter to Dr Glass, Containing a Reply to His Examination of Mr. Henry's Strictures* (London: Joseph Johnson, 1774).

Henry, Thomas, *An Account of the Medicinal Virtues of Magnesia Alba, More Particularly of Calcined Magnesia* (London: J. Johnson, 1775).

'Historical and Practical Treatise on the Venereal Disease', *Critical Review*, 20 (1797).

Ietros, 'Of Quacks and Empiricism', *MPJ*, 13 (1805), 66–75.

Laity's Directory (London: Coghlan, 1788).

'Lecture 3', *The Ghost*, 6 (1796), 21–24.

Medical Register for the year 1783 (London: Joseph Johnson, 1783).

Medicus, 'Dr Brodum's Intrigues with the College of Physicians', *The Scourge*, 2 (1811), 491–493.

Page, John, *Receipts for Preparing and Compounding the Principal Medicines Made Use of by the Late Mr Ward* (London, 1763).

Priestley, Robert, *A Few Interesting Remarks on Bilious Disorders and Yellow Fever* (Leeds: Thomas Gill, 1798).

Prosser, Thomas, *The Oeconomy of Quackery Considered* (London: J. Bew, 1777).

Public Catalogue Foundation, National Trust.

'Sketch of the Foley-House Masquerade', *Weekly Entertainer*, 39 (1802), 451–453.

Solomon, Samuel, *A Guide to Health*, 52nd edn (London, ?).

Spilsbury, Francis, *Free Thoughts on Quacks and Their Medicines* (London, 1776).

Spilsbury, Francis, *Free Thoughts on the Scurvy, Gout, Diet and Remedy* (Rochester, 1783).

Spilsbury, Francis, *Discursory Thoughts*, 2nd edn (London, 1785).

Trial of Philip Gibson, 1795, *Old Bailey Proceedings Online*, www.oldbaileyonline.org (accessed 7 April 2013).

Walker, William, *Memoirs of the Distinguished Men of Science of Great Britain: Living in the Years 1807–8* (London: W. Walker & Son, 1862).

Ward, William, *A Letter Sent to Mr Jos. Clutton* (London, 1736).

Secondary Sources

Blom, Frans, Jos Blom, Frans Korsten, and Geoffrey Scott, eds., *The Correspondence of James Peter Coghlan (1731–1800)* (Woodbridge, Suffolk: Boydell Press, 2007).

Cody, Lisa Forman, '"No Cure, No Money," or the Invisible Hand of Quackery: The Language of Commerce, Credit, and Cash in Eighteenth-Century British Medical Advertisements', *Studies in Eighteenth-Century Culture*, 28 (1999): 103–130.

Corley, T. A. B., 'Solomon, Samuel', *ODNB* (accessed 2 May 2012).

Corley, T. A. B., 'Ward, Joshua', *ODNB* (accessed 9 April 2016).

Daunton, M. J., *Progress and Poverty: An Economic and Social History of Britain 1700–1850* (Oxford: Oxford University Press, 1995).

Day, Julie, 'Elite Woman's Household Management: Yorkshire 1680–1810' (PhD thesis, University of Leeds, 2007).

Gardner, Victoria E. M., *The Business of News in England, 1760–1820* (Basingstoke: Palgrave Macmillan, 2016).

Gerhold, Dorian, 'The Development of Stage Coaching and the Impact of Turnpike Roads, 1653–1840', *Economic History Review*, 67 (2014): 818–845.

Goodwin, Gordon, 'Glass, Thomas', rev. Alick Cameron, *ODNB* (accessed 10 January 2013).

Greenaway, Frank,'Henry, Thomas', *ODNB* (accessed 10 January 2013).

Grimwade, A. G., *London Goldsmiths 1697–1837*, 3rd edn (London: Faber, 1990).
Halloran, Brian M. 'Hay, George', *ODNB* (accessed 3 November 2012).
Hancock, David, and Patrick Wallis, 'Quacking and Commerce in Seventeenth-Century London: The Proprietary Medicine Business of Anthony Daffy', *Medical History*, Supplement 25 (2005).
Helfand, William, *Quack, Quack, Quack: The Sellers of Nostrums in Prints, Posters, Ephemera and Books* (New York: Grolier Club, 2002).
Hunter, W. R., 'William Hill and the Ormskirk Medicine', *Medical History*, 12 (1968): 294–297.
Jones, Peter, *Industrial Enlightenment: Science, Technology and Culture in Birmingham and the West Midlands, 1760–1820* (Manchester: Manchester University Press, 2008).
Kirkby, William, 'Thomas Henry FRS: The Inventor of Calcined Magnesia', *Chemist and Druggist*, 121 (1934): 674–675.
Mackintosh, Alan, 'Authority and Ownership: The Growth and Wilting of Medicine Patenting in Georgian England', *British Journal for the History of Science*, 49 (2016): 541–559.
Maxted, Ian, 'Newbery, Francis', *ODNB* (accessed 10 April 2012).
'Messrs. Newbery', *Chemist and Druggist*, 15 (1874): 112–116.
Newbery, Arthur Le Blanc, *Records of the House of Newbery* (Derby: Bemrose, 1911).
Nicolson, Marjorie, 'Ward's "Pill and Drop" and Men of Letters', *Journal of the History of Ideas*, 29 (1968): 177–196.
O'Connor, Barry, 'Hill, Sir John', *ODNB* (accessed 8 March 2013).
O'Gorman, Frank, *The Long Eighteenth Century: British Political and Social History 1688–1832* (London: Hodder Arnold, 1997).
Picton, J. A., *Memorials of Liverpool, 2 Vols.*, 2nd edn (London: Longmans, Green and Co., 1875).
Pollock, Linda, *With Faith and Physic: The Life of a Tudor Gentlewoman Lady Grace Mildmay 1552–1620* (London: Collins & Brown, 1993).
Porter, Roy, *English Society in the Eighteenth Century* (London: Penguin Books, 1991).
Rawlings, F. H., 'Old Proprietary Medicines', *Pharmaceutical Historian*, 26 (1996): 4–8.
Rogers, K. H., *Medical Matters in Trowbridge*, (Trowbridge, Wiltshire: Friends of the Trowbridge Museum, 2002).
Rogers, Nicholas, *Catholics in Cambridge* (Cambridge: Gracewing, 2003).
Roscoe, S., *John Newbery and His Successors: A Bibliography* (Wormley, Herts: Five Owls Press, 1973).
Rousseau, G. S., ed., *The Letters and Papers of Sir John Hill, 1714–1775* (New York: AMS Press, 1982).

Samuel, Edgar, 'Brodum, William', *ODNB* (accessed 2 May 2012).
Slugg, J. T., *Reminiscences of Manchester Fifty Years Ago* (Manchester: J. Cornish, 1881).
Steenbrink, J. L. C., 'Coghlan, James Peter', *ODNB* (accessed 3 November 2012).
Vaisey, David, *The Diary of Thomas Turner 1754–1765* (Oxford: Oxford University Press, 1984).
Vickery, Amanda, *The Gentleman's Daughter: Women's Lives in Georgian England* (New Haven: Yale University Press, 1999).
Wagner, Peter, 'The Satire on Doctors in Hogarth's Graphic Books', in *Literature and Medicine During the Eighteenth Century*, edited by Marie Mulvey Roberts and Roy Porter (London: Routledge, 1993), 200–225.
Watson, G. M., 'Some Eighteenth-Century Trading Accounts', in *The Evolution of Pharmacy in Britain*, edited by F. N. L. Poynter (London: Pitman Medical, 1965), 45–77.
Welsh, Charles, *A Bookseller of the Last Century* (London: Griffith, Farran, Okeden & Welsh, 1885).
White, Jerry, *London in the Eighteenth Century: A Great and Monstrous Thing* (London: Vintage, 2013).
Wiles, R. M., *Freshest Advices: Early Provincial Newspapers in England* (Columbus: Ohio State University Press, 1965).
Zachs, William, *The First John Murray and the Late Eighteenth-Century London Book Trade* (Oxford: Oxford University Press, 1998).

CHAPTER 4

Connecting the Country

Nowhere in England was detached from the supply of patent medicines, and the most popular ones were available in every town. But the nature of such a medicine, made from a secret recipe by a uniform technique, required production at a single site which was often the premises of its proprietor. Thus, an efficient national distribution system was needed to transport these valuable products from their source to all parts of England, and sometimes to Scotland, Ireland, Europe and the colonies, and to get the money back.

This chapter will explore the structure of patent medicine wholesaling, particularly by the better documented larger businesses, and will seek to explain that structure. This investigation will not only expose a 'market' aspect of the Georgian medical market, a fairly rare event, but it will also illustrate the available techniques for the wholesaling of other goods in Georgian England. Medicine wholesalers can be found across the country, but medicine wholesaling was dominated by businesses in London. In the eighteenth century, several were publishing booksellers who could utilise their experience in the national distribution of books; but by the beginning of the next century, they had been superseded by medicine specialists and chemists. Arrangements varied, but much of the wholesaling was part of an established industry in that, like medicine ownership, it was stable, orderly and employed its own practices. The key point is that the wholesaling was tightly organised with each popular medicine being distributed by a very small number of wholesalers, usually only

A. Mackintosh, *The Patent Medicines Industry in Georgian England*, Medicine and Biomedical Sciences in Modern History,
https://doi.org/10.1007/978-3-319-69778-9_4

one, to recognised medicine retailers, and with each wholesaler concentrating on a limited range of medicines. This meant that, even more than ownership, wholesaling was distinct from both regular and irregular medical practice which relied on the skills of an individual operating in a single locality rather than commercial cooperation across a much wider area.

Developing these structured national markets required more than available infrastructure and business organisation: strong consumer demand was essential. In his well-known words, Adam Smith emphasised that consumption drove the market, not the other way round: 'Consumption is the sole end of all production: and the interest of the producer ought to be attended to, only so far as it may be necessary for promoting that of the consumer'.[1] As we saw in Chapter 2, the interest of the consumer in this case was a demand for more patent medicines, in terms both of their quantity and the range of products. The widespread distribution and sales of patent medicines were a response to the public seeking out the medicines they thought best for their problems, and one of the objectives of the wholesalers was to encourage that consumer demand, as well as ensuring a good supply.

In this chapter, 'wholesaling' will not be confined to the activities of dedicated self-described wholesalers; it will be used in the more general sense of linking the source of production to the retailers, including both the distribution of the medicine and the encouragement of local consumer demand. Such wholesaling could be carried out by a medicine proprietor himself or by his employed tradesman, by a national wholesaler who might specialise in medicine production and distribution, by a regional agent such as a newspaper printer, or by a larger retailer for distribution to smaller ones. First, an assessment of the different relationships between proprietor and wholesaler is required: Who was in charge? The principal London wholesalers and their stock will then be identified, together with their connections with the publishing booksellers. Why was it that some London booksellers distributed medicines early in our period, but none did so at the end? The third topic to be considered will be the methods of distribution in this pre-railway age: How were the medicines transported, and how did the wholesaler get his money? The wholesaling of patent medicines can then be placed in a wider perspective as the pioneer of some of the long-term developments in the wholesaling of other goods, particularly the switch from 'pushing' goods out from a centre, by using travelling salesmen and other agents in a method comparable to the traditional fairs, to 'pulling' goods out by peripheral demand induced by advertising and other forms of promotion.

These questions cannot be answered by recourse to a few, well-ordered, sources: no organised records from a medicine wholesaler, such as accounts or order books, are available from this period, though some snippets of accounts are available from newspaper printers who advertised or sold medicines. Most of the material for this chapter has been derived from the advertisements in newspapers and from bills, together with catalogues of medicines and other related publications, the day book of a newspaper printer, legal documents, and a surprise contribution from the records of the plates used to print the medicine excise stamps. The spine of the information again comes from the medicine advertisements in the studied newspapers from Leeds, Birmingham and Salisbury during five selected years between 1769 and 1822. Such an approach will miss some wholesalers whose advertisements happened to fall outside these periods, but it is a great deal more comprehensive than earlier accounts based on one retail locality or limited information.[2] As always, the contents of a Georgian advertisement should be analysed cautiously; much was inaccurate and all was selective. However, it is unlikely that an account of a medicine's source, which was inserted to facilitate its supply, would be deliberately misleading.

These advertisements supplied a variable degree of information on the wholesaling arrangements. A few stated clearly the names and addresses of both the proprietor and the wholesaler together with the relationship between them. More commonly, the advertisements recorded the name and address of the wholesaler without specifying his or her connection with the owner. Sometimes, the name and address of a supplier outside the catchment area of the newspaper and often in London were printed above a list of local retailers, without describing this supplier as the wholesaler; but other information in the advertisement, such as the same name printed on the excise stamp, usually confirmed that this supplier was acting as the national wholesaler and providing the medicine to local retailers. Some of these retailers, especially the newspaper printers, might also act as wholesalers for their locality.

Piecing the information together, wholesalers can be identified for most of the medicines in the studied newspapers, and their names and addresses are listed in Appendix A. A wide spectrum appears, which is not unexpected in a field of business unrestrained by legislation, or by the regulations of a guild or company. Nevertheless, we can observe, at the core, strong and stable businesses using standard skills and their capital to distribute medicines across the country, and to produce, for some, substantial profits. In other words, we see an industry at work.

Owners and Wholesalers

The most straightforward pattern of wholesaling was for the proprietors to undertake the whole process themselves. Examples from the last chapter are the market leader Francis Newbery, the tradesman owners Francis Spilsbury and Nathaniel Godbold, the elite owners Elizabeth Shackleton and James Coghlan, and the irregular owners William Brodum, Samuel Solomon and John Lignum. These irregular owners were particularly vigorous wholesalers as they needed to promote themselves as skilled medical practitioners, and to present their medicines as being extensions of their practice. The irregular owners and their medicines had to be closely linked, and indeed Solomon's name was still engraved on the excise stamp for his medicines three years after his death.[3]

As we shall see in the next chapter, provincial newspaper printers and booksellers sometimes provided a local wholesaling service, and wholesaling proprietors could enlist their assistance. One example is John Hill, who for a time used local wholesalers across the country for his range of herbal medicines. In an apparently coordinated series of announcements within medicine advertisements during January 1769, he stated in the *Leeds Intelligencer*: 'Dr Hill has appointed Griffith Wright as agent for the counties of York, Lancaster and Westmoreland'; in *Aris's Birmingham Gazette* that A. Pearson and S. Aris had been appointed wholesale and retail agents for sale in Birmingham and 'places adjacent'; and in the *Salisbury Journal* that 'I have appointed Edward Easton, bookseller, as my sole agent for Salisbury and places adjacent'.[4] Griffith Wright owned and printed the *Intelligencer*, and Pearson and Aris also did so for the *Gazette*, but Easton did not print the *Salisbury Journal*. Hill had probably assembled more formal, controllable wholesaling arrangements for each area rather than the casual provisions organised by the local newspaper printer.

If the proprietor did not want to organise distribution, he or she could use a national or regional wholesaler who commonly supplied a specified range of medicines to the local retailers. Such a wholesaler might have acquired some ownership of the medicine as part of this process, or he might have distributed it by the owner's appointment

Table 4.1 Categories of medicine wholesaling

	Brief description of category	*No. in Bacon's catalogue*
A	Small-scale wholesale and retail sales, probably without owner's specific agreement	83
B	Wholesaling of a long-established medicine of unclaimed or disputed ownership	8
C	Sole wholesaler responsibility for the medicine	8
D	Wholesaler appointed and probably controlled by the owner	9

or agreement.[5] Alternatively, he might just have sold it, wholesale or retail, without specific approval. Various combinations of these roles are apparent, but four broad categories can be discerned and are listed in Table 4.1. A good way to explain and illustrate the four categories is to explore the catalogue, from around 1790, of William Bacon, a prominent medicine wholesaler and retailer in Oxford Street (Table 4.2).[6] It will help us to answer the question: Who was in charge?

Bacon's twenty-page catalogue of 108 medicines shows that he retailed medicines, but that he had also been appointed as a wholesaler 'by special appointment of the inventors and preparers'. It described his wholesaling arrangements, such as 'orders for exportation supplied without stamps, and on the lowest terms' and 'it is requested that orders from the country may contain reference for payment in London'.[7] Bacon was also recorded in the newspapers as a wholesaler for several nationally advertised medicines in the 1790s (Appendix A.1).

For 83 of these medicines Bacon just listed their name and price at the end of the catalogue; in other words, Bacon was not actively promoting them. In this Category A, Bacon seems to have been concentrating on local retail sales, and perhaps fortuitous wholesale orders. All these medicines, which include some we have already met such as Spilsbury's Drops, Godbold's Vegetable Balsam and Dr. James's Fever Powder, were probably owned and normally distributed by others. Thus, Bacon was not seeking to be in charge of these medicines, but was selling them nevertheless.

The remaining 25 medicines all had promotional descriptions of up to one page attached to them, together with prices for different size bottles and other details. Eight of these medicines can be assigned to Category B, the wholesaling of long-established medicines; their fairly brief descriptions were followed merely by a statement of the prices. These were all old familiar medicines, such as castor oil, Dr. Dickinson's Red Drops or Switzerland Arquebusade Water, whose ownership was unknown or unenforceable. They might have been good sellers, but Bacon's control and probably his profits were limited by the ease of imitation.

The wholesaler having sole responsibility for the medicine (Category C) was exemplified by another eight medicines for which Bacon had signed his name on the medicine excise stamp, clearly seeking to be in command of it: 'Mr. Bacon's name is signed by him on every stamp, as a guard against fraud'. The 1785 Medicines Act stated that the responsibility for applying the stamp rested with the first person in the supply chain.[8] In one of his newspaper advertisements of the period, though not in this catalogue, Bacon offered a twenty guineas reward for a conviction for forging his signature on the medicine stamp or on its directions.[9] For this third category, an inventor might be named, but his or her relationship with Bacon was hidden; Bacon was claiming to be in control of a widely sold product.

A wholesaler by appointment (Category D) was represented by the nine medicines whose descriptions stated that Bacon was acting under instructions, such as 'sold, by appointment of the proprietor (T. Williams) by Mr Bacon', 'sold, by special appointment of Lady Hill,' or, more obliquely for Dr. James's Analeptic Pills, 'these Analeptic Pills are had immediately from Mr Newbery, in St Paul's Churchyard'. Bacon might be temporarily in charge, but the proprietors of these medicines in this fourth category were probably in control because they could issue instructions or change their appointment to another wholesaler.

We can find examples of all these categories in other sources. For Category A, a 1770s bill from Francis Newbery shows that he was prepared to provide small-scale retail and wholesale sales direct from his premises, without being unduly concerned about the medicines' origins.[10] For some of the 15 medicines on the bill, he was the sole wholesaler, but five were normally distributed by other wholesalers including Cluer Dicey.[11] Newbery was not being fastidious about wholesaling arrangements for these medicines. Examples of Category B arrangements for the long-standing medicines of uncertain ownership can be found in many newspaper advertisements, especially those from the Dicey family as described later in this chapter. Medicines without a recognised owner which they promoted and distributed included Squire's Elixir, Godfrey's Cordial, French Hungary Water and British Flour of Mustard.

The many proprietors who distributed their own medicines fit into Category C, control by the wholesaler; and so too do the wholesalers who acquired ownership of medicines which they had not introduced. An example of the latter was Francis Newbery, who in 1781 was only a joint wholesaler of Dr. Steers's Opodeldoc, still owned by the Steers family; but by 1807 he was advertising it as the sole proprietor and wholesaler, with his name printed on the excise stamp.[12] Another illustration was provided by Shaw and Edwards (Table 4.2) who announced in 1807 that they had acquired the ownership of the long-standing Dr. Walker's Jesuits' Drops from Joseph Wessells.[13] Examples of wholesalers being appointed by the medicine proprietor, Category D, were also common in the advertisements, one such proprietor being the 'Rev. Mr' J. Jones who carefully explained in the 1769 *Leeds Intelligencer* that his British Herb Snuff had been available from Mr. Rowley at St. Paul's Coffee House for the past four years, but was now to be had from Evans, gold-beater, in Long Acre.[14]

We can conclude that, not unexpectedly, there were different types of relationships between the owners and wholesalers of patent medicines in late Georgian England. The wholesaler could be the proprietor, an appointee, or just somebody taking advantage of a gap in the market where ownership was uncertain. Amongst these variable relationships, the important finding, as we shall see throughout this chapter, is that the widely disseminated medicines had a defined and publicised wholesaler who undertook the bulk of the distribution of that medicine. Other methods, such as salesmen travelling around the country to provide local retailers with whatever medicines the salesman chose to carry, or owners selling their medicines at fairs and markets for local distribution by others, may have occurred, but clear descriptions of them have not been found. The relationship between the owners and the wholesalers varied, but specific arrangements were made for the popular patent medicines; it was nationally organised, not a free-for-all.

London Medicine Wholesalers

Who were the principal wholesalers? The newspaper advertisements and other material indicate that they were confined to London. Provincial proprietors, such as Solomon and Lignum, did distribute their own named medicines across the country, sometimes several of them, and newspaper printers and others could act as local or regional wholesalers, but a provincial wholesaler who disseminated nationally medicines which he had not created has yet to be found. The principal London medicine wholesalers can be identified from the newspaper advertisements in the studied periods and from other sources: their names and premises are recorded in Table 4.2.

Before exploring these wholesalers, we need to avoid confusion between the principal wholesalers and the more numerous owners of medicine warehouses and medicine merchants. In this period, 'warehouse' was often used as a dignified alternative to 'shop', without any implication that the premises were used for wholesaling.[15] A wholesaler might own a warehouse or be described as a medicine merchant in addition to his national wholesaling activities, but the majority of the medicine warehouse owners and medicine merchants in London were not significantly involved in national wholesaling. Thus, thirteen tradesmen were listed as running medicine warehouses or as medicine merchants in *Wakefield's London Directory* (1794)[16]; but only four out of these can be identified as principal wholesalers in 1794, and the other nine were probably concentrating on retail sales and perhaps more local wholesaling. Another London wholesaler, John Wye, was mentioned in the advertisements but not in the directory, probably because he had only just ceased to be a partner in Dicey and Co.

Two features are immediately striking about the wholesalers listed in Table 4.2. One is their stability and continuity over time. As we shall see, the businesses mostly continued by inheritance and taking on new partners, not by purchase by outsiders or haphazard change, and this must reflect profitability. Once a business was established, it was worth continuing with it and leaving it to a son if possible, and the family businesses of the Newberys, the Barclays and the Sangers were still active in the twentieth century.[17] The other striking feature is their geographical proximity. Seven of the eight premises were in the City with only 150 Oxford Street outside, and each of these seven was no more than a short walk away from the others. Indeed, four addresses at St. Paul's Churchyard, Bow Churchyard and Cheapside were all clustered near St. Paul's Cathedral, the most popular area in London for eighteenth-century booksellers and printers.[18] This geographical concentration reflects the links between selling medicines and publishing books, which will be returned to after an exploration of these prominent wholesalers and their business structures.

Table 4.2 Principal London wholesalers as described in advertisements in the studied newspapers (1769–1822), in an 1800 handbill, and as recorded in the 1841 Post Office London Directory

	10 Bow Churchyard	*45 St. Paul's Churchyard*	*95 Fleet Market*	*14 Birchin Lane*	*150 Oxford St.*	*59 Coleman St.*	*4 Cheapside*	*66 St. Paul's Churchyard*
1769	Cluer Dicey & Co.	Newbery & Carnan[a]	Jackson & Co.					
1781	Cluer Dicey & Co.	F. Newbery jnr.	Jackson, Warter & Co.	M. & H. Wray				
1793	Dicey & Co.	F. Newbery	Jackson & Co.		W. Bacon	John Wye		
1794	Dicey & Co.	F. Newbery	J. Barclay		W. Bacon	John Wye		
1800[d]	Dicey & Beynon	F. Newbery	Barclay & Co.	H. Wray & Co.	Jeboult & Co.		Ching & Butler	(Benjamin Shaw)[e]
1807	Dicey & Sutton	F. Newbery & Sons	Barclay & Son		Bacon & Co.		R. Butler	Shaw & Edwards
1822	Sutton & Co.	F. Newbery & Sons	Barclay & Son		Sanger		Butlers	Shaw & Edwards
1841 Directory	William Sutton & Co.	F. Newbery & Sons	Barclay & Sons[b]		John Sanger		Thomas Butler	Evan Edwards?[c]

Notes
[a]65 St. Paul's Churchyard (see below)
[b]95 Farringdon St. (Fleet Market widened and renamed Farringdon St. in 1829)
[c]66 St. Paul's Churchyard now occupied by a florist, but Evan Edwards, patent medicine warehouse, next door at number 67
[d]From a dated handbill for Maredant's Antiscorbutic Drops (ECCO, British Library)
[e]74 Borough High St. (Burnby, 'Preparers', 53)

The Diceys and the Newberys

Some of the major wholesalers are rather shadowy, but we have a significant amount of material about the Diceys and the Newberys, the market-leading proprietors who were also prominent wholesalers. Both the Diceys and the Newberys were initially printers, newspaper owners and booksellers as well as being medicine wholesalers, and for this reason we know a great deal more about them than we do about other medicine wholesalers who were unrelated to the print trades and who have left only bits of historical evidence. However, we should not be unduly concerned about the possibility of being misled on wholesaling by selected material: as we shall see, the Diceys and the Newberys were also the dominant medicine wholesalers at the end of the eighteenth century.

The Dicey family business was founded by William Dicey, the son of a Yorkshire vicar, who started a printing business with Robert Raikes in St. Ives and then Northampton, launching the longstanding *Northampton Mercury* in 1720.[19] Dicey and Raikes were involved in patent medicines from the beginning of this newspaper, advertising Dr. Bateman's Drops from at least 1721 which was five years before Benjamin Okell patented this medicine.[20] Raikes went off to Gloucester in 1722 and his son, also Robert, later became well known as the leader of the Sunday School Movement. William Dicey's sister married John Cluer of Bow Churchyard, London, a well-known printer of chapbooks and ballads who had also established a new and successful system for cutting, printing and distributing popular prints.[21] After the deaths of John Cluer and his widow, William Dicey acquired John Cluer's business in 1736 and set up a partnership with his own oldest son, confusingly named Cluer Dicey, with other sons assisting.[22]

The Diceys now had an extensive business at both Bow Churchyard and in Northampton, producing prints, chapbooks and more substantial publications in London and a newspaper in Northampton, together with medicine selling at both locations. Until at least the end of the century the various Diceys split their time between the two places and the family continued to own the *Northampton Mercury* until 1885.[23] The combined business was clearly very profitable and Cluer Dicey set himself up as a country gentleman, buying Claybrooke Hall, Leicestershire, in 1765 together with other property in the same parish. Cluer's son Thomas acquired further land in the area.[24] Hannah More wrote Cluer's epitaph for his marble monument in Claybrooke Church, which still occupies a prominent position between the nave and the chancel in the church, accompanied by other memorials to his son Thomas, his grandson, grand-daughters and other descendants.[25] It is not clear whether Cluer's money came predominantly from medicines or from printing and publishing: Simmons and Burnby have both suggested that the medicines may have been more important, but agree that the evidence is inconclusive.[26] Certainly, the medicine business was important enough in 1753 to require a different partnership (William Dicey, Cluer Dicey and Elizabeth Okell) compared to that of the printing/publishing business (William Dicey, Cluer Dicey and Richard Marshall), and, in 1764, to feature in legal disputes about William Dicey's will.[27]

On his father's death in 1775, Thomas Dicey took over the combined family business, which became more specialised in medicines. He withdrew from London chapbooks and publishing, which is confirmed by the *British Book Trade Index* not listing any Diceys for London after the 1780s, and printing became confined to Northampton. Thus, Thomas's will, written in 1807, only mentions 'the craft or business of a medicinal warehouseman' in Bow Churchyard and that of a printer at Northampton, suggesting that the medicine wholesaling had been more profitable than the printing and publishing in London.[28] Several non-family partners joined the medicines side of the business at different times, including Francis Beynon, John Wye and William Sutton; John Wye set up on his own in the early 1790s.[29] By the time of Thomas's death in 1807, the medicines business was a partnership between Thomas and William Sutton, known as Dicey and Sutton. It later became Sutton and Co. when Thomas's son, Thomas Edward Dicey, a senior wrangler at Cambridge, left the partnership in 1811 and moved away from commerce in London.[30] He devoted himself to country affairs in Leicestershire and to owning the *Northampton Mercury*, without any link to patent medicine wholesaling.[31]

The medicine wholesaling activities of John and Francis Newbery have much in common with those of their contemporaries, Cluer and Thomas Dicey. As we have seen in Chapter 1, John, like Cluer, combined medicine selling with publishing and bookselling, basing his prosperity on the lucrative and best-selling Dr. James's Fever Powder. However, John's publishing was more up-market, bolstered by friendships with Samuel Johnson, Oliver Goldsmith, Tobias Smollett and many others, together with a reputation, which continues to the present day, as the father of children's literature. Like Cluer Dicey in 1775, John was wealthy at the time of his death in 1767, but in John's case it is clearer that patent medicines provided the more important source of his income.

After John's death, the combined Newbery publishing and medicines business split up around the same time as Thomas Dicey was concentrating on medicines in his London business. In 1779, twelve years after his father's death, Francis Newbery fell out with his publishing partners, Thomas Carnan (stepbrother) and another Francis Newbery (first cousin). Francis moved to new premises at 45 St. Paul's Churchyard and concentrated solely on medicine owning and wholesaling, leaving his relatives to continue with bookselling and publishing.[32] Amongst these relatives, Thomas Carnan and Elizabeth Newbery, the widow of cousin Francis, did continue their own medicine wholesaling activity for a time; for example, Elizabeth was one of the wholesalers in 1794 for a range of medicines produced by Swinfen and Son, surgeon-apothecaries in Leicester.[33] But unlike their publishing, their medicine selling does not seem to have been sustained as they are not listed as wholesalers in the studied newspapers in later years. In common with Cluer Dicey, Francis Newbery became a country gentleman, as we saw in the last chapter. One potential link between the two families is that Francis married Mary Raikes, daughter of William Dicey's original business partner Robert Raikes, in 1770.[34] But the Raikes family now lived in Gloucester and it is not clear whether this marriage reflected any social contact between the Newbery and Dicey families.

The Dicey family and successors and the Newbery business were the two largest wholesalers of patent medicines at the end of the eighteenth century and the beginning of the next, as judged by their financial success, range of medicines, and frequency of appearances in provincial newspaper advertisements. It is therefore not surprising that in 1785 Francis Spilsbury regarded them as the natural leaders of the opposition to the new medicine excise duty.[35] This pre-eminence is confirmed by an unlikely source, the plates used to print the named excise duty stamps. As we saw in Chapter 2, from 1783 an excise stamp had to be stuck on every bottle or box of patent medicines, and the larger wholesalers and prominent owners were allowed the commercial advantage of having their own names and addresses on the stamps. The named stamps were printed at the Stamp Office, and the plates required to do so were recorded in a register, together with an entry for any repairs

Table 4.3 Printing plates for medicine excise stamps registered at the Stamp Office 1795–1823, plus repairs or replacement of those plates

	Number of registered plates	*Number of repairs or replacements*
Dicey & Co.	7	9
F. Newbery	10	10
Barclay & Co.	4	nil
Wrays	3	nil
Bacon/Sanger	No plates recorded[a]	
John Wye	1	nil
Ching & Butler	3	4
Shaw & Edwards	5	2

[a]As we have seen, William Bacon *signed* his name on the stamp

or replacements. The register for 1795–1823 has been transcribed by Booth, and the entries for the eight principal wholesalers described in this chapter are summarised in Table 4.3.[36]

The number of registered plates reflects both the range and the volume of sales for each of these wholesalers. As explained in Chapter 2, stamps were required at values of 1½d, 3d, 6d, 1s and upwards, depending on the price of each medicine: so a range of medicines at different prices would require a range of stamps and their printing plates. But the volume of sales would also be reflected in the number of plates as wholesalers might require several plates of the same value to facilitate printing for their high turnover. For example, Dicey and Co. registered no fewer than six plates for a duty of 1½d, the lowest value, together with a single plate for 3d. The Diceys were clearly concentrating on a high volume of sales at the lower end of the market. By contrast, Francis Newbery had a range of plates for duties up to three shillings reflecting his rather more upmarket medicines (Dicey and Newbery stamps are pictured in Fig. 2.1). In addition, the number of repaired or replaced plates indicates the relative volume of production, since these renewals reflected the wear and tear of the plates, which in turn must have stemmed from how often these plates were used. It is possible that some wholesalers might not have been using their named stamps or that the plates may have been renewed for reasons other than wear and tear. However, I cannot see a commercial advantage in doing so, and these considerations are unlikely to explain the differences in Table 4.3.

The large number of original plates and their replacements used on behalf of the Diceys and Francis Newbery confirms that these two businesses were indeed the leaders amongst the eight prominent London wholesalers. All the Dicey plates, and six of the ten Newbery plates, were repaired or replaced at least once. The only wholesaler approaching their level of activity, at least for a time, seems to have been Ching and Butler, later Butlers, whose 3d plate had to be repaired three times between 1810 and 1816. No other wholesaler came close to the degree of printing plate usage by the two leaders. In addition, this data provides a rough impression of the volume of business handled by these prominent wholesalers in comparison to that handled by the owners who wholesaled their own medicines. Even the proprietors who distributed a substantial volume of their own medicines rarely required a repair or replacement of their plates. Amongst such proprietors mentioned in Chapter 3, Spilsbury, Godbold, Brodum, Solomon and Lignum did not need any, while William Henry, the son of Thomas Henry, had one plate repaired. So the prominent wholesalers were handling a much higher volume of medicines than the successful owners who distributed their own medicines. At the beginning of the nineteenth century, the Dicey and Newbery family businesses had left publishing in London and together they now represented a significant proportion of the whole patent medicines industry.

What types of medicines were the Diceys and the Newberys distributing? Wholesalers often repeatedly advertised a core stock, together with a variable collection of other medicines. As befits former printers of chapbooks and ballads, the Diceys' core was cheap long-standing medicines such as Daffy's Elixir, Bateman's Pectoral Drops, Anderson's Scots Pills, Squire's Elixir, Godfrey's Cordial and Radcliffe's Purging Elixir which were all priced at 1s, plus 1½d duty after 1783, for the smallest bottle (Appendix A.1). As a result, the Diceys required a large number of 1½d excise stamps, as we have seen. In contrast the Newberys' core medicines were more varied in price, more expensive and often relatively more recent, and so they were analogous to their more upmarket books: examples included Dr. James's Fever Powder (2s 6d), Dr. James's Analeptic Powder (4s 6d), Dr. Steers's Opodeldoc (2s) and Essence of Coltsfoot (3s 6d). Thus, the Diceys' and the Newberys' core medicines each reflected the market positions of their families' printed publications.

Other Prominent Wholesalers

The information on the Diceys and the Newberys is relatively plentiful, but if we turn to other wholesalers who were not booksellers, for example Thomas Jackson succeeded by the Barclays, the material becomes sparse and largely confined to their advertisements. The Jackson/Barclay business was probably started by Thomas Jackson, a chemist who patented a lotion in 1761 (Appendix B). It was recorded as Jackson and Co. in 1769, and continued into the mid-nineteenth century as Barclay and Sons (Table 4.2). A London directory for 1792 listed Jackson and Barclay as partners at 95 Fleet Market, so we can consider it as a single business, and a Barclay pharmaceutical business was still trading after the Second World War, nearly two centuries after its original foundation.[37] By 1794, James Barclay was proclaiming that he was the successor to the late Thomas Jackson and the sole proprietor of the business, and by 1807 Barclay and Son was advertising a wide range of medicines.[38] Further details on Thomas Jackson, James Barclay, or members of their families, are meagre: no family members can be identified amongst these common surnames in the *Oxford Dictionary of National Biography* or the *British Biographical Archive*, and the partnership arrangements do not seem to have been chronicled in the *London Gazette*.

Similarly, only a little detail is available for the remaining five wholesaling firms described in Table 4.2. Hilton Wray was mentioned in Chapter 3, and John Wye was an ex-partner of Thomas Dicey. We can trace some of the origins of Ching and Butler, later Butlers, at 4 Cheapside through newspaper advertisements. John Ching was a chemist and apothecary in Launceston, Cornwall, who patented a worm medicine in 1796 (Appendices 1A and 2). In an advertisement of that year he had premises in Launceston and also a medicine warehouse in London at Gould Square.[39] Another advertisement in 1798 stated that George Dixon of 4 Cheapside, the proprietor of Dixon's Antibilious Pills, 'has relinquished the business in favour of Mr John Ching'.[40] By 1800, the worm lozenges could be obtained from Ching's Medicinal Warehouse in Cheapside, and by 1802 from Ching and Butler in Cheapside.[41] By contrast, the origins of Shaw and Edwards remain obscure, though we know that in 1802 Benjamin Shaw bought the popular range of medicines introduced by John Hill sixty years earlier, and that the business moved from Borough High Street to St. Paul's Churchyard around the same time.[42] In 1807, they claimed to be 'the successors' of Joseph Wessels, the proprietor of the successful Dr. Walker's Jesuits' Drops; but this description in an advertisement may only refer to ownership of the medicine, rather than a continuation of the existing business.[43] Four of these prominent wholesaling businesses (Jackson/Barclay, Wray, Wye, and Ching and Butler) had their origins in chemistry and pharmacy in contrast to the Diceys' and Newberys' roots in bookselling, printing and publishing.

Thomas Jackson's core stock was different to that of the Diceys and the Newberys, consisting of medicines sold under his own name: Jackson's Ointment for the Itch, Jackson's Tincture, and Jackson's British Tooth Powder were frequently advertised within a range of other named products. Like the Dicey medicines, the Jackson medicines were relatively cheap with the cost of the smallest bottle or box lying between 1s and 1s 6d plus duty. From 1794, several of the Jackson medicines were replaced by Barclay medicines, reflecting the name of their new owner: but as the titles were similar, this may have been a renaming exercise rather than a change in the recipes.

The newspaper advertisements suggest that some, though not all, of the prominent London wholesalers broadened their range of medicines from the beginning of the nineteenth century, partly to accommodate provincial proprietors. Francis Newbery and his sons largely retained their existing products, whereas the Diceys and their successors added other medicines to their continuing core stock. Barclay and Son were only intermittently advertising their core stock of their own medicines, but were now distributing more medicines from provincial owners. For example, advertisements in the *Leeds Mercury* in the first half of 1822 show Barclay and Son linked with proprietors in Manchester, York, Loughborough and Penrith. One of the newer wholesalers, Shaw and Edwards, was also associated with provincial proprietors such as Elliott, a druggist in Huddersfield with three medicines, and Hallam, a surgeon in Bury St. Edmunds.[44] This tendency for London wholesalers to be increasingly allied with provincial owners was taken a stage further as some of these owners acquired multiple London agents. Advertisements in 1822 do not show how these arrangements worked, but they do reveal that, for example, Carrington's Life Pills from Barry and Son in Bristol could be obtained from four wholesalers in London, while two medicines from Dr. Roberts of Bridport, Pilulae Antiscrophulae and the Poor Man's Friend, were distributed by five (Appendices 1A and 1B).[45] Many of the old favourite medicines were still available from the usual sources, but overall the market seems to have been opening out, with more links between provincial owners and the London wholesalers. Butlers illustrated this expansive process by advertising in 1822 that they were not only at the same address on Cheapside, but now also had their own premises in the West End, Edinburgh and Dublin. We can conclude that whatever their origins, the prominent wholesalers were following broadly similar evolving practices, and a single owner could work with several of them.

Booksellers as Medicine Wholesalers

As the industry developed in the mid-eighteenth century, much of the national medicine wholesaling was being carried out by booksellers, especially when the proprietor was not distributing his own medicine. As described above, the two largest wholesalers in the late eighteenth century, the Diceys and Newberys, were originally successful publishing booksellers. Several other booksellers acted as wholesalers for a medicine, and some can be identified in the studied medicine advertisements with the help of the *British Book Trade Index*. Examples during 1769 include William Nicoll at 51 St. Paul's Churchyard distributing Beaume de Vie, Isaac Fell at 14 Paternoster Row distributing Norris's Antimonial Drops, and W. Harris at 70 St. Paul's Churchyard wholesaling the Hooping Cough Medicine.[46] Moving on to 1781, Joseph Johnson at 72 St. Paul's Churchyard was responsible for Henry's Calcined Magnesia, Lee Roe in Silver Street off Fleet Street distributed English Coffee, and John Bew at 28 Paternoster Row was the wholesaler for Beaume de Sante.[47] The premises of five of these six booksellers were, in common with four of the principal medicine wholesalers, clustered around St. Paul's Cathedral, the traditional centre of London booksellers. Most of these six booksellers were only mentioned for a single medicine in the newspaper advertisements, but they sometimes dealt with a greater number. For example, Harris gave evidence at a murder trial about his provision of another product, Dodd's Rheumatic Tincture, and William Nicoll was a wholesaler for seven medicines in the studied 1781 advertisements.[48]

Why were booksellers, including some major publishing booksellers such as John Newbery, Joseph Johnson and the first John Murray, heavily involved in wholesaling medicines in the second half of the eighteenth century? Cox and Dannehl's assessment of the development as being 'pure serendipity' is clearly inadequate.[49] A better initial response might be: Why not? Although print historians have sometimes been reluctant to acknowledge it, the booksellers of the period could indulge in a variety of additional occupations, other than selling patent medicines, to improve their income.[50] At various times, the first John Murray traded in beer for export to India, lottery tickets, Irish linen, cast reproductions of gems and seals, and game birds, as well as acting as the sole London wholesaler for the Edinburgh Febrifuge Powder.[51] Another example was John Newbery, who in 1764 was a joint patentee of a machine for printing onto a variety of fabrics; while James Coghlan provided an agency for travel to and from France.[52] We should also recall that medicine wholesaling could be very profitable and it could be a desirable occupation for many Londoners with the appropriate capital. So the question needs to be rephrased: Why did medicine wholesaling fit so well with bookselling that booksellers were more successful in this field than other tradesmen? A number of responses to this question can be put forward, but the least convincing is the traditional one that the publishing booksellers could conveniently promote and distribute their medicines alongside their books.[53] As we shall see later in this chapter and in the next, little or no evidence exists that the bookselling wholesalers did indeed send out their medicines alongside their books, and it is also hard to find examples of medicines and books being advertised together, apart from supporting treatises for medicines or addenda and puffs within the wholesalers' own publications.

The prominence of the booksellers in medicine wholesaling in comparison to other non-pharmaceutical trades, was more a matter of expertise than of convenience. Books and patent medicines were the only two widely consumed Georgian goods which *had* to be made at a single site and delivered to retailers across the country as a finished product. Other goods could be prepared, or at least processed, by the shopkeeper, who often sold a mixture of goods made on the premises and those bought in.[54] The wholesaling of books and that of patent medicines had several features in common. First, publishing books was one of the few occupations which always required management of capital: the books had to be printed, or at least bought, weeks or months before they achieved any significant revenue, and this revenue could be unpredictable. Similarly, medicine wholesalers also needed proficiency in capital management, as they had to invest in their medicines without being sure of the eventual income. So the financial skills for these two types of goods overlapped, and were sharpened in this era by the possibility of destitution and/or prison for failure. Second, the major booksellers knew how to distribute and promote goods across the country. In the late eighteenth century, only a smattering of books was published outside the capital, and the London publishers knew how to send their goods and who might be interested in selling them. This knowledge would give them a head start in medicine wholesaling, and they would also be proficient in the crucial skills, for both activities, of assessing credit worthiness and getting back any money owed. Third, the booksellers were obviously more experienced than other trades in using the printed word. Unlike orthodox therapy which relied heavily on oral communication, patent medicines needed the printed word for promotion and effectiveness, as discussed in Chapters 7 and 8. Bookselling medicine wholesalers would in general have better access to high-quality printed material of all kinds than other tradesmen, including chemists and apothecaries.

However, the picture changed over time, with medicine wholesaling by booksellers diminishing and then dying out. As we have seen, the Diceys' and Newberys' medicine interests in London had split from the family printing/bookselling businesses by the 1790s, and by this decade far fewer booksellers were involved with medicines. Joseph Johnson continued both to publish for Thomas Henry and to distribute his Calcined Magnesia, perhaps because both of them were Unitarians, and James Coghlan, the prominent Catholic publishing bookseller who already sold his own medicines, was still being approached to sell those of other Catholics, as we have seen: but Johnson died in 1809 after several years of inactivity, without apparently passing on a medicines business, and Coghlan died in 1800.[55] Coghlan's successors continued with the medicine business alongside bookselling and publishing for a short time, but it is difficult to find other examples of booksellers acting as medicine wholesalers after the turn of the century. Conversely, the prominent medicine wholesalers no longer seem to have had any connection with the print trades; with the exception of Francis Newbery who had long since given up bookselling, none of the wholesalers mentioned in Table 4.2 for the years from 1800 can be identified in the *British Book Trade Index*. By the late Georgian period, the link between bookselling and national medicine *wholesaling* was broken.

The newer wholesalers predominantly had their origins in chemistry and pharmacy. This was probably due to the expansion of druggists across the country, and an increase in their expertise, in the late eighteenth and early nineteenth centuries.[56] As a result of these changes, specific chemical and medical knowledge would be more important for the competing patent medicine wholesalers than the general skills in capital management, distribution and the printed word. In addition, the publishing booksellers probably wished to enhance their reputations amongst authors by avoiding association with other trades, especially one which was coming under increasing attack from the medical professions.

The part-time wholesaling of medicines by some of the London publishers in the eighteenth century has wider implications than just an exploration of the patent medicines industry, particularly as it has largely escaped the attention of print-culture historians.[57] As income from this source could have been substantial, it should be part of discussions on the economics of publishing in the period. It might have impacted on the booksellers' decisions about what to publish: for example, John Newbery's speculative entry into the new field of publishing children's books might have been made possible by his secure income from Dr. James's Fever Powder and other medicines. Historians have been too ready to assume that the preserved publishing records represent the totality of a bookseller's revenue.

Methods of Distribution

Medicine wholesaling in Georgian England adopted some of the distribution skills of booksellers, but there is no clear evidence that medicines were normally supplied *alongside* books in our period. Feather claimed that the medicines and books were distributed together and that this was part of the reason why local booksellers often sold patent medicines.[58] However, Feather's own case study showed that John Clay, a bookseller in Daventry who sold medicines in the 1760s and 1770s, used several national and regional medicine wholesalers, at least four of whom, including Francis Newbery, did not supply him with anything else: one stationery supplier from Gloucestershire did provide Clay with Cheltenham Water as well.[59] I have yet to find an example of a publishing bookseller sending a significant quantity of medicines with his books, though Coghlan did sometimes add a small quantity of a medicine to a single customer's order for books.[60]

Exploring the distribution of the medicines and the associated financial arrangements relies on piecemeal sources, and the practical details are often unclear. At the outset, we should not underestimate the need for robust methods to carry the medicines across the country. Most patent medicines were liquids in thick glass bottles which could be up to two pints in size: transporting the bottles would have had more in common with modern methods of delivering wine than distributing our current, easily transported, pharmaceutical pills. One useful source is the day books, for five years from 1799, of John Ware, the owner and printer of the *Cumberland Pacquet* in Whitehaven.[61] These books contain both the records of individual transactions of his printing, newspaper, bookselling and medicines business for the later making-up of accounts which have been lost, and the totting-up of repeated transactions such as the amount owing for advertising by medicine wholesalers and the cost of advertising excise duty. They are not a complete record of his business, or even of his daily transactions, but they do document many of his dealings with medicine wholesalers. The most striking finding is that, even in far off Cumberland, Ware was dealing directly with the London wholesalers without the involvement of intermediaries. In the three years 1800–1802, Ware received substantial quantities of medicines from seven London wholesalers, as well as from Solomon in Liverpool, Lignum in Manchester, and the Henry family magnesia business also in Manchester. One of the wholesalers, Dicey and Sutton, delivered medicines worth a total of £200 in this period. The day books only occasionally reveal the quantity of medicines, but one entry for Dicey and Sutton shows a delivery of 499 bottles and 268 boxes at a cost £43.18s. If this delivery was typical, Dicey and Sutton were sending an average of around 1150 items a year to Whitehaven. This despatch of large quantities of medicines directly to John Ware is unlikely to be exceptional: Whitehaven was about 310 miles from London, one of the most distant towns in England from the capital, so any arrangements for a newspaper printer in this remote, medium-sized port, were probably in place elsewhere.

It is easy to assume that the wholesalers were in vigorous competition all the time, but the Ware day books show some cooperation between them. On several occasions, John Wye and Dicey and Sutton shared deliveries and helped each other in settling their accounts; this is not entirely unexpected as John Wye was a partner in the Dicey family business until the early 1790s. More surprisingly, Dicey and Sutton were assisting the Newberys, their presumed rivals. Francis Newbery did not send any medicines, and did not advertise in the *Cumberland Pacquet*, during the period of the day books; but Ware was selling his Dalby's Carminative.[62] Broken bottles of the Carminative were returned and charged to Dicey and Sutton in July 1803, strongly implying that Dicey and Sutton had delivered them in the first place.

The clearest accounts of the mechanisms of distribution, though unfortunately only for a few medicines, can be found in the *Proceedings of the Old Bailey*, where the processes are described in detail by witnesses for the benefit of the court. One such report was provided in 1795 by the successful prosecution of Philip Gibson for stealing a consignment of Spilsbury's Antiscorbutic Drops.[63] At that time, the Drops were made and distributed by Spilsbury's widow Dorothy at Soho Square, London. Thirty-nine bottles of the Drops were packed in a deal box on Dorothy's premises before being carried by a footman to an inn on Ludgate Hill for despatch on the Newbury coach to Mr. Fuller, a printer and bookseller in that town.[64] Another Old Bailey example of medicine distribution comes from the trial and death sentence in 1812 of Thomas Collicott for forging medicine excise stamps.[65] Collicott was a medicine vendor in Oxford Street, London who supplied Dr. Jebb's Antibilious Pills to a shop in the Royal Exchange, and also a mixed box of medicines to Wood and Cunningham, medicine vendors, in Bath. The pills were carried to the Royal Exchange by hand and the box was sent to Bath by Roger's waggon which set out from an inn in Aldersgate Street. Once again, these medicines mentioned in the court proceedings were transported separately and directly to their destinations.

These Old Bailey reports describe transport by coach, by wagon, and on foot: in the market as a whole the manner of transporting the medicines was also variable, perhaps determined by the retailer rather than the wholesaler. It could be simple and direct, such as Thomas Curtis of Covent Garden being prepared to dispatch three packets of his Mucilage of Marshmallows for urinary gravel to anywhere in the country, 'carefully sent by coach', in exchange for a one pound note.[66] In contrast, bulk transport could be involved; for example, Sims and Ansell in Stockport asked Howards of Stratford, Essex, to deliver the medicines to Manchester by canal and the empty bottles would be returned by the same means.[67] William Jones, who, as we saw in Chapter 3, manufactured and distributed a mixture of regular and patent medicines, dispatched them in 1781 'by the very first coach', 'the Cranfield Carrier', or 'by the return of the wagon' at the specific request of his customers.[68] The Ware day books document the involvement of local carriers in the transport, but generally they do not reveal whether this was by ship, coach or wagon;[69] and they also indicate that the wholesalers paid at least some of the costs of the transport. Overall, these examples of varied transport methods had one thing in common: the medicines were not being sent with books.

Details of methods of payment to the wholesalers for medicines delivered are sparse, but they fitted into the normal pattern of trust and credit which drove Georgian commerce.[70] Ware used the wholesalers' advertisements to offset his payments for their medicines. His day books show that the cost of the advertisements in the *Cumberland Pacquet* was credited against the cost of the medicines that Ware received, with Ware then making a balancing payment through an intermediary to settle the account. For example, payments of about £34, £41 and £74 were made in July 1801, July 1802 and July 1804 to Mr. Bell to settle the Dicey and Sutton account, and £63 was paid to an unknown intermediary (Mr. Bell?) for the same purpose in July 1803.[71] Using the medicine advertisements to pay for, at least partly, the delivered medicines seems to have been routine in the industry: giving evidence in a legal case in 1805, the owner of the *Cornwall Gazette* confirmed that this was the usual practice.[72] The incomplete information in Ware's day books suggests that he reconciled many accounts annually, but some settlements might be several years in arrears: for example, a totting-up of Brodum's advertisements in July 1803 went right back to 1798, and a payment to Ching and Butler in August 1802 seems to have been based on the previous two or three years.

Did the medicine retailers have to pay eventually for everything they had ordered, or did they receive their medicines sale or return? Examples of both can be found. Sale or return was described in 1764 by Cluer Dicey during a legal dispute about his father's will, when he reported that country shopkeepers only had to pay for the Dicey medicines once they had been sold: a company 'rider' normally visited each shopkeeper every year to review the stock and to receive the money.[73] In contrast, Wood, one of the Bath retailers for Thomas Collicott in 1812, asserted in court that he had to pay for the medicines with no credit for any unsold, and he even had to pay the carriage down from London.[74] Ware received some of his medicines on a sale or return basis, for example in February 1801 he sent back to Ching and Butler 27 boxes of an expensive medicine for preventing yellow fever with a total value of nearly £41. However, he returned few medicines unless the bottles had been broken in transit, and some of the delivered medicines were recorded as 'on return' suggesting that this was not the standard practice. Elsewhere, William Singleton seems to have required payment *in advance* from retailers for his Dr. Johnson's Yellow Ointment, but this was probably rare.[75] The limited evidence indicates that sale or return was the exception around the beginning of the nineteenth century, but it may have been common earlier.

Medicine Wholesaling in the Georgian Context

This chapter has assessed the wholesaling of patent medicines with the help of a diversity of sources which are often incomplete and of variable quality; the gaps are larger than the substance. Yet, as noticed elsewhere in this book, the erratic nature of these sources also strengthens the conclusions. Accounts of selling and distributing goods in Georgian England have usually depended on the records of the successful few, such as Wedgwood or Boulton. The uncoordinated and, in some cases, random records of selling patent medicines, usually preserved by chance and not by the intentions of descendants, enable us, paradoxically, to build up a more representative picture of their wholesaling compared to that of other commodities that have relied on a smaller selection of more complete archives. Patent medicines are much easier to find in imperfect records than most other goods, as they must be publicised with identifiable names. So the use of these sources, especially the newspaper advertisements, allows us to know more about wholesaling patent medicines, such as the names and addresses of the principal London distributors, than can be discovered for most other movable goods in late Georgian England.

The wholesaling of patent medicines employed an assortment of procedures, but a commonality can be discerned for the most popular nationally distributed medicines. Thus, each medicine was transferred from a central source by a small number of wholesalers, often a single one. These wholesalers became increasingly committed to the industry during the period, in the sense of normally being the owner of the medicine or a distributor with recognised medicine-related skills. Their range of medicines was also focussed as they dealt in a relatively small collection, or perhaps a single agent, rather than a general mixture of remedies. The medicines were heavily advertised, as anybody who has studied eighteenth-century newspapers can testify, and Chapter 6 will show that the advertising of nationally available medicines was largely controlled by the wholesalers. Advertising techniques will also be discussed in that chapter, but it is worth noting here the sheer volume of advertising. For example, the *Leeds Mercury* publicised Lignum's Pills for venereal disease 18 times in the 26 weeks of the first half of 1794, while, at the top of the town, the *Leeds Intelligencer*, carried 27 advertisements for Spilsbury's Antiscorbutic Drops during the same period. Such frequency was unusual, but many medicines were advertised once a month or more, on average, in a particular newspaper. Many of the advertisements were lengthy, commonly 15–40 lines of print and occasionally filling a whole column of 150–160 lines.

Furthermore, the wholesalers were not only advertising heavily in the provincial press, they were also paying for other types of local promotion. Ware's day books show that most of his medicine wholesalers paid for the distribution of bills and other forms of printed publicity in the Whitehaven area. For example, in January 1800, Dr. Arnold, who wholesaled his own pills, was charged 2s for 'distributing handbills'; in the same month, Mr. Dickinson, who was responsible for Gowland's Lotion, was charged 10s for 'distributing books, etc.'; and in March of the same year, Ching and Butler were charged 2s 6d for 'postering, bills etc.'. The publicity-conscious irregular practitioners paid for more local promotion than the other wholesalers, with Solomon being charged on ten occasions in 1800–1802 for a variety of activities including 'distributing pamphlets, sticking up posting bills, etc.' and 'distributing books', and Brodum paying 3s for 'distributing Dr Brodum's pamphlets'. Ware was paying his journeymen between 12 and 15 shillings a week, so these payments for local publicity represent more than a trivial amount of work.

Medicine wholesalers may not have read Adam Smith, but they were nevertheless following his advice of concentrating on the needs of the consumer. They were aiming to supply national markets by creating demand through advertising and other publicity aimed at consumers. Thus, the standard pattern for disseminating patent medicines was fulfilling orders which had been induced by that local demand from customers, not by sending out salesmen or visiting fairs to create a market for a medicine. An important exception was the travelling irregulars such as Brodum, Solomon and Lignum who traded their services and their medicines together as they journeyed around the country. But these three irregulars still probably sold more of their medicines by fulfilling orders from a central base, in a similar fashion to other wholesalers. Certainly, Solomon employed two assistants in Liverpool to prepare and distribute his medicines when he was away travelling.[76] As already described, Cluer Dicey might also seem to have been an exception as his representatives regularly visited his retailers; but this was to settle accounts and review sales rather than to distribute the medicines. Wholesale quantities of patent medicines were heavy, and speculative transportation would probably not have been worthwhile.

The wholesalers were also trying to create demand from retailers. This is demonstrated by the printing of the name and address of the wholesaler in many of their advertisements: after all, the consumers only needed to know the name and address of a *local retailer*, with details of the wholesaler having little practical relevance to them. This may have been done to enhance the branding by clearly defining the medicine and making imitation more difficult, especially if the name of the wholesaler was also on the excise stamp attached to each bottle or box. But this is not a complete explanation as it does not explain why the wholesaler's precise address was needed, or why the wholesaler's name and address were sometimes provided in addition to those of the proprietor; the latter should have been enough for the branding.[77] The conclusion must be that many of the advertisements were aimed both at the consumers and at the actual or potential local retailers. Retailers would need to know the details of the wholesaler, and indeed a few advertisements did summarise the arrangements that a local retailer might expect from the wholesaler. Advertising aimed to recruit and encourage local retailers, who would need to know the name and address of the wholesaler, in addition to persuading consumers of the virtues of the medicine.

We can conclude that wholesaling for the nationally distributed patent medicines depended on a 'pull' from the periphery which had been induced by the promotion controlled by the centre. How does this compare with the wholesaling of other Georgian consumer goods? The national wholesaling of other goods in the period has not been extensively researched, partly because they are more difficult to identify than the branded patent medicines. The available evidence demonstrates the expected wide range of techniques, but the dominant impression is that manufacturers concentrated on moving their goods to local wholesalers and retailers by employing salesmen and methods derived from the disappearing annual fairs, rather than by utilising advertising and branding to appeal directly to the consumers.[78] For example, the Coalbrookdale iron producers dispatched salesmen to supply their consumer products to shops in the West Midlands and Manchester, and they still also used fairs in market towns.[79] The rapidly expanding London porter trade of the mid-eighteenth century relied on general merchants operating in a limited area, while the London wholesalers of wool textiles had their own travellers and local factors.[80] Producers seeking to sell in a wider area had to work hard to do so, and they might spend part of each year travelling around touting their wares.[81] Most wholesaling of the period relied on a 'push' from the centre, rather than the 'pull' induced by the branding and advertising of patent medicines across the country.

So the patent medicines industry was using wholesaling techniques which have not yet been demonstrated for most other goods in the period. The obvious exception to this individuality was bookselling, which also relied on a similar 'pull' from peripheral demand and required comparable skills, at least early in our period, as discussed above. Tea was another product which was derived from a central source and then distributed right across England, because until 1833 all legal tea in England came from the East India Company's auctions in London. The wholesaling of the tea developed some similarities to that of patent medicines during our period, shifting towards a greater degree of 'pull'. In the mid-eighteenth century, the tea was diffused through the country by a pyramidal system of several layers of dealers, starting with a cartel of London tea brokers and running down to the licensed tea vendors, who packaged the tea and determined the final price.[82] Some attempts were made to simplify tea distribution in the late eighteenth century, but the main change to the market was induced in 1818 by Frederick Gye, who used the capital from a £30,000 lottery win to set up a national distribution system which supplied fixed-price pre-packaged tea directly to local agents.[83] Interestingly, as a printer, Gye was originally a member of the print trades in common with several eighteenth-century medicine wholesalers. Others soon followed his successful example, tea advertising developed, and the wholesaling of tea became closer in structure to that of patent medicines.

Considering the bigger picture, patent medicine wholesaling can be integrated into developments in the wholesaling of movable consumer goods in general. In the seventeenth century, bookselling required the prototype national distribution system due to the legal restrictions on printing outside London, and advertising helped to create a peripheral demand for books. Like books, patent medicines were also produced at a single site, and this type of distribution with a peripheral 'pull' was transferred to medicines, sometimes by those already working in the bookselling trade. The techniques were eventually imitated for other goods, such as tea, though the timing of the changes is disputed.[84] Using advertising to generate demand was potentially quicker, though perhaps not cheaper, than setting up nationwide travellers and agents to push the product out across the country. The eventual result was a Victorian flowering of wholesaling with the same fixed-price branded goods becoming available all over the country.

Conclusion

Patent medicine wholesaling was the section of the industry which encouraged a national market and thereby accomplished the substantial sales, with a good profit for some. The arrangements for it varied, but the important point is that they were defined and organised, with most medicines having a well-publicised distributor, sometimes the owner, who provided a clear source for regional or national sales. Wholesalers could be found all over the country, but the major London wholesalers provided a stable profitable core which dominated the industry. Of course, wholesalers who were briefly active before collapsing in bankruptcy will remain obscure, and there undoubtedly were some. Nevertheless, patent medicine wholesaling in late Georgian England had a firm centre of normal successful trading, which was distant from quackery and not comparable to orthodox medicine.

With its distinct methods of creating demand and distributing patent medicines across the country to supply that demand, patent medicine wholesaling was at the front edge of the development of national markets for consumer goods in general. This wholesaling was mostly carried out by established tradesmen with the appropriate skills, not by irregular practitioners. The next chapter will explore medicine retailing and reveal more details on how the consumers' demand for medicines was met. The specialised wholesalers required accomplished local tradesmen to sell the medicines, not 'quacks'.

Notes

1. Smith, *Wealth of Nations*, vol. 2, 515.
2. Brown, 'Medicines Advertised'; Burnby, 'Preparers'.
3. *ABG*, 25 March 1822.
4. *LI*, 3 January 1769; *ABG*, 2 January 1769; *SJ*, 23 January 1769.
5. Women owners are apparent throughout the Georgian period and they sometimes distributed their own medicines; but otherwise almost all the wholesalers were men.

6. Bacon, *Patent Medicines.*
7. Patent medicines destined for export did not require an excise stamp.
8. Kearsley, *Tax Tables 1787*, 88.
9. Samaritan Water, *Swinney's Birmingham Chronicle*, 24 January 1793.
10. John Johnson Collection, Patent Medicines, 14 (41).
11. The five medicines were British Oil, Dr. Hooper's Female Pills, Dr. Bateman's Pectoral Drops, Daffy's Elixir and Dr. Anderson's Scots Pills.
12. *LI*, 16 January 1781; *ABG*, 16 February 1807.
13. *LI*, 26 January 1807.
14. *LI*, 31 January 1769.
15. 'Warehouse', *OED*, accessed 3 June 2017.
16. *Wakefield's Directory.*
17. Newbery, *Records*, 62; Burnby, 'Preparers', 53–54.
18. Raven, *Business of Books*, 168–185.
19. Simmons, *Catalogue*, 2.
20. Burnby, 'Printer's Ink', 163. The Diceys' successors, Sutton and Co., were still selling Bateman's Drops in 1822 (*SWJ*, 14 January 1822).
21. Duval, 'Diceys Revisited'; Harris, 'Scratching the Surface', 100.
22. Simmons, *Catalogue*, 2. Even more confusingly, John Cluer's son who died in infancy was Dicey Cluer.
23. Simmons, *Catalogue*, 5.
24. Macaulay, *History*, 11, 28, 38 and 53.
25. Macaulay, *History*, 65.
26. Burnby, 'Printer's Ink', 163; Simmons, *Catalogue*, 2.
27. Simmons, *Catalogue*, 3.
28. National Archives, Will of Thomas Dicey.
29. *London Gazette*, 21 December 1790 and 5 August 1800.
30. *London Gazette*, 12 January 1813.
31. Burnby, 'Printer's Ink', 163.
32. Roscoe, *John Newbery*, 18.
33. *ABG*, 21 April 1794. A rare example of a woman wholesaling the medicines of others.
34. Newbery, *Records*, 44. Two of their eight sons, John and William, succeeded to Francis's medicines business.
35. Spilsbury, *Discursory Thoughts*, 17.
36. Booth, *Catalogue*, vol. 2, A139-45.
37. *London Directory 1792*; J. Burnby, 'Preparers', 53 (footnote).
38. For example, Jackson's Patent Medicines, *ABG*, 13 January 1794. The 1807 advertisements were particularly prominent in the two Leeds newspapers.
39. Ching's Patent Worm Lozenges, *Sherborne Mercury*, 5 December 1796.
40. Dixon's Antibilious Pills, *Northampton Mercury*, 17 March 1798.

41. Ching's Patent Worm Lozenges, *LI*, 6 January 1800 and 1 November 1802.
42. Rousseau, *Letters*, xxxiii; Burnby, 'Preparers', 53.
43. Dr. Walker's Jesuits' Drops, *LI*, 26 January 1807. A St. Paul's Churchyard address for Joseph Wessells has not been found.
44. *LM*, 10 January and 17 January 1807.
45. *SWJ*, 14 January and 6 May 1822.
46. *LI*, 3 January 1769, *ABG*, 20 February 1769, *SJ*, 9 January 1769.
47. *LI*, 9 January 1781, *ABG*, 1 January 1781 and 8 January 1781.
48. Gurney and Blanchard, *Trial*, 29; *SWJ*, 1 January 1781; *ABG*, 8 January 1781; *SWJ*, 18 June 1781; *ABG* 16 April 1781. Nicoll's seven medicines were Glass's Magnesia, Dr. Henry's Chemical and Nervous Drops, Beaume de Vie and four Swinfen medicines (see Appendix A).
49. Cox and Dannehl, *Perceptions of Retailing*, 137.
50. Raven, *Business of Books*, 240.
51. Zachs, *John Murray*, 43–50.
52. Woodcroft, *Alphabetical Index; Laity's Directory*, Appendix, 43.
53. Feather, *Book Trade*, 84; Zachs, *John Murray*, 46.
54. Mitchell, *Tradition*, 38; Riello, *Foot in the Past*, 92.
55. Braithwaite, *Romanticism*, 61 and 179; Blom et al., *Correspondence*, xxxii.
56. Holloway, *Pharmaceutical Society*, 42–50; Marland, 'Medical Activities', 419.
57. For example, James Raven, *Business of Books*, the most authoritative recent account of London publishing, only briefly mentions medicine retailing by booksellers, not their wholesaling (240–241).
58. Feather, *Book Trade*, 84.
59. Feather, 'John Clay', 205–206.
60. Lancashire Records Office, Correspondence of James Coghlan, RCBu/14/56 and RCBu/14/81.
61. Cumbria Archives, Day Books of John Ware.
62. 'Cumberland Pacquet', 1800–1802.
63. 'Trial of Philip Gibson'.
64. With the help of an accomplice, Gibson stole the box of medicines from the coach shortly after it had set off.
65. 'Trial of Thomas Collicott'. The sentence was later commuted to transportation.
66. John Johnson Collection, Patent Medicines 2 (38).
67. J. Burnby, 'Preparers', 54.
68. Watson, 'Trading Accounts', 51–52.
69. An exception was Dr. Arnold being charged six shillings by Ware for the delivery of Arnold's medicines by coach in 1796.

70. For an analysis of the importance of credit in this period see Muldrew, *Economy of Obligation*.
71. The identity of Mr. Bell and the method of getting the money to Dicey and Sutton are unknown.
72. Gardner, *Business of News*, 53.
73. Simmons, *Catalogue*, 3.
74. 'Trial of Thomas Collicott'.
75. Bently, 'Trademark Case' 1012.
76. Denizen, 'Solomon', 295.
77. For example, an advertisement for three medicines owned and prepared by G. Ramsay in Penrith also mentioned the name and address of the wholesaler, Barclay and Son (*LI*, 2 February 1807).
78. Styles, 'Manufacturing', 542; Chartres, *Internal Trade*, 50; Daunton, *Progress*, 320–325; Mui, and Mui, *Shops*, 16.
79. Black, *Eighteenth-Century Britain*, 74.
80. Mathias, *Brewing Industry*, 148; Smail, *Merchants*, 71.
81. Daunton, *Progress*, 321.
82. Mui and Mui, *Shops*, 252.
83. Mui and Mui, *Shops*, 274–278.
84. Jefferys, *Retail Trading*, 6; Mitchell, *Tradition*, 11; Mui and Mui, *Shops*, 232.

Bibliography

Archives

Cumbria Archives, Whitehaven, Day Books of John Ware, DA276A.
John Johnson Collection, Bodleian Library, Oxford.
Lancashire Records Office, Preston, Correspondence of James Coghlan, RCBu/14.
National Archives, Kew, Will of Thomas Dicey, Prob. 11/1477/322.

Newspapers and Journals

Aris's Birmingham Gazette (ABG).
Cumberland Pacquet.
Leeds Intelligencer (LI).
Leeds Mercury (LM).
London Gazette.
Northampton Mercury.
Salisbury and Winchester Journal (SWJ).

Salisbury Journal (SJ).
Sherborne Mercury.
Swinney's Birmingham Chronicle.

Printed Primary Sources

Bacon, W., *An Account of Several Valuable and Genuine Patent and Public Medicines* (London, 1790?).
'Cumberland Pacquet', 1800–1802, *British Newspaper Archive*, http://www.britishnewspaperarchive.co.uk (accessed 16 March 2017).
Denizen of Liverpool, 'Samuel Solomon M D', *The Scourge*, 2 (1811), 287–303.
Gurney, Joseph, and William Blanchard, *Trial of Jane Butterfield* (London: Owen and Kearsley, 1775).
Kearsley, George, *Kearsley's Tax Tables 1787* (London, 1787).
Laity's Directory (London: Coghlan, 1788).
London Directory for the Year 1792 (London, 1792).
Macaulay, Aulay, *The History and Antiquities of Claybrook* (London, 1791).
Smith, Adam, *Wealth of Nations*, 3rd edn, 3 vols. (London: Strahan and Cadell, 1784).
Spilsbury, Francis, *Discursory Thoughts*, 2nd edn (London, 1785).
'Trial of Philip Gibson', 1795, *Old Bailey Proceedings Online*, www.oldbaileyonline.org (accessed 7 April 2013).
'Trial of Thomas Collicott', 1812, *Old Bailey Proceedings Online*, www.oldbaileyonline.org (accessed 2 June 2017).
Wakefield's Merchant and Tradesman's General Directory for London,...,for the Year 1794 (London, 1793?).

Secondary Sources

Bently, Lionel, 'The First Trademark Case at Common Law? The Story of Singleton v. Bolton (1783)', *UC Davis Law Review*, 47 (2014): 969–1014.
Black, Jeremy, *Eighteenth-Century Britain, 1688–1783*, 2nd edn (Basingstoke: Palgrave Macmillan, 2008).
Blom, Frans, Jos Blom, Frans Korsten, and Geoffrey Scott, eds., *The Correspondence of James Peter Coghlan (1731–1800)* (Woodbridge, Suffolk: Boydell Press, 2007).
Booth, R. G., *A Catalogue of the Revenue Stamps of the U K, Isle of Man, Channel Islands and Eire*, 2 vols. (Hexham, Northumberland: Tom Clutterbuck, 1982).
Braithwaite, Helen, *Romanticism, Publishing and Dissent: Joseph Johnson and the Cause of Liberty* (Basingstoke: Palgrave Macmillan, 2003).

British Book Trade Index, http://bbti.bodleian.ox.ac.uk (accessed March–July 2012).
Brown, P. S., 'Medicines Advertised in Eighteenth-Century Bath Newspapers', *Medical History*, 20 (1976): 152–168.
Burnby, Juanita, 'Printer's Ink and Patent Medicines', *Pharmaceutical Journal*, 229 (1982): 162–163.
Burnby, J. G. L., 'The Preparers and Distributors of English Proprietory Medicines', *Dental Historian*, 32 (1997): 47–55.
Chartres, J. A., *Internal Trade in England 1500–1700* (London: Macmillan Press, 1977).
Cox, Nancy, and Karin Dannehl, *Perceptions of Retailing in Early Modern England* (Aldershot: Ashgate, 2007).
Daunton, M. J., *Progress and Poverty: An Economic and Social History of Britain 1700–1850* (Oxford: Oxford University Press, 1995).
Duval, Gilles, 'The Diceys Revisited', *Factotum*, 35 (1992): 9–11.
Feather, John, 'John Clay of Daventry: The Business of an Eighteenth-Century Stationer', *Studies in Bibliography*, 37 (1984): 198–209.
Feather, John, *The Provincial Book Trade in Eighteenth-Century England* (Cambridge: Cambridge University Press, 1985).
Gardner, Victoria E. M., *The Business of News in England, 1760–1820* (Basingstoke: Palgrave Macmillan, 2016).
Harris, Michael, 'Scratching the Surface: Engravers, Printsellers and the London Book Trade in the Mid-18th Century', in *The Book Trade and Its Customers 1450–1900*, edited by Arnold Hunt, Giles Mandelbrote and Alison Shell (Winchester: St. Paul's Bibliographies, 1997), 95–114.
Holloway, S. W. F., *Royal Pharmaceutical Society of Great Britain 1841–1991: A Political and Social History* (London: The Pharmaceutical Press, 1991).
Jefferys, James B., *Retail Trading in Britain 1850–1950* (Cambridge: Cambridge University Press, 1954).
Marland, Hilary, 'The Medical Activities of Mid-Nineteenth-Century Chemists and Druggists, with Special Reference to Wakefield and Huddersfield', *Medical History*, 31 (1987): 415–439.
Mathias, Peter, *The Brewing Industry in England, 1700–1830* (Cambridge: Cambridge University Press, 1959).
Mitchell, Ian, *Tradition and Innovation in English Retailing, 1700–1850: Narratives of Consumption* (Farnham, Surrey: Ashgate, 2014).
Mui, Hoh-Cheung, and Lorna Mui, *Shops and Shopkeeping in Eighteenth-Century England* (London: Routledge, 1989).
Muldrew, Craig, *The Economy of Obligation: The Culture of Credit and Social Relations in Early Modern England* (Basingstoke: Macmillan, 1998).
Newbery, Arthur Le Blanc, *Records of the House of Newbery* (Derby: Bemrose, 1911).

Raven, James, *The Business of Books: Booksellers and the English Book Trade, 1450–1850* (New Haven: Yale University Press, 2007).
Riello, Giorgio, *A Foot in the Past: Consumers, Producers, and Footwear in the Long Eighteenth Century* (Oxford: Oxford University Press, 2003).
Roscoe, S., *John Newbery and His Successors: A Bibliography* (Wormley, Herts: Five Owls Press, 1973).
Rousseau, G. S., ed., *The Letters and Papers of Sir John Hill, 1714–1775* (New York: AMS Press, 1982).
Simmons, R. C., *The Dicey and Marshall Catalogue* (Birmingham: University of Birmingham, n.d.).
Smail, John, *Merchants, Markets and Manufacture: The English Wool Textile Industry in the Eighteenth Century* (Basingstoke: Macmillan Press, 1999).
Styles, John, 'Manufacturing, Consumption and Design in Eighteenth-Century England', in *Consumption and the World of Goods*, edited by John Brewer and Roy Porter (London: Routledge, 1994), 527–554.
Watson, G. M., 'Some Eighteenth-Century Trading Accounts', in *The Evolution of Pharmacy in Britain*, edited by F. N. L. Poynter (London: Pitman Medical, 1965), 45–77.
Woodcroft, Bennet, *Alphabetical Index of Patentees of Inventions* (London, 1854, reprinted London: Evelyn, Adams & Mackay, 1969).
Zachs, William, *The First John Murray and the Late Eighteenth-Century London Book Trade* (Oxford: Oxford University Press, 1998).

CHAPTER 5

Supplying the Consumer

The investigation of selling patent medicines to the many consumers across the country might be expected to return us to more traditional territory: irregular practitioners selling their medicines and the well-documented examples of printers, booksellers and other tradesmen supplying small quantities of the popular products in a fairly random fashion. Surely the organised industry of owners and wholesalers must give way to 'quackery' and to some shopkeepers making some money on the side by providing a range of goods and services which could include medicines. This is the impression gained from historians of commerce, print and medicine who may have come across the sale of patent medicines when investigating other topics. However, when the sale of patent medicines is investigated as an entity a different picture emerges: a coherent, and to some extent organised, configuration, looser than ownership or distribution, but still a significant component of the overall industry.

This explainable structure means that patent medicine retailing should be regarded as a distinct and autonomous element of the local medical market, not an amorphous feature of the varied irregular practice and shopkeeping within it. Two themes will emerge from the exploration of patent medicine retailing. One is that it had a recognisable structure and established practices, with the national wholesalers having a substantial influence on the vending arrangements. The second is that many printers and booksellers did not sell medicines as a small, almost accidental, sideline, as assumed by many writers; they became involved and successful

A. Mackintosh, *The Patent Medicines Industry in Georgian England*, Medicine and Biomedical Sciences in Modern History,
https://doi.org/10.1007/978-3-319-69778-9_5

in medicine retailing for good reasons, and for some it was a substantial part of their business. The next chapter will then explore how the medicines were advertised to ensure good local sales.

Who *did* sell patent medicines and how did this change over time? General traders sold medicines in rural areas, but in towns the sale became concentrated at the larger volume medicine retailers, often printers, booksellers, stationers or bookbinders. These members of the print trades formed local retailing networks, and in the second half of the eighteenth century they were even more dominant than suggested in the literature, before being joined after about 1800 by druggists. Members of the print trades had the advantage over other shopkeepers of being more experienced in the manipulation of the printed word, and being part of retailing networks. The newspaper printers also had the benefit that they had a two-way financial relationship with the wholesalers who paid to advertise in their papers, and that they were at the centre of a regional network of agents, who collected advertisements and notices for the newspaper and distributed the newspaper within their locality. The prominent role of many booksellers and printers in supplying medicines hints at medical knowledge and guidance, and it raises the question whether we should identify them as medical practitioners.

The Sellers of Patent Medicines

Getting some information about local retailers is easy. Many of the numerous newspaper advertisements and printed bills finished with a list of local retailers for that medicine. Sometimes the main occupation of the retailer was conveniently included, and for the others it can often be found in local trade directories which became general, if incomplete, from the 1780s.[1] So the inspection of a few advertisements can provide a deluge of names and their occupations: but many historians have not attempted to go much further in exploring the participants and mechanisms of retailing.

Porter's scrutiny of unspecified handbills from the early eighteenth century led him to the conclusion that the retailers belonged to a broad range of occupations: 'an oilman here, a cheesemonger there, this stationer, that coffee house, or simply Mr So-and-So at the Duck and Drake'.[2] But this may represent retailing before the establishment of the national industry, and he and his wife also recognised the importance of printers and booksellers in selling medicines.[3] Several authors have commented on the sale of medicines by members of the print trades (printers, booksellers, stationers and bookbinders), an activity which is apparent in the seventeenth century.[4] However, the reasons why the members should be successful in this trade, involving bottles of liquid, pills, ointments and other items which were physically unrelated to books and newspapers, have not received serious attention. Little attempt has been made to delve into their mechanisms of retailing, their degree of specialisation and their importance relative to other retailers, especially the druggists.

Like ownership and wholesaling, the techniques of medicine retailing were variable, but clear patterns emerge with closer inspection. To begin with the simplest method of retailing, a patent medicine could be sold from the owner's own premises, and this was probably the only source of the medicine from the local owners who just printed their own address in their advertisements. Joseph Wright's Medicine for the Bite of a Mad Dog was apparently only available at Wortley Windmill near Leeds, where he was the miller, and Mrs. Walter's Recipe for Pulmonary Complaints had to be bought from her niece Miss Hall, who was residing with Mrs. Pinkett at Oldbury-on-the-Hill, near Tetbury, Gloucestershire.[5] However, this solitary retailing was unusual for medicines advertised in newspapers, and most of the local owners would list a few other local retailers or make a non-specific claim for a wider distribution, such as from 'most respectable medicine venders'.[6] A modified form of local retailing was the owner of a nationally available medicine selling it from his or her own premises, in addition to the wholesaling or appointing another wholesaler. For example, in 1781, the range of herbal medicines invented by John Hill was made and sold at 29 St. James's Place, London by his widow, who was also responsible for the wholesaling, and the late Dr. Steers's Opodeldoc was available from his son's house in London with Francis Newbery running the national distribution.[7]

Away from the owner's own premises, medicines could be sold in the many general shops which provided a wide range of merchandise to a small local population.[8] A characteristic example would be Abraham Dent of Kirkby Stephen in Westmoreland, whose day book of transactions from 1756 to 1777 has survived.[9] Dent was a grocer, mercer, stationer and bookseller, who brewed beer and sold wine and gunpowder: he also sold non-proprietorial medicines and two patent medicines, Anderson's Scots Pills and Daffy's Elixir.[10] A two-page bill from around 1763 found in his day book indicates that the Elixir was distributed from

London by Thomas Jackson, and that the Kendal booksellers Thomas and then James Ashburnar were probably Dent's suppliers. Edward Harrison's correspondents in 1806 confirm this pattern of medicine sales by general shopkeepers in rural areas. For instance, a meeting of physicians and surgeons in Northumberland reported that in their area 'every common shopkeeper vends drugs as articles of commerce', and a correspondent from Cambridgeshire wrote that 'the grocers in villages sell drugs, which are always bad'.[11] Almost any shopkeeper could sell patent medicines as a small sideline.

According to Hogarth, Richard Rock sold patent medicines at Covent Garden in the 1730s, but selling patent, or indeed any, medicines in a public space or door-to-door seems to have become unusual towards the end of the eighteenth century, as we saw in Chapter 3.[12] Itinerant medicine selling did still occur, but it had become less acceptable as suggested by the tone of an advertisement in 1807 offering a one guinea reward for the apprehension of a travelling rag collector and 'vender of quack medicines' who had poisoned a child and 'several others have been thrown dangerously ill'.[13] A few colourful irregular owners such as Brodum, Solomon and Lignum sold their medicines as they travelled the country (Chapter 3). But, as mentioned before, they imitated regular medicine as much as possible, and they incorporated selling their medicines into their consultations, which took place indoors at a previously announced address. Brodum himself made this clear when advertising his visit to Leeds in 1793. 'Dr Brodum is not a person who goes from house to house vending medicines and calling for papers. His method is to prescribe and furnish such medicines only as are necessary for patients under his direction'.[14] A description of anybody selling *patent* medicines in public or door-to-door in late Georgian England has not yet been found; though it would be surprising if it did not happen at all. Indeed, this Brodum quotation implies that such door-to-door vending was carried out by others.

Although almost any shopkeeper could sell patent medicines, the major players, especially in the towns, were the members of the print trades and the druggists. Several historians have recognised that printers and booksellers often sold medicines, but they often felt that this was a minor sideline. John Feather claimed that 'almost all booksellers seem to have sold medicines', and Peter Isaac wrote that 'it seems to have been universal that members of the book trade also dealt in nostrums'.[15] In her unsympathetic account of patent medicines, Lisa Cody referred to 'book and print sellers who sold quack or patent medicines on the side'[16]: while Hannah Barker described 'the army of booksellers, printers and other small traders who sold medicines in provincial towns'.[17] However, none of these historians attempted to analyse in any detail the importance of the print trades in the retailing of the medicines or, conversely, the importance of patent medicines to the booksellers and printers. Brown provided more specific information for Bath with a sample of newspaper advertisements from the 1790s showing that 12 of the 36 named retailers were members of the print trades and five were apothecaries, chemists or druggists.[18] However, health was Bath's main industry, so it was a far from typical English town. My analysis of the advertised retailers in the studied newspapers and other contemporary reports shows that the members of the print trades were even more dominant than previously suggested in selling patent medicines in the late eighteenth century, before developing a shared role with the druggists at the beginning of the next century.

Figure 5.1 shows the predominance of the print trades in the late eighteenth century. Insufficient data is available on the retailers' occupations in 1769 for an assessment, but in all four newspapers during 1781 and 1794, most of the advertised retailers were members of the print trades, with a minority of chemists/druggists, grocers and other trades. This predominance is particularly striking in the Leeds newspapers, which circulated in a relatively compact area in Yorkshire based on the towns of Leeds, Bradford, Huddersfield, Wakefield, Halifax and Barnsley together with the southern Dales; few occupations other than the print trades were mentioned in 1794. The predominance was somewhat lower both in the Birmingham newspaper area, which included much of the Midlands, and in the Salisbury area, which stretched from Hampshire to Devon and Bristol; though the print trades still provided 63% and 68% respectively of the retailers with identifiable occupations who were advertised in these two newspapers in 1794. The explanation for this difference in emphasis between west Yorkshire and the other two areas is not clear, but it may have been related to the compactness and greater urbanisation of the Leeds area compared to the other two. These factors might have enabled the medicine retailing to be concentrated on the print trades amongst the greater range of shopkeepers in the towns in the Leeds area, while in parts of the other two areas the choice of local traders might have been limited. Nevertheless, in all three areas during these two years, the print trades provided the majority of advertised patent medicine retailers.

Leeds Intelligencer

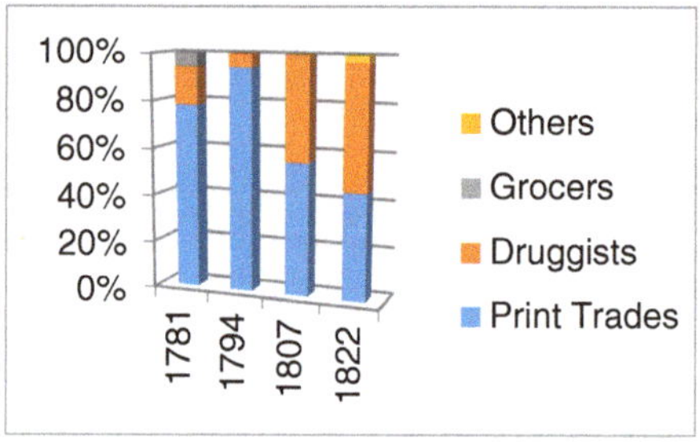

Leeds Mercury

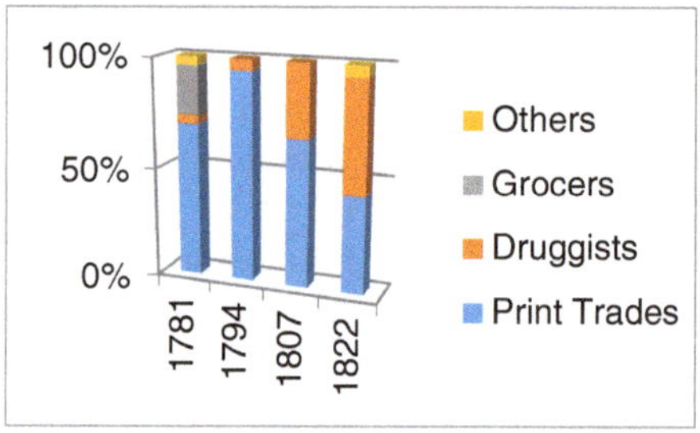

Aris's Birmingham Gazette

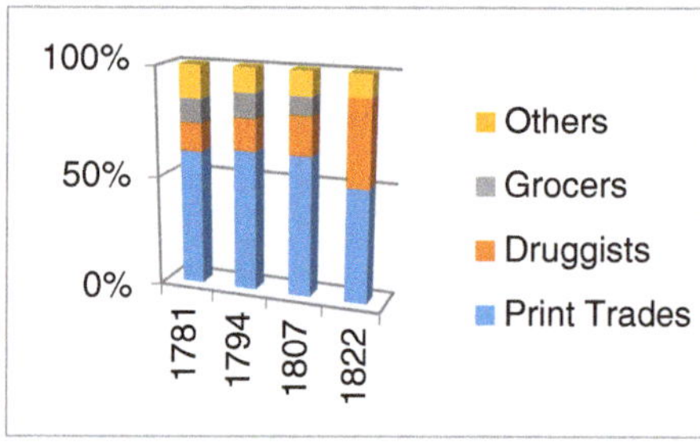

Salisbury & Winchester Journal

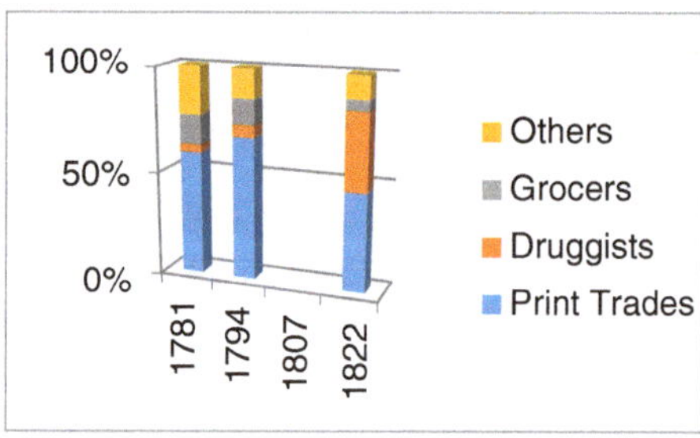

◂ **Fig. 5.1** Incidence of different trades amongst advertised medicine retailers in the four studied newspapers. No directories were available for all three areas in 1769: in 1807, no directory was available for the Salisbury area and the directory for the Leeds area was for that town only

The dominance of the print trades is rendered even more striking when we consider that booksellers and printers were not abundant in English towns. Grocers, tailors and shoemakers were all more numerous than booksellers in the 1770s and 1780s.[19] In particular, we might expect grocers to appear more frequently amongst the medicine vendors, given the overlap between medicines and food as explained in Chapter 1; yet only a few grocers were advertised as such, in spite of grocers being at least four times commoner than booksellers in English towns in the 1790s.[20]

In the last two years covered by the studied newspapers the picture altered. In all three areas, the advertised retailing of medicines was now shared between the print trades and the druggists, with few other trades involved. The timing of this change was similar to the switch in *wholesaling* away from booksellers and towards medicine specialists and chemists/druggists which we saw in Chapter 4. Indeed, some of the reasons for the change in retailing were similar to those promoting the switch in wholesaling. The expertise of the druggists was probably growing in importance, while the general promotional and retailing skills of the printers and booksellers were less sufficient on their own for the successful selling of patent medicines. Later in this chapter, the degree of medical expertise which the print trade members might have acquired will be discussed.

The alterations in the medicine excise duty arrangements provide an additional, more specific, reason for the greater role of the druggists in selling medicines in 1807 and 1822: druggists could now freely advertise their provision of patent medicines. As discussed in Chapter 2, the 1783 and 1785 Medicines Acts, which introduced the medicine excise duty, exempted regular druggists from the requirement to take out a licence to sell medicines. But if a druggist clearly sold patent medicines, he would have to pay to take out a licence and more of his stock would probably be subject to the excise duty.[21] The 1802 Medicines Act abolished this exemption for druggists, and, as they were now all required to take out a licence, they had no reason to avoid selling patent medicines and be identified as doing so.[22] Druggists could now compete freely with the members of the print trades in the sale of advertised medicines.

So, the members of the print trades had a dominant role in selling the patent medicines advertised in these studied newspapers in the late eighteenth century, and a major one afterwards. Can we generalise this role to the sale of all patent medicines across the country? The answer would seem to be yes as writers in the late eighteenth century often assumed that nearly all patent medicine vendors would be members of the print trades. When John Hunter was advising Edward Jenner on the sale of Jenner's proposed patent medicine, a tartar emetic, he suggested: 'Had you not better let a bookseller have it to sell, as Glass of Oxford did his magnesia?'.[23] Francis Spilsbury wrote a polemic against the workings of the 1783 Medicines Act, and he started it with: 'To the Booksellers of Great Britain' on the apparent assumption that the booksellers encompassed most patent medicine vendors at that time.[24] As one of the prominent medicine owners who organised his own wholesaling across the country, Spilsbury would have had a clear impression of the national market. In addition, newspaper advertisements sometimes reflected a belief that unidentified medicine retailers were likely to be members of the print trades. For example, a 1769 advertisement for Beaume de Vie named the newspaper printer as a retailer and then added 'all county booksellers', while a 1794 advertisement for Hill's Pectoral Balsam of Honey only mentioned the 'printer of this paper' and 'all booksellers and stationers in the circuit thereof'.[25] Later writers might have recognised a shared role for provincial booksellers and druggists in selling patent medicines. Thus, one of Harrison's respondents, an anonymous Suffolk physician, wrote about inappropriate prescribing by druggists in 1806 and concluded that 'he [the druggist] and the booksellers are generally the venders of nostrums, patent or not patent, which deluge this country'.[26] An 1822 advertisement in *Aris's Birmingham Gazette* listed 'all respectable medicine venders, booksellers and druggists'.[27]

In summary, the participation of booksellers and printers in selling patent medicines has often been recognised, though it has usually been considered that many other types of tradesmen were also involved. Early in the Georgian era, a reasonably wide range of tradesmen did sell medicines, but we can now see that members of print trades were the majority of patent medicine vendors in the second half of the eighteenth century, dominating this area of retailing. The druggists, freed from the constraints of the early Medicines Acts, joined in after the 1802 Medicines Act; but the members of the print trades still had a major role.

Newspaper Printers as Medicine Retailers

The extensive participation of members of the print trades in selling patent medicines does not mean that all those members were equally involved. For a few it was a substantial part of their business, for many it was a small sideline, and some chose not to be included at all. The newspaper printers, who were commonly also the newspaper owners, were normally in the first category. They were frequently the most prominent retailers advertised in the local newspapers and, as we saw in the last chapter, they could also be local wholesalers. But there were some exceptions. Looking at the studied newspapers, the printer of the *Leeds Mercury* in 1769, James Bowling, seems to have had no involvement with patent medicines: only a single advertisement for one of them appeared on three occasions in his newspaper during the first half of the year and his name was not mentioned in it.[28] Elsewhere, the account book of John Fletcher at the *Chester Chronicle* shows that he advertised, but did not sell, patent medicines for several years after buying the newspaper in 1783; though he did sell them in later years.[29] In contrast, in 1769, Benjamin Collins, the printer of the *Salisbury Journal*, was heavily engaged in medicine selling: in the first six months of the year, the *Salisbury Journal* advertised 38 medicines and Collins was printed as a retailer for 30 of them.

An inspection of provincial newspapers quickly shows that James Bowling and John Fletcher were uncharacteristic newspaper printers when they did not participate in medicine selling. The other printers in the remaining studied years, including James Bowling himself in 1781 and 1794, were selling at least half of the medicines advertised in their newspapers, sometimes all of them (Fig. 5.2). One apparent exception was the *Leeds Mercury* in 1807; but this is misleading. As we shall see later, Edward Baines, the printer of the *Leeds Mercury*, had been barred from selling medicines himself when he bought the newspaper in 1801. In 1807 he installed his brother John as the medicine retailer on the newspaper premises: but this arrangement only started in March, producing an artificially low figure for the whole six months. Elsewhere, John Ware was listed as a retailer for 18 of the 25 medicines he advertised in Whitehaven during 1800. This data from across England demonstrates that most newspaper printers were selling a substantial quantity of

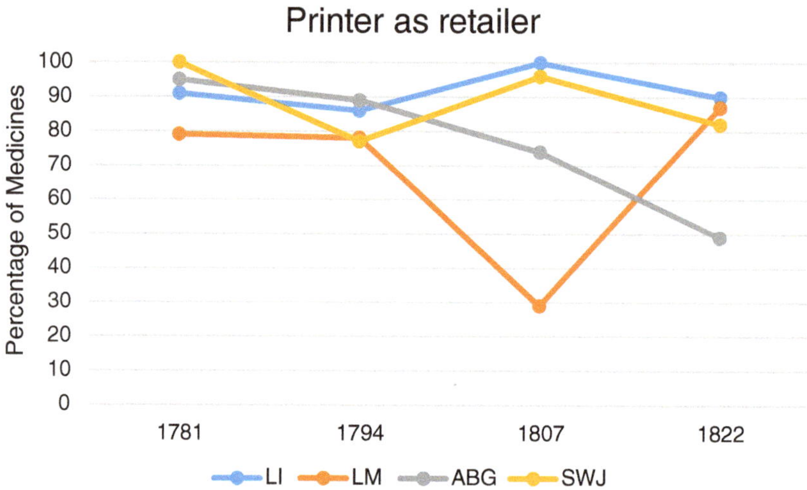

Fig. 5.2 Percentage of advertised medicines in the studied newspapers with the newspaper printer named as a retailer

patent medicines, and they were doing so throughout the later Georgian period.

Away from the medicine advertisements, other evidence confirms the impression that medicine vending was a significant part of the businesses of the majority of newspaper printers. The printers of *Aris's Birmingham Gazette*, Pearson and Rollason, were described as retailers for 72 of the 76 medicines advertised in their newspaper in the first half of 1781; but this was just part of their stock. Their handbill, probably from 1782, listed their 161 available medicines together with veterinary products, inks and cleaning materials.[30] Indeed, Joseph Greene, a Stratford parson, described in 1778 how he stopped for breakfast in Birmingham while returning from Lichfield, and took the opportunity to go next door to Pearson and Rollason, where he bought a book, checked on a previously ordered subscription, and purchased a rat killer called Poultey's Paste on Rollason's recommendation.[31] Rat killers were often sold by patent medicine vendors. Another illustration of the importance of medicine selling to the business of a printer is provided by a non-medical advertisement from 1769 in the *Salisbury Journal*, printed by Benjamin Collins. The advertisement offered a job for 'a sober industrious man, who has been accustomed to serve and make up medicines in an apothecary's shop' or perhaps a promising apprentice, with both enquiries and applications to be made to the newspaper office.[32] Now it is possible that Collins was advertising a job for somebody else, but he did not normally do this, and it is probable that he was seeking his own assistant to manage his substantial medicines business.

By coincidence, the owners of both the *Leeds Mercury* and *Aris's Birmingham Gazette* sold a proportion of their businesses in February 1801, and the legal documents for the sales provide further compelling evidence of the importance of selling medicines to the business of a newspaper printer. John Binns, bookseller and medicine vendor, and George Brown, bookbinder, had bought the *Leeds Mercury* from James Bowling in 1794. Binns died in 1796, and his widow and Brown sold the newspaper to Edward Baines in 1801.[33] In the Articles of Agreement for the sale, Mrs. Binns and Brown agreed not to publish a newspaper in the Leeds area, and in return Baines agreed that he 'shall not nor will at anytime hereafter vend or sell any medicine or medicines'.[34] The next line is written in a different ink between the existing lines: 'during the time he occupies the above premises'. It seems that their medicines business had been sufficiently large for Mrs. Binns and Brown to seek to retain it without any competition from the new owner. Baines probably then realised that the Articles of Agreement as they stood amounted to a life ban on selling patent medicines, and so he arranged for the qualifying phrase to be added after the original document had been written out. He would not have wanted to be excluded permanently from this profitable activity. The document contains no other additions. Baines appears to have stuck to the agreement by advertising medicines in his *Leeds Mercury*, but not listing himself as retailer. In 1807 he got round it in two ways. He moved the printing office up Briggate to different premises, and he installed his younger brother John as stationer and medicine vendor in the front of his new office.[35] In an advertisement in the *Mercury*, repeated two days later in the *Intelligencer*, John Baines announced the opening of his shop, restrictions on credit arrangements to keep prices low and a large stock of stationery, and he finished with 'he has also received supplies, and will regularly keep all the prevailing patent medicines, from the warehouses of the patentees, warranted genuine'.[36] Medicine selling was important to the Baines family.

Whereas Mrs. Binns and Brown sold their newspaper while retaining the medicines business, Thomas Pearson, the owner of *Aris's Birmingham Gazette*, sold the bookselling, stationery and medicines side of his business on a twenty-one year lease to Jonathan Knott and Robert Lloyd, while retaining the printing side and the newspaper.[37] Pearson agreed to pass on 'all his stock of books papers medicines and stationery articles of every sort and kind', and he undertook not to engage 'in the trade or business of a stationer, bookseller or vendor of medicines' within thirty miles of Birmingham. Knott and Lloyd made similar assurances about printing and producing a newspaper. Selling medicines was the only business activity unrelated to print mentioned in these Articles of Agreement, and clearly it was a sufficiently large part of Knott and Lloyd's future income to require a specific arrangement. Knott and Lloyd then acquired the newspaper and the printing business in December 1803 after Pearson's death. Elsewhere, John Fletcher's lack of involvement in retailing patent medicines after his hurried purchase of the *Chester Chronicle* in 1783 is probably explained by a similar type of agreement with John Poole, a local printer and the dominant partner in the group which sold the newspaper to Fletcher.[38] Although a legal document has not survived, a retention of the patent medicines side of the business by Poole is suggested by his payment for medicine advertisements in the *Chester Chronicle* and his frequent listing as a medicine retailer. In common with most newspaper printers, selling medicines was a substantial part of the income of these proprietors, not a minor sideline.

To what extent were the printers who did not publish newspapers engaged in selling patent medicines? This of course varied from printer to printer, but none in our studied areas was as heavily involved as the newspaper printers. For example, Leeds in the mid-1790s had three printers.[39] Two, Thomas Wright and James Bowling, published newspapers and were frequently named as medicine retailers in the runs of the two newspapers in 1794 (Fig. 5.2), whereas the third, Thomas Gill, was not named at all. Gill might have been selling medicines without seeking to be named as a retailer, and he certainly printed handbills and treatises for those who were actively engaged in the market.[40] However, Gill was not seeking to develop a patent medicines trade, in contrast to the newspaper printers, and any such activity would have been small. Printers who did not produce newspapers were named as retailers in all three of the studied areas, but none appeared as frequently as the newspaper printers, and others did not appear at all.

So why were the newspaper printers so prominent as patent medicine retailers? There are several possible answers to this question; but they all need to reflect the fact that the patent medicines industry could be very profitable and many tradesmen might have wanted to engage in it. So we need to consider why the newspaper printers were in a strong position in the market, not why they sold medicines at all. One initial possibility is that the newspaper printers might seem to have an advantage in being able to advertise medicines for free in their own publications. However, there was no such thing as a free newspaper advertisement after the introduction of a one-shilling duty on every newspaper advertisement in 1712.[41] From 1757, two shillings were collected for every advertisement, and the duty increased in stages up to 3s 6d. Also, we have no evidence that the printers generally allowed medicine advertisements to appear in their newspapers without the normal fees being paid. Indeed, the opposite is demonstrated in the advertising account books for the *Hampshire Chronicle*, which listed who was paying for each individual advertisement.[42] An inspection of the books for 1781 reveals that 277 advertisements for patent medicines were published in that year, and *all* were charged to individuals or companies other than the printer John Wilkes. This happened even when John Wilkes was mentioned in the heading, which implied that he had some responsibility for the advertisement. The incomplete evidence from the day books of the *Cumberland Pacquet* also suggests that the vast majority, perhaps all, of the medicine advertisements during 1800–1802 were paid for.[43] Printers could not advertise in their own newspaper without paying the excise duty, and, as advertising was a major source of income, they normally expected someone to pay for each one.

Another potential explanation for the prominence of newspaper printers may have some merit. This reason is that the newspaper printers had the advantage of being able to use their own system of newsmen and other local distributors to deliver the patent medicines within their immediate locality. However, the practicality of this type of distribution is unclear, especially as we do not know whether the newsmen normally walked or rode their weekly route.[44] As mentioned before, patent medicines were mostly liquids in bottles of commonly half or one pint, and so they would have been much heavier to carry than the single sheet, four-page, newspapers of the period. So if the newspaper distribution system in the immediate locality was used extensively to deliver medicines, it would be more a question of carrying the newspapers with the medicines

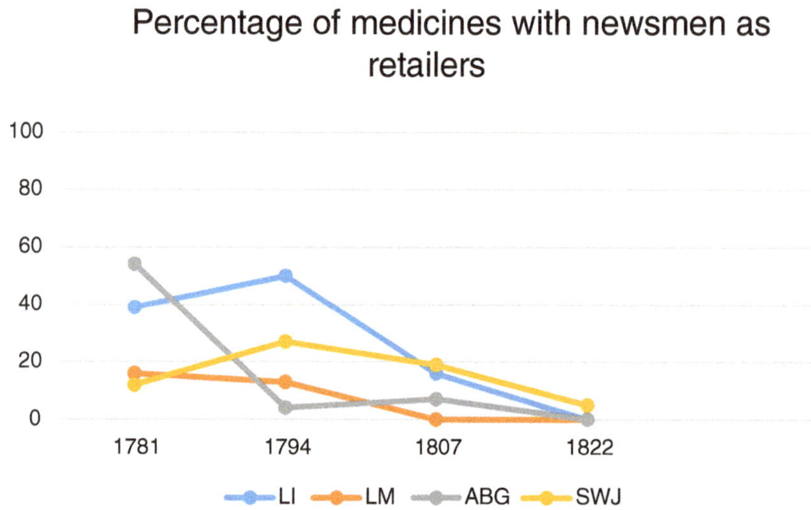

Fig. 5.3 Percentage of advertised medicines with the newspaper newsmen or distributors mentioned as retailers

rather than the other way round. Distributing patent medicines in this way did occur and it could be mentioned in the newspaper advertisements.[45] For the studied newspapers, this occurred relatively frequently in the eighteenth century (Fig. 5.3); but it was mentioned less often in these newspapers in 1807 and 1822; for example, only four times for 88 medicines in the *Salisbury and Winchester Journal* in 1822. Whatever effectiveness in disseminating medicines can be attributed to the newspaper distribution system in the immediate locality, it was diminishing with time. This distribution system probably played only a small part in the prominence of newspaper printers in medicine vending.

A more persuasive explanation for the importance of the printers was their position at the centre of a *regional* network of agents. All provincial newspapers required a system of regional shopkeepers, often booksellers and stationers, who would sell the paper in their shops, supervise the distribution of the newspaper in their area, receive orders for books and other items which the newspaper printer might be able to supply, and, most importantly, receive and take payment for all types of advertisements. The newspapers could not survive without such a network, and the one for the *Salisbury and Winchester Journal* has been well documented by Christine Ferdinand.[46] The lists of medicine retailers printed in the advertisements often had many names in common with the published list of regional agents for that newspaper. For example, 22 of the 35 booksellers who were agents for the *Salisbury and Winchester Journal* in 1781 were also printed in advertisements as medicine retailers.[47] Not all the agents in a newspaper's regional network were advertised as selling patent medicines, but the newspaper printer had easy access to the many who did. He could then act as the local wholesaler, simplifying national distribution, and, as we shall see, having a strong influence on medicine advertising. With this central position in the regional market, medicine selling by the newspaper printer would be both inevitable and profitable.

Channelling the medicines through the newspaper printers also had major financial advantages for both the wholesalers and the printers. As we saw in the last chapter, the cost of the medicines supplied to the printer was often balanced against the charges to the wholesalers for their advertisements in the newspapers. So the amount of money which had to be transferred across the country to pay for the medicines or to pay for the advertisements was sharply reduced. For some wholesalers delivering to John Ware in Whitehaven, the debit for their advertisements almost balanced the credit for their delivered medicines; though for others, a substantial sum still had to be transferred. For example, only £2 2s 3d was required to settle for the delivered Spilsbury's drops at the end of 1799, while we saw in the last chapter that Ware paid £34–£74 annually to Dicey and Sutton from 1800 to 1803. Whatever the final balance, dealing with a single newspaper printer would have helped both the wholesaler and the printer by ensuring that payments were simplified and by limiting the amount of credit which built up. Another possible reason for the newspaper printer selling medicines was that it could encourage the wholesalers to place lucrative medicine advertisements in the newspaper. But there is little to back up this suggestion, with Edward Baines and John Fletcher advertising medicines without selling them and John Ware selling medicines which had not been advertised in his newspaper.

This central position in the regional market would have encouraged the newspaper printer to develop expertise in the use of print to promote the medicines and in the medical knowledge to support their use, reinforcing his or her dominant role.[48] The importance of print in encouraging the sale of patent medicines and boosting their effectiveness will be discussed in Chapter 7: the newspaper printer would be the local expert on how to use print in this manner. The degree of medical knowledge of printers and other print trade members who sold medicines will be discussed below, but we can already see that many printers carried large stocks of medicines and they would have known something about their indications and use. This double expertise in medical therapy and the use of print gave the newspaper printers a continuing advantage over other medicine vendors.

Booksellers as Medicine Retailers

Although the newspaper printers were important in medicine retailing, they were few in number: most members of the print trades who sold medicines were booksellers, stationers and occasionally bookbinders. The term 'bookseller' was often used to refer to both the vendors of books and the vendors of stationery. In this period, those who sold books also usually sold stationery and vice versa; so they were often described as booksellers regardless of which activity was predominant. This eighteenth-century usage will be followed here, allowing 'bookseller' to refer to a variable mixture of both occupations. We should also appreciate that most newspaper printers were also booksellers, but this section is largely concerned with the much greater number of booksellers who did not publish a newspaper.

This section will show that these print trade members did not show a uniform involvement in providing medicines for the community. Some specialised in medicine vending and became important in the local provision of healthcare, while others only dabbled in selling medicines, or perhaps avoided it altogether. Some of the ways in which booksellers were appointed and remunerated for selling patent medicines will then be assessed, and the section will finish by answering the question: Why were so many patent medicines sold by booksellers?

As we saw earlier, in rural areas without a local bookseller almost any shopkeeper could sell medicines. A single bookseller in a smaller town often sold medicines, but, according to the newspaper advertisements, such a bookseller commonly sold only a small number of them; and the few surviving records of provincial booksellers from this period also demonstrate only a modest engagement. John Cheney founded a long-lasting family printing and bookselling business in Banbury in 1767, and his advertising handbill from around 1788 listed 46 categories of items for sale, including garters, laces and bodkins, and miniature paintings by his son, without any medicines being mentioned.[49] But in 1809–1811, his successors purchased medicines from a London wholesaler for £7 5s 9d.[50] John Feather studied the records of John Clay, bookseller in Daventry and surrounding towns, and concluded that Clay sold medicines in all his shops without selling very many.[51] Feather's analysis of Clay's accounts has been questioned by Fergus and Portner, but all their figures show a minimal role for medicines in the business finances. For example, in October 1768, income from medicines was 3s 2d out of a total income of £18 19s 7d, and for a three-month period in 1779, the income was coincidentally 3s 2d out of a total of £154 14s 0d.[52] It seems that selling medicines was undertaken by many booksellers in small towns, but it was often only a trivial sideline to their business.

In larger towns with more booksellers, some would provide an extensive range of patent medicines, while others would have little or no involvement in medicine vending, as we can see with the booksellers named as retailers in the newspaper advertisements in Leeds (Table 5.1). If we look at the major towns in the distribution area of the two studied Leeds newspapers, the medicine-selling booksellers became a minority of all the booksellers when six or more booksellers were available in the town.

Table 5.1 The total number of booksellers in each town in contemporary directories, and the number of booksellers named as retailers in 1794 and 1822 medicine advertisements in either of the two Leeds newspapers. Some of the booksellers were also non-newspaper printers. *Sources*: *Leeds Intelligencer*; *Leeds Mercury*; Barfoot and Wilkes, *Universal British Directory*; Baines, *Baines's Yorkshire*

	1794		*1822*	
	Total no. booksellers	*Named in medicine adverts*	*Total no. booksellers*	*Named in medicine adverts*
Leeds	7	2	30	7
Bradford	4	2	7	2
Wakefield	3	2	6	2
Halifax	3	2	14	2
Huddersfield	1	1	7	2
Barnsley	2	2	3	3

Also, the named booksellers in the two Leeds newspapers varied in their enthusiasm for medicine selling. Some were listed in advertisements as agents for a wide range of medicines while others were only mentioned for one or two. Thus, John Heaton, the most prominent bookseller in Leeds in 1822, was listed for 37 medicines, while two other booksellers were only printed in the advertisements for one medicine each. The presence in Leeds of two newspaper printers with an active interest in medicines did not prevent a bookseller such as Heaton from having a substantial involvement in medicine vending. A bookseller's participation could vary, and in the account book of the Leeds printer Thomas Gill we can see a growing attention to medicine selling by one bookseller, Edward Greenwood, who was also a library owner.[53] The book recorded the orders of many tradesmen, including booksellers and druggists, for advertising bills and labels. In 1791 and 1792, Greenwood ordered stationery and printed bills, but none had anything to do with medicines. Greenwood did not appear in a Leeds newspaper advertisement for medicines in the first half of 1793. However, in the first half of 1794, he was listed as an agent for three medicines from Thomas Jackson. Then in March 1795, he ordered 200 printed bills for 'Balsam of Tolu' from Gill, amongst a similar quantity of other materials. Between February 1796 and December 1797, Greenwood placed five smaller orders, the first and last of which contained no medicine-related items, but the second in July 1796 included 100 labels for 'American Syrup'. In the third order in October 1796, he ordered 210 labels for 'American Vegetable Syrup', and the fourth in July 1797 included 230 labels for 'American Vegetable Syrup'. The intended use of the labels was not defined; but, assuming that they were bottle labels and that these syrups were the same, Greenwood probably distributed around 300 bottles in a year before ordering the third batch, a reasonable sale for a trader who does not seem to have devoted much earlier attention to medicines. Three other booksellers had no medicine-related items amongst their repeated orders from Thomas Gill. It seems that Greenwood, unlike several other booksellers, was pursuing a growing interest in medicine selling.

The greater involvement of some booksellers in medicine vending compared to others is also evident in the Birmingham area, although the data are less clear cut due to the more diffuse nature of the catchment area of *Aris's Birmingham Gazette*, which overlapped those of several other Midland newspapers, obscuring the total number of medicines sold by each bookseller. One example was Thomas Wood, who was listed for 15 medicines in the *Gazette* in the first half of 1822. His combined bookselling and patent medicines shop survived under different owners till the beginning of the twentieth century, retaining its original features from 1800 (Fig. 5.4).[54] An eyewitness account of Birmingham in the 1820s described Thomas Wood wearing a wig with a hat on top and standing on the steps of this shop, which 'was well patronised for stationery and patent medicines'.[55] Those steps were probably the same ones shown in this photograph from the end of the nineteenth century. Outside Birmingham, the Leicester bookseller Hursley was named for 21 medicines in 1781, while Gregory, another bookseller in the town, was only mentioned for two.

Further south in the catchment area of the *Salisbury and Winchester Journal* a different picture can be seen, with some booksellers having a significant interest in medicine selling early in our period, but this specialisation becoming less apparent later. For example, Thomas Baker, a Southampton bookseller, acted as an agent for London medicine wholesalers in the 1780s as we shall see in the next chapter, and he was named in advertisements for sixteen medicines in 1781. Over time, the specialisation in medicine vending diminished amongst booksellers in the area, with none being named in more than three medicine advertisements in this newspaper during the first half of both 1807 and 1822. This later absence of specialisation in medicine vending amongst the booksellers might have been secondary to the area having few large towns and a modest population growth. Salisbury itself had a stable population of about 6500–7000 throughout the eighteenth century, only gradually rising to 9531 in the 1841 census.[56] The lack of urbanisation in comparison to the growing towns in the Leeds and Birmingham areas might have meant that the population in each town was insufficient to support substantial medicine vending by both booksellers and druggists, once the latter were competing actively in the market.

The specialised medicine-selling role of some of the booksellers was recognised in their publicity. In 1807, John Heaton, who had a stock of at least 43 medicines according to the advertisements in the studied newspapers, made it clear that medicine vending was a substantial part of his business when he headed an advertisement as 'John Heaton,

Fig. 5.4 Britten's book and medicines shop, 78 High Street, Birmingham, *c.* 1895 (reproduced with permission of the Library of Birmingham). In 1822, a previous owner of this shop, Thomas Wood, was named as a retailer for 15 patent medicines in advertisements in *Aris's Birmingham Gazette.* The sign over the door is a model of a bible, symbolising a bookseller, while medicines are displayed in the windows

Bookseller, Stationer, and Vender of all the most approved Patent Medicines, exc'.[57] Other booksellers were described as patent medicine 'venders', alongside their other business interests, in trade directories. For example, the Huddersfield booksellers and printers Joseph Brook and Thomas Smart were both designated as 'venders of patent medicines' in *Baines's Directory* (1822).[58] For some of these booksellers, as for newspaper printers, medicine retailing was an important part of their income.

How did booksellers manage to get named as advertised retailers for medicines? For most medicines, probably any shopkeeper could become a retailer and be named in the advertisements, but a few wholesalers sought to have some influence on who sold their medicines locally. As we saw in the last section, many booksellers were advertised as medicine vendors because they were local agents for the newspaper printer. Also, a bookseller or other shopkeeper could apply to the newspaper printer for his or her name to be added to the list of local retailers: as an advertisement for the Pectoral Lozenges of Tolu put it, 'Any shop-keeper of credit, in any of the towns thro' which this paper circulates, who are inclinable to deal with the above articles, may on application to the printers, have their names added to this advertisement'.[59] A few wholesalers tried to select their retailers, or at least limit their number, and, as we have often observed in this book, Francis Newbery had a sophisticated system. Retailing his medicines required his annual approval with a legally valid certificate, and he seems to have restricted sales to a single agent in most towns.[60]

What were the financial arrangements for medicine retailers? A striking finding is that patent medicines were sold to the customer at a fixed publicised price which was normally the same in Leeds, Birmingham, Salisbury or anywhere else. The price often remained unchanged over several decades, as demonstrated in Appendix A.1. Thus, in 1781, Leake's Pills were advertised in Leeds, Birmingham and Salisbury for 2s 6d, and Norton's Maredant's Drops were advertised for 6s 0d in the same towns in both 1769 and 1781. In 1807, Leake's Pills were being advertised in Leeds for 2s 9d, which was the same price as 1781 plus the 3d excise duty introduced in 1783. These fixed prices were in use earlier in the Georgian period: for instance, Dr. Henry's Chemical Nervous Medicine was sold at 7s 0d in 1748 as well as in 1769 and 1781, and Turlington's Balsam was sold at 1s 9d in the same years.[61] The price of a few medicines crept up over the years (for example, Spilsbury's Antiscorbutic Drops, which increased in price from 4s to 4s 6d [before tax] between 1781 and 1794), but for the majority the price before tax remained unchanged from decade to decade. Omission of a price from any advertisement consisting of more than a few lines was uncommon, and, with an occasional exception, the minimum price for a bottle or box of a patent medicine was one shilling plus any duty. Of course, the retailers could still have offered a reduced price or other forms of discount in a variety of ways, but evidence of this occurring is not available and contemporary accounts seem to assume that the retail cost of a medicine was fixed when they refer to the usual prices. In other words, the retailers competed with each other by providing a good range of fresh products accompanied by relevant information, not by offering the lowest price.

The few surviving examples of the financial arrangements between wholesalers and retailers demonstrate a range of profit margins for the retailers. William Jones supplied his Tincture of Peruvian Bark to retailers at 30s per 'dozen', with a 'dozen' containing 14 bottles: at a retail price of 3s 6d, this gave the retailer nearly 40% of the amount paid by the consumer.[62] This margin was higher than usual: the Collins family were accused in an anonymous pamphlet of taking a profit of 25% on their patent medicines, and the vendors of the Edinburgh Febrifuge Powder had a modest profit margin of about 15% which was only paid if they ordered a minimum of two dozen packets.[63]

The day books of the *Cumberland Pacquet* reveal rather more on the profit margins of John Ware. The wholesale price of nine of his medicines are provided by the amount of credit he received for returning broken bottles to the wholesalers, and when compared with the retail prices they reveal the available retailers' profits (Table 5.2). Around a quarter

Table 5.2 Percentage of retail price retained by the retailer(s) in Cumberland. *Source* Cumbria Archives, Day Books of John Ware. Gribble was the second husband of Dorothy Spilsbury

Medicine	*Wholesaler*	*Wholesale price*	*Retail price*	*% of retail price to retailer(s)*
Essence of Mustard	Johnston	1s 10.5d	2s 9d	32
Velno's Syrup	Swainson	8s 3d	11s 6d	28
Godfrey's Cordial	Dicey & Sutton	6d	8d	25
Wheatley's Ointment	John Wye	1s 4d	1s 9d	24
Greenough's Tincture	Dicey & Sutton	10.5d	1s 1.5d	22
Hooper's Pills	Dicey & Sutton	10.5d	1s 1.5d	22
Spilsbury's Drops	Gribble	4s 0d	5s 0d	20
Dixon's Pills	Ching & Butler	2s 3d	2s 9d	18
Robberd's Cough Drops	Robberds/Forbes	2s 3d	2s 6d	10

of the price paid by the consumer was retained in Cumberland; but not necessarily all by Ware as the day books also provide a few examples of his supply of small quantities of medicines to local medicine vendors at less than the retail price, sharing the overall profit. For example, in April 1800 he supplied Mr. Scott, a Carlisle printer, with four bottles of Essence of Mustard at 2s 6d each, 3d less than the retail price, and around the same time he supplied Mr. Walker, a Cockermouth bookseller, with Spilsbury's Drops at 4s 6d, 6d less than the retail price. Confusingly, the day books also provide examples of Ware sending medicines to local retailers at the full price, but these are not account books and the complete financial arrangements between Ware and the local medicine retailers remain hidden.

The reasons why booksellers often sold medicines, and sometimes made this a substantial part of their business, were similar to the reasons for the involvement of London booksellers and provincial newspaper printers in the wholesaling and retailing. Vending medicines and vending books could utilise common skills, including obtaining a quick and reliable supply from a central source, selling at a fixed price determined by others, and participating in nationally based promotion. Many booksellers were already part of regional networks for the distribution of the local newspaper and books from London. In addition, like the other engaged members of the print trades, the provincial booksellers were more familiar and skilled in the use of the printed word, the essential ingredient for promoting patent medicines, than other potential rivals such as grocers and druggists. As mentioned already, the booksellers were also helped by the apparent reluctance of the druggists to engage in publicised patent medicine selling before the 1802 Medicines Act.

Ware's day books demonstrate that he was supplying medicines to a regional network of retailers in Cumberland and the Isle of Man, but they provide very little detail on how the network worked. As we have seen, the books document the arrival of thousands of bottles and boxes from the wholesalers, but only a small fraction of these medicines are recorded in the books as being sold directly to consumers or being passed on to local retailers in the network. The supply of medicines to local retailers must have been recorded elsewhere. The handful of entries in the day books dealing with transactions with local agents over the five years show medicines being sent to Ware, as well as distribution from Ware. For instance, in June 1802, Mr. Crosthwaite, a Whitehaven bookseller, was credited for sending nine bottles of the Balm of Gilead to Ware, and then debited in the following month for nine bottles of the same medicine 'returned to him' (i.e. Crosthwaite). The probability is that the day books were only used for extraordinary arrangements with local retailers, with the routine supply being documented by other means.

One speculative additional reason for the extensive participation of booksellers in medicine retailing would be the favourable ambience of bookshops. Bookshops could be sites for gossip, news and discussion, providing an environment for decisions to be made at leisure.[64] Many contained libraries for books to be browsed and borrowed, they advertised reading rooms, and illustrations show advice being given.[65] *Real Life in London* provides a contemporary fictional account of a Georgian bookshop and library, describing the range of advice given to customers while the narrator was 'lounging away an hour in digesting the politics and news of the day'.[66] The decision to buy a patent medicine was considerably more difficult than buying other movable goods, requiring time, information and, sometimes, the opinions of others. The leisured supportive atmosphere of a large bookshop might have been a good place to make such a decision, in comparison with the potentially more 'take it or leave it' feeling of a grocer's or mercer's shop, or even a druggist's. In the 1810s and 1820s, John Heaton's Leeds bookshop was 'the chief bookselling business in Yorkshire', and it was also the most prominent source of patent medicines in the town apart from the newspaper printers; though it retained the appearance of a private house with no shop front.[67] Booksellers' shops, and particularly Heaton's shop, were noted as centres for dialogue and information:

> Whilst the taverns furnished a rendezvous for the local politicians, the booksellers' shops were the recognised gathering-places of those who were inclined towards literature. It was here that the clergy met on Monday morning, to discuss together perchance the sermons of the previous day, more probably the latest pamphlet from London, or the contents of the new number of 'The Gentleman's Magazine'.[68]

Following the normal practice of Victorian writers, this account of Heaton's shop makes no mention at all that medicines were sold. However, we can conjecture that the discussions on a Monday, and any other day, might also have included the virtues or otherwise of patent medicines.

Printers and Booksellers as Irregular Medical Practitioners?

The extensive involvement of many newspaper printers and booksellers in the sale of patent medicines raises the question as to whether they should be regarded as irregular medical practitioners. This is a difficult question to answer about an era when medical knowledge was not confined to those who had received formal or informal medical training. Some understanding of medicine was part of a good education, and the lady of the house was expected to know recipes for medicines as well as recipes for food. Lower down the social scale, domestically prepared remedies were a large part of healthcare, and the popularity of the medical advice in John Wesley's *Primitive Physic* made the book a bestseller for half a century. So a degree of medical knowledge, or at least easy access to it, was universal, and the questions need to be more precise. Did these members of the print trades use knowledge not generally available in the community to advise consumers on their health problems? Did any such advice extend beyond the use of particular patent medicines to other forms of therapy?

The information does not exist to answer these questions directly. The indirect evidence suggests that at least some of the provincial booksellers and printers possessed specialist medical knowledge which they passed on to the consumers. However, this knowledge was probably confined to the products they were selling, and there is no good evidence that they extended their advice to a wider range of therapies or provided other forms of treatment. One strand of the indirect evidence is that some booksellers and printers ran large stocks of patent medicines, often with several different medicines available for the same condition. So they probably needed to give advice as to which would suit a particular customer. As we saw earlier in the chapter, Pearson and Rollason, the printers of *Aris's Birmingham Gazette*, published a stock of 161 medicines in the early 1780s, and it is likely they provided advice to a consumer when he or she was faced with this extensive choice. Indeed, the Rev. Greene's account from 1778 of getting advice from Rollason himself on the best rat poison was described earlier, and Rollason probably would not have confined his recommendations to non-medical products. John Ware in Whitehaven also had a large stock of 87 'genuine medicines', which he listed in an advertisement.[69] Other major medicine-vending printers and booksellers may have held fewer medicines in their shops; but the local newspaper advertisements often indicate that they had at least 25–50 available, probably with others which were unadvertised. Amongst this number of medicines, there were several for a single condition. For example, when John Baines advertised thirty-two medicines available from his newly opened stationery and patent medicines business in 1807, four of them were primarily indicated for venereal disease.[70] Booksellers such as John Baines would probably have offered advice on which medicine to take, particularly when more than one was available in the shop for a specific condition.

Another type of indirect evidence for some specialist knowledge amongst the booksellers and printers is the family links with regular medical practitioners, which suggest a background awareness of medicine and the possibility of good access to medical knowledge if required. The studied newspapers provide two examples. Thomas Pearson part-owned and printed *Aris's Birmingham Gazette* from 1780 to 1802, and had an extensive interest in patent medicines as already described. His younger brother, Richard, was a well-known physician in Birmingham and London with an interest in therapeutics, succeeding William Withering on the staff of Birmingham General Hospital, and becoming in 1825 the first lecturer in materia medica and medical botany at the new Birmingham Medical School.[71] In a similar fashion, William Bird Brodie, a grandson of Benjamin Collins, was part-owner of the *Salisbury and Winchester Journal* for 40 years from 1808 and was able to supply at least 72 patent medicines in the first half of 1822 according to the advertisements in his paper.[72] His younger brother was Benjamin Collins Brodie, who was already a distinguished London surgeon in 1822 and was later to be President of both the Royal College of Surgeons and the Royal Society, receiving a baronetcy in 1834.[73] Both these younger brothers would have been able to provide advice on medical therapy and perhaps could have suggested other medical contacts when required. A rather different medical link involved John Heaton, the prominent Leeds bookseller and medicine vendor. He apprenticed his son, John Deakin Heaton, to a local surgeon at the age of sixteen and the young man later became a well-known Leeds physician.[74]

In addition, apprenticeships and the retention of a newspaper business within the family meant that knowledge in both the techniques of medicine selling and the medicines themselves could be passed down the generations. Both the *Leeds Intelligencer* and the *Salisbury and Winchester Journal* were owned by three generations of the same family, the Wrights in Leeds and the Collins/Brodie family in Salisbury. Apprenticeships link another four of the prominent medicine sellers in the Leeds area. Nathaniel Binns was a successful bookseller and medicine vendor in Halifax in the mid-eighteenth century. His son, John, was apprenticed to both his father and a London bookseller, and then came to Leeds in the mid-1760s.[75] He became a prominent bookseller, printer of the *Leeds Mercury* and a medicine vendor. One of his apprentices was John Heaton, mentioned at several points in this chapter, who became Binns's shop manager before setting up on his own, and another was Edward Baines who, as we have seen, bought the *Mercury* from Binns's widow and George Brown.[76] So John Binns, John Heaton and Edward Baines would all have gained experience and knowledge in medicine vending, and probably the medicines themselves, during their apprenticeships.

So far, the evidence presented has shown that the prominent vendors had a potential need and an opportunity to learn more about their medicines, but it has not demonstrated that they actually provided medical advice. A suggestion that they did so can be found in the advertisement testimonials, which describe the circumstances which induced the beneficiary to take the medicine. Barker has concluded that the members of print trades who sold medicines were trusted for their standing in the community, not for their medical expertise.[77] However, the testimonials suggest that a degree of knowledge about a medicine and its indications could enhance this trust. Of course, testimonials might not have been truthful and many of the stories were probably exaggerated; but the core of the story in the testimonial was probably correct, particularly, as we shall see in Chapter 8, many derived from an identifiable person. Even if a testimonial was entirely fabricated, it still had to present a plausible account, and so it should indicate how the medicine vendors often behaved.

The testimonials can show booksellers and printers using their experience on the efficacy of treatment to make recommendations: this indicates medical knowledge, albeit not acquired from an orthodox source. For example, a Liverpool bookseller demonstrated the advantage of experience when he reported that he had sold 150 doses of a medicine for the bite of a mad dog with no known failures.[78] Booksellers and printers could give direct advice on the best therapy. Sutton, a Nottingham bookseller and printer, provided a testimonial for Elliott's Restorative Tincture which described how he had recommended the medicine for one of his workmen who was dangerously ill with cough and asthma: 'Knowing that your tincture had been useful in similar complaints, I gave him a bottle, and prevailed upon him to make a trial of it'.[79] A former editor of the *Carlisle Journal* described how the subject of the testimonial 'exhibited every symptom of a person in the last stage of consumption', and this ex-editor 'prevailed upon him at length to make a trial of your Cordial Balm of Gilead.[80] In a widely used testimonial for Hickman's Pills, John Bent, a Barnsley bookseller and medicine vendor, described how he had been suffering from abdominal pain which had been unresponsive to medicines from an apothecary.[81] Bent's daughter had explained her father's problems to Griffith Wright, the printer of the *Leeds Intelligencer* and medicine vendor, when she had been visiting Leeds, and Wright had recommended and supplied the Hickman's Pills to Bent's daughter, curing her father's abdominal pain.

The printers and booksellers may have acquired a degree of medical knowledge: but the key observation is that contemporaries did *not* regard them as irregular practitioners selling patent medicines. This is particularly clear in the responses to Harrison's 1806 circular on medical reform. The correspondents often strongly criticised irregular practitioners of all sorts, druggists and the patent medicines themselves as being a threat to the health of the public and their own income, but they did not regard the booksellers who sold the medicines as being a problem. For example, a committee of regular practitioners in Nottingham described and counted various discreditable groups which hindered their work including druggists, irregulars of both sexes, quacks and uninstructed midwives; but booksellers or printers were not mentioned.[82] The report of a meeting of 'medical gentlemen' at Market Bosworth went further and added to the list 'ignorant mechanics', 'ignorant inoculators' and druggists who were only trained as apprentices to grocers and tea dealers, but again not print trade members.[83] This group of correspondents recognised that booksellers sold the patent medicines which they often deplored: however, they did not regard the booksellers as irregular practitioners or a threat in themselves to the regular profession.

So, although the degree of medical knowledge of the medicine-selling booksellers and printers remains unclear, circumstances suggest that they often would have been aware of the indications and potential benefits of their medicines, if only to advise the customers and to maximise their sales. Yet, however well-informed they were about the medicines, we have nothing to indicate that their medical advice extended beyond the use of these medicines. The contemporary regular practitioners took a similar view when they did not see the members of the print trades who sold the medicines as a threat, unlike the patent medicines themselves. We should regard these medicine vendors as probably being informed about their medicines and the remediable diseases, but they were not irregulars, quacks or empirics.

Conclusion

The main finding in this chapter is that the retailing of patent medicines had an organised and coherent structure whose main features were largely consistent across the country. Before this research, an initial glance at the retailing might have seemed to show that patent medicines were sold in a disorganised fashion by many different types of vendors, often as a minor sideline. This style of vending was more apparent at the beginning of the Georgian period, and this is as far as several historians, including Porter, have penetrated in their exploration of medicine selling. However, when more investigation is undertaken, controlled networks for the promotion and sale of patent medicines can be observed as the industry developed in the mid-eighteenth century. The earlier picture of the sale of patent medicines as part of a free-for-all Georgian medical market, with the participants entering and leaving the market and with the medicines being provided at the whim of unspecified local demand, is incorrect, particularly for the many nationally available medicines. The major wholesalers influenced the retailing of these medicines, which were sold at the same fixed price across the country and at a broadly similar local profit margin. The members of the print trades had the necessary skills, the business premises and the local contacts to be successful in selling patent medicines, and some took the opportunity to become major vendors, deriving a significant part of their income from medicine selling and continuing to do so for many years. The most prominent vendors were often the provincial newspaper printers who could act as local wholesalers by both providing the medicines and organising their distribution. Some other printers and booksellers were only involved in this trade to a minor extent or not at all.

Thus, medicine vending was often a stable and structured business which sold its pre-packaged fixed-price goods through specialised retailers. With its durable organisation and specialised participants, we must regard the retailing of patent medicines as being part of the established industry, alongside ownership and wholesaling. The national structure of this retailing is more apparent than that of other forms of retailing in Georgian England, and it was probably more precisely configured than the retailing of any other class of consumer goods in this period, apart from books. This emergence of a stable defined business, more systematic than either orthodox medical practice or quackery, does not mean that selling patent medicines was no longer part of the uninhibited medical market. Patent medicines were still in it, but as a separate entity not as an amorphous ingredient of an uncontrolled free-for-all. Within this entity, many of the medicine vendors were experienced in the techniques required and knowledgeable about their products; but they were not regarded as irregular medical practitioners by themselves or their contemporaries. Patent medicine vending was a distinct component of the local medical market with its own rules, personnel and organisation. It competed vigorously with both orthodox therapy and the treatments of irregular practitioners, but it did so by its own methods. Central to these methods was local publicity through the use of the printed word, and the next chapter will explore how this was organised.

Notes

1. Norton, *National Directories.*
2. Porter, *Health*, 113.
3. Porter and Porter, *Patient's Progress*, 98.
4. Brown, 'Venders', 360; Feather, *Book Trade*, 83; Isaac, 'Pills'; Cody, 'No Cure', 106; Barker, 'Medical Advertising', 396; Hancock and Wallis, 'Quacking', 19.
5. *LM*, 4 May 1793; *SWJ*, 8 April 1822.
6. Randall's Elixir, *SWJ*, 21 January 1822.
7. *ABG*, 5 February 1781 and 8 January 1781.
8. In the 1750s there was an average of one shop for about every 40 people in England (Mitchell, *Tradition*, 77).
9. Willan, *Shopkeeper.*
10. Willan, *Shopkeeper*, 6–18.
11. *MCR*, 13 (1806), xlv and xliii.
12. Wagner, 'Satire', 204.
13. *LI*, 18 January 1807.
14. *LM*, 22 June 1793. 'Papers' were testimonials from satisfied customers.
15. Feather, *Book Trade*, 83; Isaac, 'Pills', 41.
16. Cody, 'No Cure', 106.
17. Barker, 'Medical Advertising', 397.

18. Brown, 'Venders', 357.
19. Mitchell, *Tradition*, 38.
20. Mitchell, *Tradition*, 39.
21. Kearsley, *Tax Tables 1786*, 88–93.
22. Kearsley, *Tax Tables 1808*, 119.
23. Paget, *John Hunter*, 165. See Chapter 3 for more details of Samuel Glass, his magnesia and the dispute with Thomas Henry.
24. Spilsbury, *Power of Gold*, 2.
25. *ABG*, 13 February 1769; *SWJ*, 6 January 1794.
26. *MCR*, 12 (1806), clx.
27. Marshall's Cerate, *ABG*, 7 January 1822.
28. Hey's Medicine for the Bite of a Mad Dog, *LM*, 2 May 1769.
29. Chester City Archives, Account Book of John Fletcher; Hughes, *Chronicle of Chester*, 180.
30. Pearson and Rollason, 'Handbill'.
31. Fox, *Correspondence*, 119.
32. *SJ*, 9 January 1769.
33. Thornton, 'Biographical Study', 61–62.
34. West Yorkshire Archives, Baines Papers.
35. Thornton, *Life of Edward Baines*, 43.
36. *LM*, 14 February 1807.
37. Birmingham City Archives, Deeds for Aris's Birmingham Gazette.
38. Hughes, *Chronicle of Chester*, 30; Chester City Archives, Account Book of John Fletcher.
39. Barfoot and Wilkes, *British Directory*; Parr, 'Leeds Printers'.
40. For example, as mentioned in Chapter 3, Priestley, *Remarks*.
41. Nevett, *Advertising*, 18.
42. National Archives, Wilkes v Collins and others. For more details, see Chapter 6.
43. Cumbria Archives, Day Books of John Ware.
44. Ferdinand, 'Distribution Networks', 145.
45. For example, 'may be had, carriage free, of the newsmen' (Dr. Solomon's Cordial Balm of Gilead, *SWJ*, 2 February 1807).
46. Ferdinand, *Collins*, 104–122.
47. *SWJ*, 25 June 1781.
48. A few were women, such as Ann Rollason, owner and printer of the *Coventry Mercury* from 1813 to 1846 and sister-in-law of James Rollason, part-owner of *Aris's Birmingham Gazette*.
49. Cheney, *John Cheney*, plate 22.
50. Cheney, *John Cheney*, 12.
51. Feather, *Book Trade*, 83.
52. Feather, 'John Clay', 201; Fergus and Portner, 'Provincial Bookselling', 160.

53. Thoresby Society, A Leeds Printer's Account Book; Parr, 'Leeds Printers', 105–109.
54. Hill, *Bookmakers*, 101.
55. Hill, *Bookmakers*, 101 (footnote).
56. Chandler, *Endless Street*, 33–35.
57. *LM, 2 May 1807.*
58. Baines, *Yorkshire.*
59. *LI*, 9 January 1781.
60. Dr. James's Analeptic Pills, *ABG*, 8 January 1781; Dalby's Carminative, *LI*, 26 January 1807.
61. *GM*, 18 (1748), 348–350; Appendix A.
62. Watson, 'Trading Accounts', 75.
63. Ferdinand, *Collins*, 42; Zachs, *John Murray*, 47.
64. Including the bookshop of the medicine wholesaler and retailer, the first John Murray, see Raven, *Business of Books*, 226. Murray's shop opened for twelve hours a day, six days a week.
65. Brewer, *Pleasures*, 176; Raven, 'Promotion to Proscription', 181.
66. 'An Amateur', *Life*, vol. 1, 284.
67. Reid, *Memoir*, 50.
68. Reid, *Memoir*, 32.
69. *CP*, 6 December 1803.
70. *LI*, 9 March 1807. The four medicines for venereal disease were Lignum's Pills, Walker's Jesuits' Drops, Freeman's Gutta Salutaris and Leake's Pills.
71. Reinarz, *Health Care*, 47.
72. Richardson, 'Salisbury and Winchester Journal', 64.
73. Brock, 'Brodie, Sir Benjamin Collins'.
74. Reid, *Memoir*, 71.
75. Raven, *New Wealth*, 114.
76. Reid, *Memoir*, 30–32; Thornton, 'Biographical Study', 51.
77. Barker, 'Medical Advertising', 397.
78. *LM*, Ormskirk Medicine for the Bite of a Mad Dog, 19 June 1781.
79. *LM*, 28 March 1807.
80. *LM*, 11 April 1807.
81. *ABG*, 1 January 1781. This testimonial was also printed in the *Leeds Intelligencer* and the *Hampshire Chronicle* during 1781.
82. *MCR*, 13 (1806), ci–cii.
83. *MCR*, 13 (1806), clxxi–clxxii.

Bibliography

Archives

Birmingham City Archives, Deeds for Aris's Birmingham Gazette, MS 3700/1/32 B.
Chester City Archives, Account Book of John Fletcher, CRO D3876.
Cumbria Archives, Whitehaven, Day Books of John Ware, DA276A.
National Archives, Kew, Wilkes v Collins and others, E 140/90 and 140/91.
Thoresby Society, Leeds, A Leeds Printer's Account Book 1790–1797, MS IV/2.
West Yorkshire Archives, Leeds, Baines Papers, WYL 383/43.

Newspapers and Journals

Aris's Birmingham Gazette (ABG).
Cumberland Pacquet (CP).
Gentleman's Magazine (GM).
Hampshire Chronicle.
Leeds Intelligencer (LI).
Leeds Mercury (LM).
Medical and Chirurgical Review (MCR).
Salisbury and Winchester Journal (SWJ).
Salisbury Journal (SJ).

Printed Primary Sources

'An Amateur', *Real Life in London*, 2 vols. (London: Jones & Co., 1821).
Baines, Edward, *Baines's Yorkshire* (Leeds: Edward Baines, 1822, reprinted Newton Abbott, Devon: David & Charles, 1969).
Barfoot, Peter, and John Wilkes, *Universal British Directory, 1793–1798* (London, 1793–1798).
Kearsley, George, *Kearsley's Tax Tables 1786* (London, 1786).
Kearsley, George, *Kearsley's Tax Tables 1808* (London, 1808).
Pearson and Rollason, 'Handbill' (Birmingham: 1782?), British Library, Eighteenth Century Collections Online, http://0-find.galegroup.com, accessed 5 October 2012.
Priestley, Robert, *A Few Interesting Remarks on Bilious Disorders and Yellow Fever* (Leeds: Thomas Gill, 1798).
Spilsbury, Francis, *Power of Gold Displayed* (London, 1785?).

Secondary Sources

Barker, Hannah, 'Medical Advertising and Trust in Late Georgian England', *Urban History*, 36 (2009): 379–398.

Brewer, John, *The Pleasures of the Imagination: English Culture in the Eighteenth Century* (Chicago: University of Chicago Press, 2000).

Brock, W. H., 'Brodie, Sir Benjamin Collins', *ODNB* (accessed 10 May 2014).

Brown, P. S., 'The Venders of Medicines Advertised in Eighteenth-Century Bath Newspapers', *Medical History*, 19 (1975): 352–369.

Chandler, John, *Endless Street: A History of Salisbury and Its People* (Salisbury: Hobnob Press, 1983).

Cheney, Christopher, *John Cheney and His Descendants: Printers in Banbury Since 1767* (Banbury: Private circulation, 1936).

Cody, Lisa Forman, '"No Cure, No Money," or the Invisible Hand of Quackery: The Language of Commerce, Credit, and Cash in Eighteenth-Century British Medical Advertisements', *Studies in Eighteenth Century Culture*, 28 (1999): 103–130.

Feather, John, 'John Clay of Daventry: The Business of an Eighteenth-Century Stationer', *Studies in Bibliography*, 37 (1984): 198–209.

Feather, John, *The Provincial Book Trade in Eighteenth-Century England* (Cambridge: Cambridge University Press, 1985).

Ferdinand, C. Y., 'Local Distribution Networks in 18th-Century England', in *Spreading the Word: The Distribution Networks of Print 1550–1850*, edited by Robin Myers and Michael Harris (Winchester: St. Paul's Bibliographies, 1990), 131–149.

Ferdinand, C. Y., *Benjamin Collins and the Provincial Newspaper Trade in the Eighteenth Century* (Oxford: Clarendon, 1997).

Fergus, Jan, and Ruth Portner, 'Provincial Bookselling in Eighteenth-Century England: The Case of John Clay Reconsidered', *Studies in Bibliography*, 40 (1987): 147–163.

Fox, Levi, ed., *The Correspondence of the Reverend Joseph Greene: Parson, Schoolmaster and Antiquary 1712–1790* (London: H.M. Stationery Office, 1965).

Hancock, David, and Patrick Wallis, 'Quacking and Commerce in Seventeenth-Century London: The Proprietary Medicine Business of Anthony Daffy', *Medical History*, Supplement 25 (2005).

Hill, Joseph, *The Bookmakers of Old Birmingham: Authors, Printers and Booksellers* (Birmingham: Cornish Brothers, 1907).

Hughes, Herbert, *Chronicle of Chester: The 200 Years, 1775–1975* (London: Macdonald and Jane's, 1975).

Isaac, Peter, 'Pills and Print', in *Medicine, Mortality and the Book Trade*, edited by Robin Myers and Michael Harris (Folkestone, Kent: St. Paul's Bibliographies, 1998), 25–47.

Mitchell, Ian, *Tradition and Innovation in English Retailing, 1700–1850: Narratives of Consumption* (Farnham, Surrey: Ashgate, 2014).

Nevett, T. R., *Advertising in Britain: A History* (London: Heinemann, 1982).

Norton, Jane, *Guide to the National and Provincial Directories of England and Wales* (London: Royal Historical Society, 1984).

Paget, Stephen, *John Hunter: Man of Science and Surgeon (1728–1793)* (London: Fisher Unwin, 1897).

Parr, Elizabeth, 'Early Leeds Printers, Publishers and Booksellers' (MA thesis, University of Leeds, 1973).

Porter, Dorothy, and Roy Porter, *Patient's Progress: Doctors and Doctoring in Eighteenth-Century England* (Stanford: Stanford University Press, 1989).

Porter, Roy, *Health for Sale: Quackery in England 1660–1850* (Manchester: Manchester University Press, 1989).

Raven, James, *Judging New Wealth: Popular Publishing and Responses to Commerce in England, 1750–1800* (Oxford: Clarendon Press, 1992).

Raven, James, 'From Promotion to Proscription: Arrangements for Reading and Eighteenth-Century Libraries', in *The Practice and Representation of Reading in England*, edited by James Raven, Helen Small, and Naomi Tadmor (Cambridge: Cambridge University Press, 1996), 175–201.

Raven, James, *The Business of Books: Booksellers and the English Book Trade, 1450–1850* (New Haven: Yale University Press, 2007).

Reid, T. Wemyss, *A Memoir of John Deakin Heaton, M.D. of Leeds* (London: Longmans, Green and Co., 1883).

Reinarz, Jonathan, *Health Care in Birmingham: The Birmingham Teaching Hospitals 1779–1939* (Woodbridge, Suffolk: Boydell Press, 2009).

Richardson, Mrs Herbert, 'The Salisbury and Winchester Journal (1729 - Present Day)', *Wiltshire Archaeological and Natural History Magazine*, 41 (1920): 53–69.

Thornton, David, 'Mr Mercury: A Biographical Study of Edward Baines with Special Reference to His Role as Editor, Author, Politician' (PhD thesis, University of Leeds, 1999).

Thornton, David, *Mr Mercury: The Life of Edward Baines 1774–1848* (Chesterfield: Merton Priory Press, 2009).

Wagner, Peter, 'The Satire on Doctors in Hogarth's Graphic Books', in *Literature and Medicine During the Eighteenth Century*, edited by Marie Mulvey Roberts and Roy Porter (London: Routledge, 1993), 200–225.

Watson, G. M., 'Some Eighteenth-Century Trading Accounts', in *The Evolution of Pharmacy in Britain*, edited by F. N. L. Poynter (London: Pitman Medical, 1965), 45–77.

Willan, Thomas Stuart, *An Eighteenth-Century Shopkeeper: Abraham Dent of Kirkby Stephen* (Manchester: Manchester University Press, 1970).

Zachs, William, *The First John Murray and the Late Eighteenth-Century London Book Trade* (Oxford: Oxford University Press, 1998).

CHAPTER 6

Persuading the Poorly

Taking a patent medicine was a personal and delicate decision. Exploring Georgian healthcare as a market should not obscure the fundamental difference between medicines and other forms of movable goods. Only medicines could confer so much potential benefit for personal well-being; but at the same time carry the risks of unpleasant side effects or being the wrong treatment for a severe, or even fatal, condition. So buying a patent medicine was not the same as purchasing, say, a roll of cloth, a piece of pottery or a packet of tea. In the unregulated Georgian healthcare world, little help was available to the consumers, who had to rely on their own judgement whether to pay a significant sum of money to buy a patent medicine and take it. True, relatives, friends and neighbours often provided advice, but no mechanism existed to prevent the English buying ineffective or dangerous medicines, nor to stop anybody heavily promoting such medicines. Consumers had to weigh up carefully the verbal and printed advice they had received, and then make their own decisions, perhaps over a prolonged period. Buying and taking a patent medicine was a serious, potentially lengthy, and often expensive affair, not a sudden fancy.

How were the medicines promoted? Unlike regular physic and the treatments of irregular practitioners which both relied on verbal promotion, the printed word in newspaper advertisements and bills was the dominant means of persuading the poorly. In common with the ownership, wholesaling and retailing of medicines, this persuasion was an

A. Mackintosh, *The Patent Medicines Industry in Georgian England*, Medicine and Biomedical Sciences in Modern History,
https://doi.org/10.1007/978-3-319-69778-9_6

organised and structured part of the patent medicines industry, not predominantly the product of haphazard local decision making. To explore the advertisements, the studied newspapers and other previously identified sources have been joined by accounting records from the *Hampshire Chronicle*, a largely uninvestigated archive which has generated detailed information on the insertion of medicine advertisements and their contribution to the finances of the newspaper. Many locally printed bills have been preserved, though their origins and methods of use are more difficult to decipher. The main finding from these sources is that the major wholesalers were heavily involved in promoting their medicines, normally providing the text of the advertisements, perhaps giving specific instructions about their insertion, and often paying for them. The precise information in the multiple bills and the repeated newspaper advertisements provided the consumer with all the necessary details for medicine purchase and effectiveness, and the frequency of the bills and the advertisements ensured that these details could be assimilated over time, at an appropriate rate for each individual.

Organisation of Newspaper Advertisements

Much has already been written in this book about newspaper advertisements for patent medicines. This is nothing to apologise for, as not only are these advertisements numerous, comprehensive and unselected sources of information for the historian, but they were probably a very effective form of publicity, especially for a new product. Skilled owners/wholesalers, such as Spilsbury and Solomon, would not have spent many hundreds of pounds a year on newspaper advertisements unless they were going to get a good return.[1] The newspaper advertisements for nationally available medicines were mostly part of organised campaigns, and this section will explore who created them, who ordered them, and who controlled their insertion in the newspapers. We shall find that the advertisements were not only part of a national structure of medicine retailing, but they also played a major role in financing the provincial Georgian press. Medicine advertisements were also printed in, or circulated with, books, almanacs, periodicals, treatises or indeed almost any type of printed publication. However, these advertisements were similar in content to those in the newspapers, and they will not be considered separately. How the content of the advertisements achieved the advertisers' aims will be the subject of the following two chapters.

The major source of information for this section will be the advertisements themselves. Once again, the continuous runs of the studied newspapers provide an unbiased selection of advertisements: in particular, they allow a comparison of the publicity for the same patent medicine across the country. Another productive source for the organisation and the finances of medicine advertising is the advertising account books, some general account books, and marked-up copies of the *Hampshire Chronicle* for the period 1778–1783.[2] The background to this unique archive and the explanation for its preservation have been described by Christine Ferdinand.[3] In summary, a partnership of local booksellers and printers bought the paper in 1778 and moved it from Southampton to Winchester, where it was printed by John Wilkes. The partnership was dissolved in 1783 and the papers were preserved as evidence in a legal dispute between the ex-partners. The accounts reveal that 56% of the paper was owned by the Collins family and their Salisbury business associates, who also owned the *Salisbury and Winchester Journal*, one of the studied newspapers. This controlling interest was kept from the public, but it is relevant to us as the business practices of the two newspapers probably had much in common. The accounts describe who paid for every single advertisement in the *Hampshire Chronicle*, except the occasional one which was paid for in cash. Also, the marked-up copies reveal some of the instructions received by the newspaper compositor for the insertion of advertisements. These detailed arrangements for advertising a product in an eighteenth-century provincial newspaper have not been revealed before.

An inspection of the advertisements in the studied newspapers quickly shows that most of their content was provided centrally by the wholesaler, not by the newspaper printer or local retailers. For example, almost all the testimonials referred to patients who lived outside the newspaper printer's area, sometimes at the other end of the country. Similarly, the wholesaler must also have provided the endorsements by the nobility and gentry who were living elsewhere. Another observation which demonstrates the central origin of much of the copy is the predominant style of informing the consumer about the activities of the proprietor and the national, or London, successes of the medicine in the past, rather than mentioning local events such as how many bottles have been sold in the town, the number of successful cases in that area, or the wisdom of local retailers. Apart from the lists of nearby agents, the only common local input into the advertisements' content was starting the text with a phrase that a local printer or bookseller 'had just received a fresh supply of', or similar wording, followed by the name of the medicine. The obvious exceptions to this central provision of advertising copy were the medicines with local owners: some examples of the content of their advertisements have already been described in Chapter 3.

The clinching evidence for this central origin for the core text of advertisements is that much of it is very similar in the three studied towns, each more than a hundred miles apart. Each advertisement was formed from moveable type in separate newspaper offices and had to incorporate local retailing arrangements; so they were not identical. However, the main text of the advertisements for nationally available medicines often employed the same wording; for example, the advertisements for Dr. Steers's Opodeldoc, which were printed in the *Leeds Intelligencer* and the *Salisbury and Winchester Journal* during January 1807, used different typefaces, but contained identical wording apart from the list of local retailers. This text must have been supplied directly or indirectly by the wholesaler or the owner. In the late eighteenth/early nineteenth century, London advertising agents, such as James White and the partnership of William Taylor and Thomas Newton, replaced booksellers' agents and specialised coffee houses as the suppliers of many national advertisements to the provincial press.[4] However, as described below, the medicine wholesalers apparently did not use these London agents to implement their advertising.

Who wrote the text remains a mystery. In the early nineteenth century, leading authors did write advertising copy, such as Coleridge, Scott and Southey for Warren's Shoe Blacking and Charles Lamb for the lottery.[5] Several of the leading medicine owners/wholesalers were friendly with well-known literary figures and prominent publishers, such as the trio of Samuel Johnson, Oliver Goldsmith and Tobias Smollett (The Newberys, Robert James and Benjamin Collins), Joseph Johnson (Thomas Henry) and Hannah More (the Diceys), as well a sometimes being authors themselves. However, the advertisements, in common with much of the material in Georgian periodicals, remain anonymous. Any papers linking a prominent writer to the text of a medicine advertisement would have run the risk of being destroyed by Victorian descendants.

We have one account of a proprietor/wholesaler providing copy for a newspaper; it was described in an advertisement designed to expose a counterfeit preparation.[6] The advertisement was for a rat poison, the

Hampshire Miller's Infallible Rat Powder, not a medicine for humans; but pest killers and medicines for animals were often advertised and distributed in a similar fashion to, and sometimes alongside, patent medicines. In the advertisement, the proprietor, Thomas Raiss, published letters he had sent to, and received from, Barclay and Son, a prominent London medicine wholesaler, in which he complained that the Barclays had advertised another Infallible Rat Powder, but used one of his advertisements. The Barclays responded that they had received the medicine and advertisement in good faith from John Baxter in Edinburgh, without knowing the original source of the advertisement copy. Raiss replied to Barclays that 'this very advertisement which is now made to appear as yours, was the identical one sent him [Baxter] to insert in the Scotch newspapers'; but he accepted their explanation. Baxter claimed in a letter that it was done by accident, but Raiss did not believe this. Raiss seems to have suspected collusion amongst the London medicine wholesalers, and he announced that his powder would no longer be available wholesale in London from one of them, John Wye, but only from two booksellers in the City.

So the text of the advertisements was mostly provided from central sources, but who organised and paid for their insertion? Normally, they were not inserted solely at the whim, and perhaps the expense, of the newspaper printer. If we first consider who *paid* for the advertisements, one of the most striking findings from the *Hampshire Chronicle* records is that the advertisements were not inserted without charge or paid for by the printer. During the whole of 1781, each advertisement was identified by a number on the marked-up copies of the newspaper and then charged to an individual account in the advertising account books: all the 277 medicine advertisements of that year were charged to an external source, such as Francis Newbery, Francis Spilsbury or Jackson and Co. As shown in Table 6.1, the majority of these sources comprised London wholesalers. Now this completely external funding of the advertisements may not have been the full picture as two of the local sources, John

Table 6.1 Types of accounts charged for the 277 medicine advertisements in the *Hampshire Chronicle* during 1781

London owners/wholesalers	165
Booksellers, printers or druggists in local area	76
Uncertain	35
Paid in cash	1

Breadhower of Portsmouth and Collins and Johnson of Salisbury, were partners in the *Hampshire Chronicle*: partners in newspapers could be required to buy a number of advertisements each month on the grounds that a good show of advertisements would encourage others to contribute.[7] So perhaps not all these charges were to a source which was truly external to the newspaper. However, these partners were only involved in a small number of advertisements, and, as we shall see, it is also possible that they were paying for the advertisements as local agents for London wholesalers and not on their own initiative. The accounts do show that external payment for all medicine advertisements was the normal business practice for this newspaper, and that in Winchester in 1781 the majority of these payments were made by London wholesalers.

However, the person who *organised* the insertion of the medicine advertisement was not necessarily the person who paid for it. The *Hampshire Chronicle* records also demonstrate that many of the advertisement insertions were organised by the London wholesalers. In order to appreciate this, we need first to understand the markings on the advertisements in the preserved copies of the newspaper, as they can reveal the instructions to the printer. These brief handwritten comments, which were probably used by the compositor to set up the type for the following week, can indicate whether the printer was working to precise instructions from a wholesaler or his local agent, or perhaps using his own initiative in inserting the advertisements. The markings were not always adhered to, but most were, and they can demonstrate a pre-ordered number of advertisements to be inserted at frequent intervals, perhaps weekly, and then discontinued. On the marked copies, this arrangement was counted down in the manner of 'five more weeks', then 'four more weeks', until the last one was marked with a cross. Other markings are less specific, perhaps for the time interval between insertions in the weekly issues but without any idea of a specific number of insertions, and they include markings such as 'fortnightly', 'omit three weeks', 'until forbid', or just a cross which usually indicated that the particular advertisement would be discontinued for at least a few weeks. Some advertisements have no markings. All the markings refer to a specific advertisement, not to any advertisement for a particular medicine: one advertisement with a cross could be followed the next week by a different advertisement for the same medicine.

As might be expected from his usually prominent role in organised medicine selling, Francis Newbery provided examples of pre-ordered advertisements. In the middle of 1781 he paid for six insertions of an advertisement for Dr. James's Analeptic Pills, which were printed and counted down on the marked copies between June and September. He also paid for six pre-ordered insertions of an advertisement for Dr. Steers's Opodeldoc between June and August, and for another six inclusions of a different advertisement for the same medicine during the same period, though the last inclusion seems to have been omitted in error. Newbery paid for other pre-ordered advertisements during 1781, always in threes or sixes. Similarly, 12 insertions of an advertisement for Hickman's Pills were counted down between March and July, though with payment being charged to one of the partners, John Breadhower, not to the wholesalers W. and H. Wray. However, another series of a different advertisement for Hickman's Pills later in the year, originally for six inclusions but increased after the second insertion to 12, was charged directly to the Wrays. Breadhower was probably acting as a local agent for the Wrays in the earlier series.

Francis Spilsbury seems to have given less precise instructions when he advertised his Drops sixteen times in the newspaper during 1781. Although the text changed, these advertisements appeared at mostly three-week intervals up to early October without any marks on them. Insertions later in the year had markings such as 'omit three weeks' and 'twice more' without a clear arrangement being apparent. Similarly, the advertisements for the Tasteless Ague and Fever Drops, paid for by G. Earle, a Winchester druggist who claimed to be the main wholesaler, were intermittently marked with comments such as 'omit a week' and 'til forbid'. Some of these marked-up advertisements without a predetermined total of inclusions may have been accompanied by more precise instructions from the wholesaler which were not marked on the newspaper. For other advertisements without marked-up comments, the compositor probably did not have detailed instructions which had to be followed, and they may have appeared, at least partly, at the discretion of the printer John Wilkes. In summary, the wholesaler was largely responsible for the advertisement text and often for paying for it, but the insertion of that text depended on a variable combination of decisions by the wholesaler, his local agent or the newspaper printer.

We can discover a little more about who was ordering the insertion of the advertisements as distinct from paying for them. The initial available advertising accounts of the *Hampshire Chronicle* from 1778 provide a detailed, but only brief, glimpse of this. These first accounts under the newspaper's new owners recorded, in a separate column, who ordered the advertisement as well as who paid for it; but the name of the person who instructed the printer soon began to be omitted and was rarely written down after the initial four months. In this short period, advertisements could be directly ordered and paid for by London wholesalers, such as Thomas Jackson, the Diceys, Francis Newbery, and John Norton. They could also be requested by local retailers with the payment being the responsibility of the distant wholesaler: under these circumstances, the local retailer was presumably acting as an agent for the wholesaler. Many of the advertisements with this arrangement were ordered by Thomas Baker, a Southampton bookseller and transient early partner in the newspaper, and charged to London owners such as James Berry (Ormskirk Medicine for the Bite of a Mad Dog) and Lady Hill (John Hill's range of herbal-based medicines). Another group of advertisements was ordered *and* paid for by local retailers, particularly John Wise, also a Southampton bookseller, who did this for a number of nationally available medicines such as Glass's Magnesia, Beaume de Vie and Spilsbury's Drops. It is not clear whether the local retailers in this last group were acting as agents for a wholesaler and passing the charges on, or were just seeking to boost their own sales of these medicines.

So the responsibility and methods for inserting medicine advertisements in the *Hampshire Chronicle* varied. First, wholesalers could order advertisements directly from London either for a specific number, as Francis Newbery normally did, or with less formal arrangements which remain hidden to us. Second, the London wholesalers and owners could work through local agents, once again either for a planned number of insertions, or by less structured arrangements. Third, the limited information from 1778 on who ordered the advertisements shows that local booksellers could order and pay for advertisements themselves, but these earlier records do not document a planned number. The role of the newspaper printer John Wilkes in the insertions of each advertisement without a planned number is unclear; though he was probably responsible for deciding which advertisements would be included in a particular issue as only he would know the available space. The plausible procedure for the less clearly organised advertisements was that the wholesaler/owner provided an overall framework for the insertions and then left the details to Wilkes: it seems unlikely that a wholesaler would send along the text of the advertisement to the printer and then just wait for an unknown bill. However, evidence on this is lacking in the accounts and marked copies.

Were these arrangements for inserting medicine advertisements typical of those for all provincial newspapers at the time? They probably were, as the wholesalers who dispatched medicines and advertising copy across the country would have been unlikely to have unusual procedures for one newspaper. In addition, as already described, the owners of the *Salisbury and Winchester Journal* had a majority stake in the *Hampshire Chronicle* and so the business practices of the two newspapers would probably be similar. The medicines advertised were similar in type and sometimes the same as those advertised in the studied newspapers in 1781. The number of medicine advertisements in the *Hampshire Chronicle* in the first six months of 1781 was 133, which was within the range for the same period in the studied newspapers; the *Chronicle's* total was greater than that of the *Leeds Intelligencer* and the *Leeds Mercury* at 120 and 73 respectively, but considerably less than that of *Aris's Birmingham Gazette* and the *Salisbury and Winchester Journal* at 225 and 361 respectively. Overall, the pattern of medicine advertising in the *Chronicle* was not noticeably different from that of the other newspapers.

Do the studied newspapers support the existence of these arrangements for the ordering and funding of medicine advertisements? Evidence from the studied newspapers is more indirect, but it does demonstrate some wholesaler involvement in not only providing the text, but also in arranging the insertion of their advertisements. For example, we have seen that Francis Newbery ordered his advertisements in multiples of three for the *Hampshire Chronicle* in 1781, and the *Leeds Intelligencer* printed an advertisement for one of his medicines, Dr. Steers's Opodeldoc, on three consecutive weeks in January 1781, followed by three insertions of another advertisement for the same medicine in March/April 1781.[8] As discussed above, the text could be similar in different parts of the country, but this is not by itself evidence for the ordering of the advertisements by wholesalers: the newspaper printers might have independently decided to advertise popular medicines. But the insertion of the *same number* of a particular advertisement at different locations is more suggestive of wholesaler participation. One example was the same advertisement for Glass's Magnesia, then distributed by two London booksellers, inserted three times in the *Journal* in Salisbury starting in January 1769, and also three times in the *Gazette* in Birmingham starting in February 1769. Another was an advertisement for Dr. Steers's Opodeldoc (wholesaler Francis Newbery and Sons), which was printed four times at approximately monthly intervals both in these two newspapers and in the *Leeds Intelligencer* in the first half of 1822. When they chose to do so, wholesalers could organise the printing of the same number of advertisements in specific newspapers.

On other occasions, the advertisements in the studied newspapers seem noticeably uncoordinated from place to place, suggesting that the newspaper printers were determining their frequency and choosing the type of content without the direct intervention of the wholesaler. This can be seen in the different advertisements for the Dicey's version of Daffy's Elixir in 1794. The *Leeds Intelligencer* promoted the elixir in a detailed description, with a long list of indications and a warning about counterfeiters[9]; the advertisement was repeated four times in the six-month period. By contrast, the only advertisement for the elixir in *Aris's Birmingham Gazette* during the same six months concentrated on the use of this medicine for stone and gravel and printed a testimonial from Yarmouth.[10] The *Salisbury and Winchester Journal* took a different approach to the other two papers by advertising the elixir briefly along with twenty other Dicey medicines on four occasions, and twice more with the name of the elixir being used as the heading together with more details.[11] Now it is possible that Dicey and Co. had issued different instructions to the three newspapers, but it is more probable that they had left the insertion of their advertisements to the discretion of the newspaper printers.

At times, the newspaper printers seem to have used this discretion to fill up the space in their four closely packed pages. In the studied newspapers, the majority of the medicine advertisements appeared on Page four, often at the bottom of a column and sometimes truncated to fit the space. It seems that the much of the space in the rest of paper had already been allocated before the medicine advertisements were inserted and altered to fit the remaining space. The next most popular page was Page one, which was printed with Page four. An exception to this approach was shown by the *Leeds Intelligencer* in 1822 and the *Salisbury and Winchester Journal* in 1807 and 1822, which had a more even distribution of advertisements between the pages, but still with a preference for the bottom of a column. Further, although medicine advertisements were normally all paid for, printers might sometimes have inserted medicine advertisements without payment to fill up otherwise unused space; or they might have ignored the intended timing of pre-ordered advertisements. Almost certainly, both these manoeuvres occurred occasionally, though evidence for either is missing.

We can conclude that much of the newspaper advertising was organised across the country, with a high degree of control by the wholesalers. They directly or indirectly paid for many of the advertisements, they provided much of the text, and they could directly control the insertion of the advertisements if they wished. The *Hampshire Chronicle* accounts suggest that the financial arrangements were normally tightly controlled, with each insertion being charged to a wholesaler or local retailer. Nevertheless, some decisions about insertions were made by the newspaper printer, probably within a framework of less specific instructions from the wholesaler or a local retailer. Details of the content of the newspaper advertisements will be discussed in Chapter 8.

Finances of Medicine Advertising

It is generally agreed that a good advertising revenue was essential for the finances of Georgian newspapers, and the income derived from advertising patent medicines was a significant part of it.[12] However, these conclusions have been based on the number of advertisements, not their revenue, and detailed information on the contribution of medicine advertisements to the finances of a provincial newspaper has not been reported. The surviving accounts of the *Hampshire Chronicle* do allow a precise calculation of both the total advertising revenue and the revenue from medicine advertising, together with their relative contribution to the overall income of the newspaper. For the first time, we can see the importance of the medicine advertisements to the economics of a provincial newspaper. The revenue from advertisements for patent medicines was a significant part of the total income of the newspaper, but not necessarily the key to survival.

The cost for each insertion of a medicine advertisement in the *Chronicle* was commonly 5s–8s during 1781, including the 2s 6d excise duty; but large advertisements could be charged at up to 15s. The cost of each advertisement was directly related to its length (Fig. 6.1). Writing in 1776 when the excise duty per advertisement was 6d less at 2s, Francis Spilsbury described the typical cost of a single medicine advertisement as 4s or 6s.[13] So we can regard the charges by the *Chronicle* as being in line with other newspapers of the time. Later, in 1800, advertisements in the *Cumberland Pacquet* cost between 5s and 12s 6d, including a duty of 3s, with the cost of each advertisement again being directly related to its length.[14] Spilsbury also wrote that it

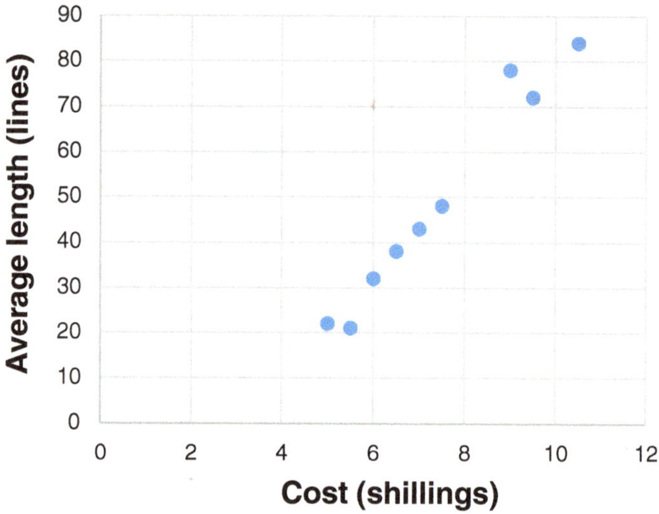

Fig. 6.1 Average length of advertisements for each cost in the *Hampshire Chronicle* during 1781. The cost included a 2s 6d advertisement excise duty

cost a minimum of £800–900 a year to advertise a medicine all over England.[15] He spent £5 8s 6d on advertising in the *Hampshire Chronicle* during 1781, Francis Newbery spent £7 15s 6d on advertisements for several medicines, and John Breadhower and the Wrays between them (see above) spent £9 13s 6d on Hickman's Pills. The more limited and later information from Ware's day books from the *Cumberland Pacquet* reveals broadly similar annual sums being paid by wholesalers for advertising their medicines: Dicey and Sutton were charged a total of £20 17s 0d for the three years 1800–1802, and advertising Spilsbury's Drops and Lignum's Drops cost their proprietors a total of £17 19s 6d and £22 10s 0d respectively for the three years 1797–1799. It has been estimated that there were about 50 provincial newspapers in England in 1782.[16] So Spilsbury's estimate of a *minimum* cost was excessive, but these figures from the *Hampshire Chronicle* and the *Cumberland Pacquet* indicate that spending several hundreds of pounds a year on advertising a single medicine across the country would not have been unusual.

A particular virtue of the accounting records of the *Hampshire Chronicle* is that the combination of the weekly *advertising* accounts and the marked-up copies of the newspaper provide a complete indication of the payment for an advertisement, even when it has been concealed in the editorial matter. However, the marked copies run only to the end of 1781. So the calendar year of 1781 was selected for a detailed analysis of the advertising revenue from medicines. This year also has the benefit of containing one of the six months used for the analysis of the studied newspapers. Looking at the *general* accounts, the total income and expenditure are summarised for each six-monthly accounting period from 1778 to 1783, the whole period of ownership by this group of partners. But the detailed weekly records of income and expenditure, which provide a breakdown of the origins of the income, only start in September 1781. So the examination of the contribution of *all* advertising to the paper's income is from September 1781 to September 1782, resulting in two twelve-month periods for analysis which overlap, but are not identical.

The income of the newspaper came from sales of the newspaper and from advertising revenue. Using the twelve months of general accounts from September 1781, the net income from sales can be determined by subtracting the stamp duty of 1½d per copy and the value of the returned copies from the gross sales. This net sales income was £359 3s 5d. A similar calculation can be made, from the same general accounts, for the net revenue from all advertising by subtracting the stamp duty of 2s 6d on each advertisement from the gross revenue. The net advertising revenue was £348 1s 10d. So the sales income and the total advertising income were equally important to the newspaper's income in that twelve-month period.

The contribution of the medicine advertisements to the overall advertising revenue was calculated for the calendar year 1781 from the advertising accounts. The paper carried 1754 advertisements during the year, of which 277 were medicine advertisements; thus, 16% of the advertisements were for medicines. However, the financial contribution of the medicine advertisements was somewhat higher as they were longer, and so generated more revenue, than the general run of advertisements. The total net income (gross income minus stamp duty) from all the advertisements during the calendar year was £311 13s 7d. The medicine advertisements provided £58 4s 6d of this, that is 19% of the advertising revenue. In addition, the newspaper printed advertisements for medical books, treatments for animals, and the availability of medical practitioners, but this calculation is confined to those for medicines.

So, about 10% of the total revenue of the *Hampshire Chronicle* during 1781–1782 was generated from advertisements for patent medicines. Are these findings in the accounts applicable to other newspapers? The accounts of any two newspapers will never be the same, but the finances of the *Chronicle* seem to be broadly in line with those suggested for other late Georgian provincial newspapers.[17] One important feature of this newspaper, which contrasts with many of the other newspapers which have been mentioned in this book, was that it was losing money. All the completed six-month accounting periods between 1778 and 1783 recorded a loss, and the partners had to put £935 into the paper over this period. Around 1100 copies of the *Chronicle* were printed each week and about 90–100 returned, so the sales were lower than those of some successful newspapers, but the circulation was greater than that of the *Chester Chronicle* in 1784, which was making at least a current account profit, and it did not preclude viability.[18] As we saw in the previous section, the number of medicine advertisements was similar to those in the studied newspapers. Overall, the *Chronicle* does not seem to have been wildly different from other newspapers of its time. Its financial problems can be attributed to an overall shortage of advertising: successful provincial newspapers often published more advertisements per year than the 1754 in the *Hampshire Chronicle*. In 1784, the *York Courant* printed 4397 advertisements in the year and the *Leeds Intelligencer* printed 2816, while the *Salisbury Journal* included over 3300 in 1770.[19]

The importance to the newspaper of the revenue from medicine advertising would obviously vary with the newspaper's financial circumstances. Although the medicine advertising provided 10% of the *Hampshire Chronicle's* income, it would not have been decisive for the newspaper's future. Even if the revenue from the patent medicines had tripled, the partners would still have been out of pocket. For some newspapers on the margin, such income might have been crucial in turning a loss into a profit, but, for most, the revenue from all the advertising would have been more important. The additional income the printer derived from the sale of the patent medicines might also have had a substantial impact on the viability of the provincial press, but no details are available. In this case, we have no information on the profitability of the separate medicines business run by John Wilkes, the printer of the *Chronicle*. Regardless of the effects on newspaper viability, the revenue from medicine advertising would still have had a substantial influence on newspaper *profits*. As there were few marginal costs of inserting extra advertisements for patent medicines apart from the excise duty, any increase in their net revenue would have fed directly through into the rewards for the newspaper owners.

Publicising Patent Medicines Outside Newspaper Advertisements

So far the discussion on the publicising of patent medicines has concentrated on advertisements in provincial newspapers. Regardless of any successes from these advertisements, wholesalers and retailers grasped other opportunities to get the virtues of their medicines better known. Word of mouth and other oral techniques of publicity are largely unknowable to the twenty-first century historian; but other methods which have left artefacts can be explored, especially the printed bills which were handed out or stuck on walls and were probably at least as important as newspaper advertisements. The range of the promotional techniques away from newspapers is not only relevant to studying patent medicines, but it also provides a glimpse of possibilities in the sale of goods in general.

Shouting in a public place, such as a street or market, more formal speeches, or just discussion in the retailer's shop might have been common point-of-sale advertising techniques, though, as we have seen, the travelling mountebank became rare as the period progressed. Unfortunately, we do not have a surviving account of these oral techniques in any detail. Much more is available from their written equivalent, the printed bills. Exactly how a particular printed bill was used is rarely clear, but they were designed to be either given out to be read by customers and others, or to be fixed to the wall in indoor or outdoor public spaces. A large collection of bills has been assembled in the John Johnson Collection at the Bodleian Library, Oxford, and other examples can be found in many archives and libraries. They were printed to provide the maximum amount of information in the available space, not to create a visual impact from a distance. Some could be headed by a symbol, such as a royal crest for those medicines which had been granted a patent, but the bills mostly consisted of dense text, designed to be read carefully and running up to 32 pages or more in length. Illustrations were rare, and large multiface type, which was popular in bills later in the nineteenth century, was used sparingly. Like the newspaper advertisements, the primary aim was to provide the consumer with information and to support him or her in coming to a decision, not to grab attention or to provoke an emotional reaction. The taking of a patent medicine was a matter for negotiation; the retailer provided the information and the consumer came to a decision, taking time if necessary and receiving advice from relatives, friends and the writers of testimonials.

The bills contained the same type of information as the newspaper advertisements; that is, a variable amount of description of the medicine's indications, its advantages and its mode of use, together with the price and availability. However, the much greater number of words printed in the bills resulted in the provision of a great deal more detail. Even a one-page bill usually contained more information than a long newspaper advertisement, with perhaps multiple testimonials, detailed instructions and a long list of retailers printed at the end.[20] So we should not think of these bills as flimsy local substitutes for the newspaper advertisements; on the contrary, they were probably the main source of the detailed information which many consumers would have required to make the important decision on which medicine to purchase, and how to use it subsequently. The newspaper advertisements can be regarded as the summaries of the printed bills, prepared for the wider dissemination of information in the weekly press.

How were the bills used? Precise information on this topic is sparse, and, unfortunately, we rarely know the origin of surviving copies. They were given out in the street: 'A person cannot walk from St. Paul's to Temple Bar without a dozen quack bills thrust in their hands'.[21] Fixing them to the wall in public spaces was common, as depicted by Hogarth in several pictures including *The March to Finchley* (1750) which shows a bill for medicines from Rock, a patent medicine owner (Appendix B), on a wall.[22] In late Georgian England, their extensive use could offend critics of patent medicines: 'The disgusting practice of having one's hands and eyes polluted in every corner of a street with the abominable bills and placards of quacking vermin, is past endurance, and loudly calls for suppression'.[23] An anonymous 1806 letter to the *Medical and Physical Journal* was more specific on the posting of medicine bills. The author, Veritas, opposed the medical reform which Edward Harrison was advocating. Firstly, Veritas commented on the inability of the College of Physicians to regulate medicine even in London, and he described an observer of the patent medicine bills:

> The inhabitants of London will bear testimony to the liberality of those gentlemen who so amply supply him with recommendatory papers in every alley about the 'Change; the walls of which buildings are ornamented with the portraits of many such illustrious personages as Drs Brodum, Sibly exc. who equally smile at the handbill of the College of Physicians and the circulating enquires of Dr Harrison.[24]

Secondly, Veritas went on to comment on a bill from Lincoln, which described an institution for mental patients part-owned by Harrison. This threatened Harrison's regular status:

> I cannot be equally tolerant on the subject of the handbill, which was obtained by me from the White Hart Inn, Lincoln, where it adorned the walls in the same manner that Dr Solomon is seen to figure in the stationers' shops.[25]

Many walls, both inside and out, seem to have been festooned with bills for patent medicines, but how systematic this was and who did the festooning usually remain unclear. But the day books of John Ware from Whitehaven are an important exception. They record that Ware arranged the sticking up of bills and then charged some of the wholesalers for the cost of doing so. During 1800–1802, the same Dr. Solomon mentioned by Veritas was charged a total of 28s for the fixing of bills on six occasions, Dr. Brodum was charged 3s 6d for 'distributing pamphlets and sticking up bills', and Ching and Butler were charged 2s 6d for 'postering, bills, etc.' As we saw in Chapter 4, most of Ware's wholesalers paid for the more general distribution of bills and other forms of printed publicity. Presumably, Ware was acting under instructions from the wholesalers, and it is unlikely that the arrangements were unique to Whitehaven. The conclusion must be that the sticking-up of bills, as well as their more general distribution, could be by the wholesalers' initiative, but the importance of this is unclear.

Details on the utilisation of the handbills made available for the customers, and perhaps for the shopkeepers themselves, are also sparse. The significant number of surviving copies suggests a widespread use, and we also know that they were printed in large runs. For example, between July and September 1796, John Newsome, a Leeds druggist, ordered the printing of 500 bills for Bateman's Drops, 400 bills for the Samaritan Balsam of Life, and 'three reams of Daffy's Bills' from

Thomas Gill in Leeds.[26] In the same town four months later, another druggist asked Gill to print four reams of Daffy's Elixir handbills. So bills appear to have been widely available. Whether they were intended to be given out at the point of sale to interested customers, or circulated freely elsewhere to drum up enthusiasm, is unclear; but both practices were likely.

Patent medicines could also be promoted by less conventional methods than printed bills or advertising in newspapers, almanacs and books. Puffs in books and in newspaper editorial matter were common. The first paragraph of Chapter 1 in the enduring children's book *Little Goody Two-Shoes* (publisher John Newbery) contains a blatant puff for Dr. James's Fever Powder (wholesaler John Newbery), which explains that the heroine's father 'died miserably' due to a lack of the Powder.[27] Macklin, the Southampton proprietor of Le Coeur's Imperial Oil, paid 6s on each occasion to have a 36-line promotion of his medicine inserted nine times into the editorial matter of the *Hampshire Chronicle* during 1781. A more inventive method of promotion was the distribution of imitation halfpennies stamped with information about a medicine. A small collection of these tokens has been preserved at the Thackray Medical Museum, Leeds, and one promoted the worm lozenges which were patented in 1796 by John Ching, a Cornish apothecary who became a major medicine wholesaler in London. On the obverse are the words 'the best medicine in the world' with details of the price and packaging, surrounding a portrait, perhaps of Ching, and the reverse has a royal crest surrounded by the words 'by every principal medicine vender in the kingdom'. Matthew Boulton produced similar tokens for Isaac Swainson, owner of De Velnos Vegetable Syrup.[28] These imitation halfpennies were produced in large quantities and were used as currency, because Georgian England was plagued with a chronic shortage of small-denomination coins: no copper coins were struck by the Royal Mint between 1775 and 1797.[29] Lord Liverpool remarked in 1805 that many manufacturers had to pay their workers with tokens instead of legal currency.[30] Used as currency, these tokens promoting medicines would pass through several hands, and even if they were not, they would be picked up and inspected.

Controlled Promotion of Medicines

In the early Georgian period, much of the promotion of patent medicines outside London seems to have been in the hands of the local retailers, though precise details are lacking. For instance, some issues of the *Stamford Mercury* in 1720 just briefly described a particular medicine which was available from the newspaper, and the *Northampton Mercury*, part owned by William Dicey, included two patent medicines in its list of items for sale in 1721.[31] These fairly random arrangements persisted with the locally owned medicines, but this chapter has shown that as the patent medicines industry developed the promotion of many medicines became structured and often controlled by the national wholesalers, who paid for the newspaper advertisements, supplied their text, and could determine their frequency. Some of the wholesalers also paid for the local distribution of bills, including sticking them on walls, and the imitation halfpennies illustrate that they were prepared to adopt less conventional forms of publicity. These findings verify that the patent medicines industry was truly national in its scope, and its financial resources are confirmed by several wholesalers spending hundreds of pounds a year on provincial newspaper advertising.

The varied methods of promotion which were employed to encourage the sale of patent medicines had two aims in common. First, the consumer could, and perhaps should, be repeatedly provided with the same information. The advertisers were willing to pay for the same advertisement to be printed frequently in a newspaper, sometimes every week, and the druggists arranged for the printing of many hundreds of copies of a bill. Most newspaper advertisements were altered from time to time, but this could be only a different testimonial, not a fundamental change. With a few exceptions, a wide range of different advertisements for a particular medicine was not attempted, and reiteration was the aim of many advertisers. Second, the bills and newspaper advertisements detailed as much information as possible in the available space, mostly avoiding large typefaces, engravings or anything to create a visual impact. Apart from a few royal crests, consumers were to be persuaded by facts and reason, not by eye-catching headlines, pictures, punchy phrases or other methods of obtaining an immediate reaction. We can see these two aims as part of a negotiating process which might be lengthy. Consumers expected to see the same information several times, interspersed with the opinions of friends and relatives and culminating in a trial of the therapy:

a medicine might seem familiar even before any of it had been taken. The important decision on whether to take a patent medicine had to be based on sufficient information; but an excessive quantity of ephemeral evidence, or an encouragement to make snap judgements, might have been counter-productive.

The facilitator for persuading the poorly by effective promotion was the printed word, not the oral negotiation typical of a market. Print was not only the main instrument of promotion, but it was essential for the correct use of the patent medicines, with the directions delivering the necessary instructions and any accompanying treatises providing the background explanations. The next two chapters will explore in more detail the crucial role of the printed text in the consumption and the effectiveness of patent medicines.

Notes

1. Spilsbury, *Free Thoughts*, xxxiii; Denizen, 'Solomon', 296.
2. National Archives, Wilkes v Collins.
3. Ferdinand, 'Distribution Networks', 131–133.
4. Gardner, *Business of News*, 54–58.
5. Gardner, *Business of News*, 64–65.
6. Hampshire Miller's Infallible Rat Powder, *LI*, 4 November 1799.
7. Ferdinand, 'Newspapers', 442.
8. *LI*, 16, 23, 30 January 1781, 27 March 1781, and 10, 17 April 1781.
9. *LI*, 10 March 1794.
10. *ABG*, 17 February 1794.
11. *SWJ*, 13 January 1794.
12. Barker, 'Medical Advertising', 384; Looney, *Advertising*, 78; Ferdinand, 'Selling', 399; Cranfield, *Press and Society*, 184.
13. Spilsbury, *Free Thoughts*, xxxiii.
14. [14] Cumbria Archives, Day Books of John Ware.
15. Spilsbury, *Free Thoughts*, xxxiii.
16. Ferdinand, 'Newspapers', 434.
17. Ferdinand felt that the *Chronicle* was representative of county newspapers of the time, but she was mainly concerned with the management arrangements (Ferdinand, *Distribution Networks*, 133).
18. Ferdinand, 'Distribution Networks', 134. Between 770 and 900 copies of the *Chester Chronicle* were printed each week during 1784 (Chester City Archives, Account Book of John Fletcher).
19. Looney, *Advertising*, 25; Ferdinand, 'Selling', 398.

20. For example, a one-page bill from *c.* 1764 for Glass's Magnesia, probably intended for fixing to a wall, contained 170 lines of print, plus marginal notes [John Johnson Collection, Patent Medicines 3 (34)]. Most newspaper advertisements had 20–60 shorter lines of print.
21. *Morning Post* 15 December 1782, cited in Burnby, 'Preparers', 49.
22. Wagner, 'Satire', 205–208.
23. *Deadly Adulteration*, 154.
24. *MPJ*, 16, (1806), 351. The 'Change' was the Royal Exchange. Sibly was an irregular physician and astrologer. Portraits on surviving handbills are rare and the concept might have been included to allow these two famous irregular medicine owners to smile metaphorically at the College. The College handbill was probably its proposals for medical reform.
25. *MPJ*, 16, (1806), 352.
26. Thoresby Society, Printer's Account Book. A ream was a somewhat variable quantity, but was probably 480 copies here.
27. *History of Little Goody Two-Shoes*, 14.
28. Waters, *Issuers*, 53.
29. Waters, *Issuers*, viii; Black, *Eighteenth-Century Britain*, 76. Waters estimated that John Ching ordered nearly 26,000 tokens (p. 16).
30. Daunton, *Progress and Poverty*, 426.
31. *Northampton Mercury*, 20 November 1721.

Bibliography

Archives

Chester City Archives, Account Book of John Fletcher, CRO D3876.
Cumbria Archives, Whitehaven, Day Books of John Ware, DA276A.
John Johnson Collection, Bodleian Library, Oxford.
National Archives, Kew, Wilkes v Collins and others, E 140/90 and 140/91.
Thoresby Society, Leeds, Printer's Account Book, MS IV/2.

Newspapers and Journals

Aris's Birmingham Gazette (ABG).
Hampshire Chronicle.
Leeds Intelligencer (LI).
Medical and Physical Journal (MPJ).
Northampton Mercury.
Salisbury and Winchester Journal (SWJ).

Printed Primary Sources

Deadly Adulteration and Slow Poisoning (London: Sherwood, Gilbert and Piper, 1830).

Denizen of Liverpool, 'Samuel Solomon M D', *The Scourge*, 2 (1811), 287–303.

Spilsbury, Francis, *Free Thoughts on Quacks and Their Medicines* (London, 1776).

The History of Little Goody Two-Shoes, 3rd edn (London: J. Newbery, 1766).

Secondary Sources

Barker, Hannah, 'Medical Advertising and Trust in Late Georgian England', *Urban History*, 36 (2009): 379–398.

Black, Jeremy, *Eighteenth-Century Britain, 1688–1783*, 2nd edn (Basingstoke: Palgrave Macmillan, 2008).

Burnby, J. G. L., 'The Preparers and Distributors of English Proprietory Medicines', *Dental Historian*, 32 (1997): 47–55.

Cranfield, Geoffrey Alan, *The Press and Society: From Caxton to Northcliffe* (London: Longman, 1978).

Daunton, M. J., *Progress and Poverty: An Economic and Social History of Britain 1700–1850* (Oxford: Oxford University Press, 1995).

Ferdinand, C. Y., 'Local Distribution Networks in 18th-Century England', in *Spreading the Word: The Distribution Networks of Print 1550–1850*, edited by Robin Myers and Michael Harris (Winchester: St. Paul's Bibliographies, 1990), 131–149.

Ferdinand, C. Y., 'Selling It to the Provinces: News and Commerce Round Eighteenth-Century Salisbury', in *Consumption and the World of Goods*, edited by John Brewer and Roy Porter (London: Routledge, 1994), 393–411.

Ferdinand, C. Y., 'Newspapers and the Sale of Books in the Provinces', in *The Cambridge History of the Book in Britain, Vol. 5 1695–1830*, edited by Michael Suarez and Michael Turner (Cambridge: Cambridge University Press, 2009), 434–447.

Gardner, Victoria E. M., *The Business of News in England, 1760–1820* (Basingstoke: Palgrave Macmillan, 2016).

Looney, John Jefferson, *Advertising and Society in England, 1720–1820: A Statistical Analysis of Yorkshire Newspaper Advertisements* (University Microfilms International, 1984).

Wagner, Peter, 'The Satire on Doctors in Hogarth's Graphic Books', in *Literature and Medicine During the Eighteenth Century*, edited by Marie Mulvey Roberts and Roy Porter (London: Routledge, 1993), 200–225.

Waters, Arthur, *Issuers of Eighteenth Century Tokens Struck for the County of Middlesex* (Leamington Spa: Simmons and Waters, 1906).

CHAPTER 7

Utilising the Imagination as Therapy

As has been mentioned before, assessing patent medicines by the twenty-first century views on human physiology, pathology and therapeutics tells us only a modest amount about their efficacy and very little about the reasons why consumers chose to take them. Key to understanding why the medicines were taken in large quantities across society are the opinions and actions of Georgian contemporaries. These contemporaries often recognised that a medical therapy, whether pharmacological or physical, seemed to be more effective than the sum of its pharmacological constituents, or in the case of a mechanical, electrical or magnetic technique, its demonstrable physical processes. Some experts, particularly physicians, attempted to explain the mechanisms responsible for this extra benefit, perhaps with the help of experimental observations, and as a result they exposed their views, directly or indirectly, on the workings of patent medicines. Others, particularly owners, wholesalers and advertisers in the patent medicines industry, did not indulge in such theory, but their promotional techniques often followed the principles laid down by the physicians for obtaining this extra benefit. In their search for sales, the promoters followed the advertising techniques which not only drew the consumers' attention, but also rendered the medicines more effective for those consumers. In the language of the period, the vehicle which made patent medicines more effective than the aggregate pharmacology of their constituents was the 'imagination' of the patient.

A. Mackintosh, *The Patent Medicines Industry in Georgian England*, Medicine and Biomedical Sciences in Modern History,
https://doi.org/10.1007/978-3-319-69778-9_7

Historians have concentrated on the important role of the imagination in the fine arts and literature of the time, and have been less active in exploring its effects in healthcare.[1] Joseph Addison, who introduced the phrase 'pleasures of the imagination', recognised in the early eighteenth century that changing the imagination had general health benefits:

> Delightful scenes, whether in nature, painting or poetry, have a kindly influence on the body, as well as the mind, and not only serve to clear and brighten the imagination, but are able to disperse grief and melancholy, and to set the animal spirits in pleasing and agreeable motions.[2]

As the century progressed, physicians increasingly ascribed a more specific role to the imagination in improving or maintaining health. For us in the twenty-first century, 'imagination' is a relatively passive faculty, referring to a mental capacity to form internal images or ideas of objects and situations which are not actually present to the senses:[3] but in the past, the imagination was a more active instrument which could create, worsen or ameliorate disease, as well as overall wellbeing. Many in the sixteenth and seventeenth centuries felt that the imagination was a direct cause of disease, a force which could act at a distance and which was not 'imaginary', to use our connotation of this adjective.[4] For Paracelsus, the imagination was a supernatural force which could act directly on the material world.[5] For Van Helmont, the imagination could cause disease in any organ but particularly in the spleen (especially hypochondria) and the uterus (especially hysteria).[6] Galenists were less convinced by the material force of the imagination, but they recognised that it affected the body by changes in the spirits, humours and the movement of the heart.[7] Francis Bacon summarised the common opinion when he wrote that the imagination 'altereth the bodie proper of the imaginant', and that it had the power to both hurt and heal.[8]

By the early eighteenth century, a separation between mind and body was more clearly recognised, and the imagination was rarely considered as the direct cause of structural changes in the body, with the controversial exception of congenital abnormalities created by the mother's imagination.[9] It was now regarded more clearly as a mental process which could influence the health of the body, perhaps by altering its involuntary functions and its state of arousal.[10] Robert James, of the eponymous fever powder, devoted three-and-a-half pages of his medical dictionary to the imagination, and he provided an exuberant illustration of its benefits, writing that it could 'cause the eyes and countenance to sparkle, while the hand and every member exult for joy'.[11] He also recognised that the imagination could have a negative influence, and 'hence the countenance is dejected and the limbs enfeebled'. John Gregory, Professor of Physic at Edinburgh and one of the founders of the study of medical ethics, was more precise when he emphasised that the imagination influenced both the mind and the physiology of the body.[12]

Although the influence of the imagination on health, and many other aspects of human life, was well recognised, the sense contained in the term was difficult to define and writers of the era struggled to do so: as Addison put it, 'there are few words in the English language which are employed in a more loose and uncircumscribed sense'.[13] The common theme was that the imagination formed the bridge between the body, especially the senses, and the mental or extra-corporeal concepts variously described as the soul, the mind, or the moral system. For Mark Akenside, author of the book-length poem *The Pleasures of Imagination*, the imagination was 'a middle place between the organs of bodily sense and the faculties of moral perception', while Nathan Bailey wrote in his dictionary: 'it is a power or faculty of the Soul, by which it conceives and forms ideas of things, by means of certain traces and impressions that had been before made on the brain by sensation'.[14] Ephraim Chambers's *Cyclopaedia* plagiarised Bailey's description of fifty years earlier, and then went on to speculate on the neuro-anatomical site for the imagination.[15] Edmund Burke was less specific, but he emphasised its central role in human life, writing that 'the imagination is the most extensive province of pleasure and pain, as it is the region of our fears and hopes, and of all our passions that are connected with them'.[16] Samuel Johnson took a simpler, descriptive, approach in his dictionary with three meanings: first 'fancy; the power of forming ideal pictures; the power of representing things absent to one's self or others'; second 'conception; image in the mind; idea'; and third 'contrivance; scheme'.[17]

Confronted by the difficulties in arriving at a precise description of the imagination, most compilers of eighteenth-century *medical* dictionaries were content to accept its medical role without struggling to define it. The imagination just could be categorised as one of the five 'internal' senses, alongside memory, conscience, affections of the mind, and reason, without further explanation.[18] Barrow briefly described it as a means of picturing 'corporeal substances in the mind', and Quincy could only provide an example of the imagination over-riding the evidence of the eyes.[19] Motherby just referred to the controversy of the mother's imagination producing abnormalities in her baby.[20] James went further than most with a narrow mechanical explanation which is difficult to understand and seemingly at odds with his illustrations quoted earlier: 'A fancy, once excited by the appearance of an agreeable object, presently raises desire, and produces local motion, either to approach or avoid the object according to its different circumstances'.[21] Perhaps reflecting these uncertainties, some physicians, such as William Falconer, John Coakley Lettsom and John Haygarth, often preferred to express the concept of altering the imagination as the more specific 'passions'.[22] In this period, 'passions' could refer vaguely to a broad range of impulses and feelings, but these physicians used it for specific states of mind, such as joy, pleasure, love, anger, grief, hope and fear.[23] 'Imagination' also had a more specialised medical use as a term for diseases considered to be predominantly affecting the mind, such as melancholy or hypochondriasis: disorders of the imagination.[24]

Rather than debating the exact nature of the imagination, physicians were normally more concerned about its effects on health and how they were achieved. The royal physician and experimental chemist Peter Shaw was convinced of the general therapeutic powers of the imagination: 'And as we see by experience, what extraordinary efficiency the imagination has in the cure of diseases'.[25] Shaw attributed the beneficial powers of the imagination to confidence: 'A quack or a farrier, in whom a patient places great confidence is, in my opinion, a better physician for that patient, than a graduate doctor, from whom he has no expectations'.[26] At the end of the eighteenth century, the probing Bath physician John Haygarth linked confidence more specifically to the success of drug therapy, writing that medical therapy had one highly important rule: 'In the best manner possible a patient ought to be always inspired with confidence in any remedy which is to be administered'.[27]

Haygarth was not alone in his views on the benefits of confidence and the imagination in medical therapy. A detailed analysis was provided when the recently formed Medical Society of London instituted in 1786 an annual competitive essay in memory of Dr. John Fothergill, and the title for the first competition was 'What diseases may be mitigated or cured, by exciting particular affections or passions of the mind?'[28] The choice of this title by a society seeking to establish itself as a prestigious unifier of the branches of the medical profession shows that the topic was current and important. The prize was won by the physician William Falconer, one of the early proponents, alongside his friends Thomas Percival and John Haygarth of the use of statistics in public health.[29] Falconer was a thoroughly orthodox practitioner with MDs from both Edinburgh and Leiden, and a fellowship of the Royal Society, who moved from Chester to Bath in 1784. In his essay, Falconer provided numerous examples of the influence of the passions on the treatment of illnesses, both in the past and in his own time, concluding that successful treatment required not just the correct medicines but also the 'calling into our assistance the strong powers of the imagination'.[30] This assistance was, for example, crucial in the management of intermittent fevers, important in treating typhus, and helpful in the control of haemorrhage.[31]

The importance of the imagination in medical therapy was mentioned throughout the Georgian era, but opinions changed on whether a practitioner should seek to alter its state. Early in the eighteenth century, orthodox opinion maintained a distance from the powers of the imagination. John Quincy recognised in his best-selling *Pharmacopoia Officinalis* that the imagination was responsible for popularity of the touch of the hand of a recently executed prisoner as a treatment for scrophula; 'but such a practice as this is too whimsical to be countenanced by any of worth in the Profession of Physic'.[32] The surgeon James Handley also felt that the imagination was responsible for any success with the Royal Touch for the same condition; but 'you might have a cure as well, by rubbing the part with a broomstick'.[33] James gave numerous examples in his medical dictionary of the effects of the imagination, particularly in inducing foetal abnormalities, but he did not suggest that physicians should try and alter it. This reluctance to investigate and utilise the powers of the imagination was partly due to concerns that the imagination could have both harmful and beneficial effects depending on the circumstances of the patient.[34] It was also probably secondary to a need for physicians in this period to avoid anything which might hint at supernatural practices or folk remedies.

Later in the century, physicians were more prepared to influence the state of the imagination for the patient's benefit: as already described, Shaw, Falconer, Lettsom and Haygarth felt that the imagination was intrinsic to therapy. Falconer went further than most when he recommended unorthodox treatments for their effects on the imagination. These included charms for typhus if the patient had confidence in them, and a magnet in moderate toothache for its influence on the patient's imagination, not because of any animal magnetism, which he did not believe in.[35] Other physicians stuck to recognised therapy, but they regarded the imagination as an important and adjustable ally in the management of disease.

Falconer also commented that the powers of the imagination and the passions were not confined to prescribed therapy and they extended to 'empirical remedies', including patent medicines:

> Hence it is, that the same remedy will not always produce the same effect, even in the same person, and that common remedies often prove wonderfully successful in the hands of bold quacks, but do not answer the purpose in a timorous and distrustful patient.[36]

Other practitioners agreed with Falconer that the imagination rendered patent medicines more effective than just the sum of the pharmaceutical consequences of the recipe, and they suggested some of the mechanisms involved. Gregory wrote that mystery was important: 'When a nostrum is once divulged and sold for a trifle, all its wonderful qualities immediately vanish, and in a few months it is utterly forgot'.[37] Haygarth emphasised that confidence created the favourable effects of the imagination with patent medicines:

> On this principle we may account for the marvellous recoveries frequently ascribed to empirical remedies, which are commonly inert drugs, and generally applied by the ignorant patient to disorders totally different from what the quack himself pretends they can cure. Magnificent and unqualified promises inspire weak minds with implicit confidence.[38]

Furthermore, other writers concluded that this crucial boost from the imagination could result in the patent medicines being more effective than the same medicines being administered by a regular practitioner. Duncan Forbes, an Edinburgh physician, commented in a long paper on empiricism:

> Mankind are fond of mystery; and it is more congenial with a sick man's mind to expect relief from the occult qualities of a medicine, than from its sensible virtues. Hence, in a great measure, arises the success of many boasted secret remedies, which, when compounded in an apothecary's shop, instantly lose their efficacy.[39]

An anonymous Edinburgh physician was sharply critical of patent medicines and their owners in a long letter to a medical journal, but nevertheless had a grudging admiration for their superiority over regular therapy under certain circumstances:

> Even the inexhaustible impudence and monstrous lies, and confident assurances, of the quacks afford more relief and comfort to many thousands of patients than any of our regular Faculty could give them, even if they were administering to them the same medicines that the quacks employ.[40]

Contemporary physicians were also exploring the imagination in experiments, using techniques that anticipate twentieth-century evidence-based medicine. A remarkable example was the French government's Royal Commission of 1784 into Animal Magnetism, which had appeared to be of considerable benefit in some medical conditions; its inventor, Anton Mesmer was practising in Paris.[41] Animal Magnetism depended on the transfer of an undetectable magnetic fluid between the operator and the patient, or from magnetised objects to the patient: speech, gestures and touch by an operator were required, perhaps accompanied by a 'magnetic' object such as an iron rod, a bucket or a magnetised tree. The appointed commission was chaired by Benjamin Franklin, the US ambassador, and consisted of four well-known Paris physicians and four prominent natural philosophers: Dr. Guillotin was one of the physicians, and Lavoisier and Bailly were amongst the philosophers.[42] A range of experiments, devised by Lavoisier to prove that the magnetic fluid did not exist, was performed under variable and controlled conditions, such as blindfolding the patient, forbidding additional verbal communication, varying any instruments, and altering the size of the audience. Several experiments showed that blindfolded patients benefited from verbal and other forms of communication even when no 'magnetism' should have been present.

The commissioners concluded that 'the imagination is the true cause of the effects attributed to the magnetism'.[43] These alterations in the imagination were induced 'by solemn preparations, by extraordinary proceedings, by the confidence and enthusiasm inspired by magnificent promises'.[44] The medical benefits resulted from stimulation of the nerves which produced demonstrable effects on the bowel and other organs by unique unknown mechanisms.[45] They found that the imagination was altered by a less powerful stimulation on a second occasion: in other words, familiarity increased the chance of successful animal magnetism.[46] The response was better in a public group than in private, which is a form of peer pressure, and that the lower classes were easier to treat than the higher levels of society, a reflection of authority. In addition, the commissioners confirmed their views on the power of the imagination by submitting a secret supplementary report on the risk of 'magnetised' female patients becoming erotically attracted to male magnetisers.[47]

The action of the imagination in medical care was also explored by John Haygarth and others when they argued that it was responsible for the benefits of the transiently popular Perkins's Tractors (Fig. 7.1). These tractors were a combination of metals made into tapered rods which had been introduced by Dr. Elisha Perkins in Connecticut and were patented in London in 1798 by his son Benjamin Perkins. When passed over the surface of the body, they relieved pain, paralysis and other manifestations of a variety of 'topical' diseases. Sets of tractors were

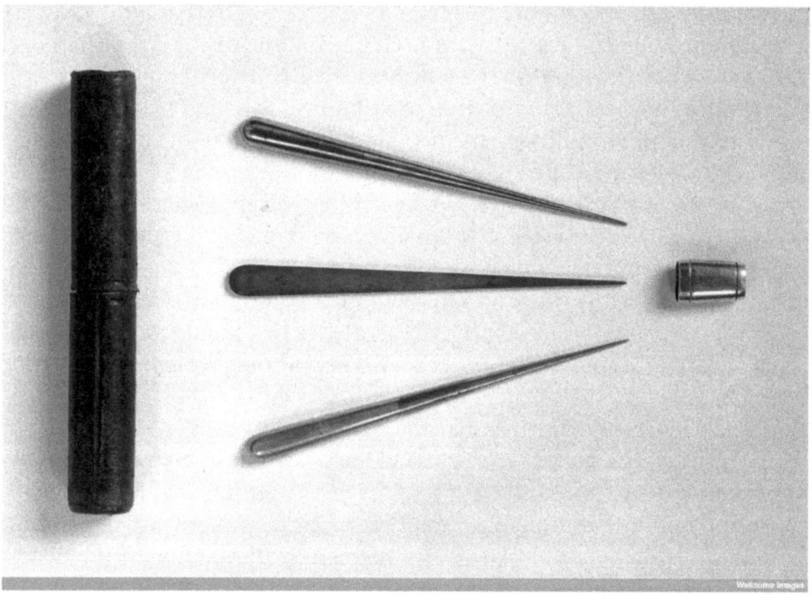

Fig. 7.1 Perkins's tractors with their box (courtesy of The Wellcome Library, London. Creative Commons License CC-BY 4.0)

imported from the USA in batches of 200 at five guineas a set, and they were used by some regulars and by a variety of irregular practitioners, including two London booksellers, David Ogilvy and James Matthews, who provided long supporting certificates describing their positive experience.[48] An account of their application by the clergy in Lancashire indicates that they were popular across the country.[49]

To explore the action of the Tractors, Haygarth used similar shaped pieces of wood painted grey to resemble the genuine metal version.[50] Without the patient being informed of the change, he tried them on five hospital patients in Bath, and they produced the same benefits as the real tractors did the following day. He then sent a pair of the false tractors to Dr. Richard Smith at Bristol Royal Infirmary who showed that they were effective in many, though not all, patients with a wide range of conditions. A bone tractor and one made from a common nail also worked. Haygarth realised that these findings were more important than just a refutation of 'Perkinism'. For him, these experiments 'clearly prove what wonderful effects the passions of hope and faith, excited by mere imagination, can produce upon diseases'.[51] Hope and confidence could be as powerful as the medicinal qualities of a drug, and a physician should confidently express his genuine appreciation of the virtues of a prescribed medicine.[52]

Haygarth also made a point which is very relevant to our understanding of the production and retailing of patent medicines: the advocates of Perkins's Tractors were not fraudulent.[53] The tractors did indeed relieve symptoms in many patients: though it turned out that cheap simple materials could produce similar benefits with the help of the imagination. Similarly, the French Royal Commission had concluded that the effects of Animal Magnetism on the body were genuine, but they were due solely to the imagination and not to magnetic fluids.[54] The proponents of these therapies correctly believed that the treatments were effective, though they misunderstood the reasons why. However, the general public did not recognise this distinction between ignorance and dishonesty: Mesmer disappeared from Paris soon after the publications of the report, and Animal Magnetism was extensively ridiculed and dismissed as fraudulent on both sides of the Channel after the publication of the Commission's report.[55] Perkins's Tractors vanished in the early years of the nineteenth century. Likewise, the advocates of patent medicines could be honest in attesting to their effectiveness even when this was due to the powers of the imagination rather than the medicines' pharmaceutical components: but contemporary and later critics of patent medicines were reluctant to appreciate this point of view.

Thus, the general remarks of Georgian physicians on the benefits of the imagination and the passions were largely in agreement with the more specific conclusions from the experiments in Paris on Animal Magnetism and in England on Perkins's Tractors. Indeed, Haygarth regarded his experimental findings as being an extension of these earlier opinions.[56] To summarise the opinions and the experimental results, the most important promoter of the beneficial alteration in the imagination was the consumers' confidence, which in turn was inspired by 'confident assurances', 'magnificent promises', familiarity, the attitude of the physician, solemnity, peer pressure and authority. In addition, the benefits of confidence were boosted by 'mystery', which could be produced by the secrecy of a recipe, or by 'extraordinary proceedings'.

The Imagination and the Placebo Effect

Nowadays, a beneficial result from drug therapy which cannot be attributed to the pharmacology of the agent is commonly referred to as a 'placebo effect'. This effect can be induced by many types of recognised therapy and should be part of routine medical practice; it can also be induced by a placebo, but today this would usually be unethical outside a therapeutic trial.[57] Was this understanding of the imagination as a therapeutic agent in the Georgian era a precursor to the modern concept of a placebo effect? Obviously, the culture of the eighteenth century was very different from that of today, and we should be very cautious about any links as we know that the occurrence of a placebo effect is culture-specific.[58] In addition, a historical discontinuity exists between the understanding of the effects of the imagination and the idea of a placebo response. Some would say that we should not even pose this question; but it has repeatedly been asked in discussions on this research, and it is difficult to ignore.

During the nineteenth century, developing medical theories and observations eclipsed the appreciation of the powers of the imagination. The ambition to explain all diseases in terms of physical changes in organs resulted in the rejection of the imagination as a positive or negative influence on health, except for the specifically female disease of hysteria.[59] By the end of the nineteenth century, mental processes were largely separated from illness of the body, and the imagination had ceased to be mentioned in orthodox medicine rather earlier.[60] The placebo effect was a late arrival on the medical scene, only becoming clear with the routine adoption of placebo-controlled drug trials in the mid-twentieth century. Up to the end of the eighteenth century, 'placebo' had the non-medical meaning of a flatterer or sycophant.[61] Its medical use was originally confined to the sense of a trivial therapy of little benefit, and then it began to refer to a harmless substitute for conventional therapy. The word did not appear in the medical dictionary written by Robert James; nor did the first edition of Motherby's medical dictionary include it in 1775; but the second edition in 1785 did, defining it simply as 'a common place method or medicine'.[62] By the fourth edition ten years later, the definition had expanded to 'a common-place method or medicine calculated to amuse for a time, rather than for any other purpose'.[63]

During the nineteenth century, the placebo remained outside orthodox medicine with little or no understanding of the placebo effect as an agent which could improve the receiver's health.[64] Raieck and colleagues digitally searched the archives of the *British Medical Journal* from 1840 to 1899 and found diverse reasons for the use of a placebo.[65] These citations could refer to an ineffective treatment, a means of allowing the natural history of the disease to unfold, a therapy given solely to fulfil the patient's expectations, something to buy time, or a way of financially benefiting the doctor. Only one of the 71 citations implied that the placebo had a clinical effect. The first analysis of the term 'placebo effect' seems to have been by Stewart Wolf in 1950.[66] So the general recognition of the beneficial placebo effect was not a restatement of the powers of the imagination for a later era: it was a fresh, mid-twentieth-century, discovery.

Over the last few decades, the placebo effect has been frequently reported, especially in placebo-controlled drug trials, but it has proved resistant to detailed exploration. For example, Richard Kradin has recently described the placebo effect as a nervous system mechanism for dealing with symptoms of distress, which is mediated by an adult version of the normal childhood attachment to a caregiver: but this is a plausible theory rather than a scientific explanation, and it is very culture-specific.[67] Cecil Helman has attempted to study the placebo response outside its cultural context, but with only limited success.[68] So any direct and detailed comparison of the modern placebo response with the powers of the imagination from 200 years ago would be invalid. But we can see whether they have anything in common, and one shared feature is that both require the confidence of the patient. The Georgian experts regarded this confidence as the key to the therapeutic powers of the imagination, and a placebo stimulus must be credible to the patient.[69] Similarly, the next chapter will show that Georgian medicine advertisers were reluctant to risk reducing confidence by exaggerating the benefits of their medicines, while David Morris has described how a disbelief in a therapy will actively counteract any placebo benefits.[70]

However, differences emerge once we compare the more detailed initiators of the placebo response to the methods used by Georgians to stimulate the imagination. Howard Brody summarised the factors which promote a placebo response in today's Western culture as an understandable and satisfactory explanation of the illness and its treatment, a demonstration of care and concern, and an enhanced promise of symptom control.[71] The first factor is not shared with the powers of the Georgian imagination, which were heightened by mystery and 'extraordinary proceedings', while the second was not prominent in the descriptions of two centuries ago and in any case would have had little relevance to the promotion of medicines by print. Only the assurance of improved symptoms has been designated as a necessity for both forms of therapy.

The conclusion must be that the therapeutic effects of the imagination were not a precursor to the placebo effect. They were formulated in eras with different cultures and different understandings of human physiology and pathology, and the imagination had disappeared from orthodox medicine before the placebo effect was appreciated. From the limited information available, they also appear to be initiated by different circumstances. This separation does not exclude the possibility that, whatever their culture-based drivers, the effects of the imagination and the placebo response employ some common physiological mechanisms in the brain and the peripheral nervous system: human culture and human pathology are constantly changing, but it is reasonable to assume that human physiology has not altered over this period of time. But any physiological commonality will remain hidden: Georgians expressed their understanding of the imagination within a system of medicine which is radically different from that of today, and, in any case, the understanding of physiological mechanisms of the modern placebo effect remains very limited.[72] Changes in the imagination were a recognised therapeutic instrument in the long eighteenth century, and they should be studied on this basis, not as a potential antecedent to something else.

The Imagination and the Printed Word

The beneficial effects of altering the imagination has a historical application beyond an understanding of Georgian medical therapy and the effectiveness of patent medicines. Therapies, such as Animal Magnetism and Perkins's Tractors, which stood outside both the contemporary theories of orthodox medicine and our current science-based knowledge, did relieve symptoms and improve the subject's health. As Haygarth recognised, the advocacy of these therapies was often not fraudulent, contrary to the beliefs of most writers in later periods: the enthusiasts could be honest, but misguided. This conclusion is applicable to other eras, and the common assumption that alternative therapies in the past were normally based on deceit and avarice is ripe for revision.[73] The 'quack' might have genuinely believed in the effectiveness of his or her remedy, and, in common with orthodox practitioners and their treatments, he was entitled to earn money from it.

How did the owners and advertisers in the patent medicines industry harness the imagination, as contemporaries recognised they did, to maximise the therapeutic response? Their approach was comparable to that of the regular practitioners, but used a different medium. The practitioners usually described the medical effects of the imagination in the context of a direct medical consultation, whether a regular or irregular practitioner was involved, and this would be conducted as a two-way verbal communication. For patent medicines, the benefits of the imagination were obtained by a different route: the popular medicines were largely promoted by the printed word in newspaper advertisements and printed bills, and the communication was one-way. For most patent medicines, the imagination was altered directly by the printed word when read by the consumer, or indirectly when the print was read by friends and relatives, or the medicine retailer, who then advised the consumer. Consequently, the printed word introduced the powers of the imagination and so it should be regarded as a therapeutic agent, alongside the included pharmaceuticals. Indeed, print might have contributed more to the beneficial effects of a medicine than the pharmaceuticals. The manufacturers did not insert this agent with a spoon or a measuring glass; they used the printing press to add it in.

The advertisers have not left accounts of the principles behind their use of the printed word in this way, but we do have copious quantities of their promotional material in the form of newspaper advertisements, bills and supporting treatises, and this material demonstrates how the imagination was altered. Some owners spent several hundred pounds a year on advertisements in provincial newspapers to ensure that their medicines were often purchased, but the advertising techniques were probably based on feedback from sales rather than advice from Falconer, Haygarth and other experts on the therapeutic powers of the imagination. Using print to alter favourably the imagination would increase effectiveness and consumer satisfaction, enhancing sales. So exploring the promotional techniques of the advertisers reveals the role of the printed word in altering the imagination. What were the predominant methods by which print altered the imagination and so added potency to the medicines? The next chapter will answer this question by demonstrating the common techniques with the help of a systematic study of newspaper advertisements.

Notes

1. For example, Brewer, *Pleasures of the Imagination*; Hoock, *Empires of the Imagination*.
2. *Spectator*, No. 411, 21 June 1712, 87.
3. 'Imagination', *OED* (accessed 9 June 2017).
4. McMahon, 'Pre-Cartesian History'; Fischer-Homberger, 'Medical History'; Kirmayer, 'Medicine'.
5. Kirmayer, 'Medicine', 585.
6. Fischer-Homberger, 'Medical History', 620.
7. Pender, 'Rhetoric', 60; McMahon, 'Pre-Cartesian History', 181.
8. Bacon, *Proficience*, 80.
9. McMahon and Hastrup, 'Post-Cartesian History', 206; Wilson, 'Out of Sight'; Motherby, *Medical Dictionary* (1775), entry for 'imaginatio'; Fischer-Homberger, 'Medical History', 624.
10. Kirmayer, 'Medicine', 585; McMahon and Hastrup, 'Post-Cartesian History', 206–208; Fischer-Homberger, 'Medical History', 622; Hartley, *Observations*, vol. 1, i–iii.
11. James, *Medicinal Dictionary*, vol. 2, entry for 'imagination'.
12. Gregory, *Lectures*, 79, 110.
13. *Spectator*, No. 411, 21 June 1712, 84.
14. Akenside, *Pleasures of Imagination*, 5; Bailey, *Dictionarium Britannicum*, entry for 'imagination'.
15. Chambers, *Cyclopaedia*, vol. 2, entry for 'imagination'.
16. Burke, *Philosophical Enquiry*, 15.

17. Johnson, *Dictionary*, entry for 'imagination'.
18. Hooper, *Medical Dictionary*, entry for 'senses'.
19. Barrow, *Dictionarium Medicum*, entry for 'imaginatio'; Quincy, *Lexicon Physico-Medicum*, 225.
20. Motherby, *Medical Dictionary* (1775), entry for 'imaginatio'.
21. James, *Medicinal Dictionary*, vol. 2, entry for 'imagination'.
22. Falconer, *Passions*, 8–21; Lettsom described the effects of the passions in the introduction to Falconer, *Passions*; Haygarth, *Imagination*, 25.
23. Dixon, *Passions to Emotions*, 5.
24. Gregory, *Lectures*, 24; Cullen, *Lectures*, 61.
25. Shaw, *Reflector*, 229.
26. Shaw, *Reflector*, 229.
27. Haygarth, *Imagination*, 28.
28. Falconer, *Passions*, 1.
29. Borsay, 'Falconer, William'.
30. Falconer, *Passions*, 88.
31. Falconer, *Passions*, 23, 27 and 52.
32. Quincy, *Pharmacopoia*, 221; Davies and Matteoni, 'Virtue'.
33. Handley, *Essays*, 420.
34. James, *Medicinal Dictionary*, vol. 2, entry for 'imagination'; Falconer, *Passions*, 9; Fischer-Homberger, 'Medical History', 621–625
35. Falconer, *Passions*, 28 and 46.
36. Falconer, *Passions*, 88.
37. Gregory, *Observations*, 55.
38. Haygarth, *Imagination*, 29.
39. Forbes, 'Empiricism', 369.
40. *MCR*, 12 (1806), clviii.
41. Darnton, *Mesmerism*, 61–64; Ellenberger, *Discovery*, 64–67.
42. *Report of Dr. Franklin*.
43. *Report of Dr. Franklin*, 78.
44. *Report of Dr. Franklin*, 10.
45. The increased tone of the sensible and nervous fibres generated 'their impulse towards certain organs, and to excite in them evacuations or excretions, without these resulting in any additions to the sciences, either philosophy or medicine' (*Report of Dr. Franklin*, 10).
46. *Report of Dr. Franklin*, 96.
47. Ellenberger, *Discovery*, 65.
48. Perkins, *Directions; Quack Doctors*, 24; Perkins, *Efficacy*, 38–45.
49. *MPJ*, 16 (1806), 457, footnote.
50. Haygarth, *Imagination*, 2–25.
51. Haygarth, *Imagination*, 29.
52. Haygarth, *Imagination*, 30.

53. Haygarth, *Imagination*, 16.
54. *Report of Dr. Franklin*, 88.
55. Darnton, *Mesmerism*, 65; Ellenberger, *Discovery*, 67; Winter, *Mesmerized*, 99.
56. Haygarth, *Imagination*, 25.
57. Brody, 'Therapeutic Agent', 78.
58. Brewer, 'Placebo'; Morris, 'Placebo', 192.
59. Fischer-Homberger, 'Medical History', 625; Morris, 'Placebo', 187.
60. McMahon and Hastrup, 'Post-Cartesian History', 209; Ellenberger, *Discovery*, 149.
61. 'Placebo', *OED* (accessed 4 June 2016).
62. Motherby, *Medical Dictionary*, (1785), entry for 'placebo'.
63. Motherby, *Medical Dictionary*, (1795), 593.
64. Kaptchuk, 'Powerful Placebo'.
65. Raieck, Stone and Kaptchuk, 'Placebos'.
66. Wolf, 'Suggestion'.
67. Kradin, 'Placebo Response'.
68. Helman, 'Placebos and Nocebos'.
69. Shapiro, Arthur, and Elaine Shapiro, 'Placebo', 30; Morris, 'Placebo', 189.
70. Morris, 'Placebo', 189.
71. Brody, 'Therapeutic Agent', 79.
72. Shapiro, Howard, 'Clinical Reflections', 40; Ernst, 'Scientific Understanding'; Kradin, 'Placebo Response', 439.
73. Two recent examples of this assumption are Rance, *Quack Doctor* and Strathern, *Quacks*.

Bibliography

Newspapers and Journals

Medical and Chirurgical Review (MCR).
Medical and Physical Journal (MPJ).
The Spectator.

Printed Primary Sources

Akenside, Mark, *The Pleasures of Imagination* (London: R. Dodsley, 1744).
Bacon, Francis, *Of the proficience and advancement of learning, divine and humane* (London, 1605), quoted in Richard Hunter and Ida MacAlpine, *Three Hundred Years of Psychiatry 1535–1860* (London: Oxford University Press, 1963), 80.
Bailey, Nathan, *Dictionarium Britannicum* (London, 1730).

Barrow, John, *Dictionarium Medicum Universale* (London, 1749).
Burke, Edmund, *A Philosophical Enquiry into the Origin of Our Ideas of the Sublime and Beautiful*, new edn (London, 1798).
Chambers, Ephraim, *Cyclopaedia: Or, an Universal Dictionary of Arts and Sciences*, 5 vols (London, 1778–1788).
Cullen, William, *Clinical Lectures, Delivered in the Years 1765 and 1766* (London, 1797).
Falconer, William, *A Dissertation on the Influence of Passions Upon the Disorders of the Body* (London: C Dilly, J Phillips, 1788).
Forbes, Duncan, 'On the Origins and Progress of Empiricism', *MPJ*, 16 (1806), 362–370.
Gregory, John, *Observations on the Duties and Offices of a Physician* (London: Strahan & Cadell, 1770).
Gregory, John, *Lectures on the Duties and Qualifications of a Physician*, new edn (London, 1772).
Handley, James, *Mechanical Essays on the Animal Oeconomy* (London, 1721).
Hartley, David, *Observations on Man, His Frame, His Duty and His Expectations*, 2 vols (London, 1749).
Haygarth, John, *Of the Imagination* (London: Cadell and Davies, 1800).
Hooper, Robert, *A Compendious Medical Dictionary* (London, 1798).
James, Robert, *A Medicinal Dictionary*, 3 vols (London, 1743–1745).
Johnson, Samuel, *A Dictionary of the English Language*, 2nd edn, (London, 1756).
Motherby, George, *A New Medical Dictionary* (London: Joseph Johnson, 1775).
Motherby, George, *A New Medical Dictionary*, 2nd edn (London, 1785).
Motherby, George, *A New Medical Dictionary*, 4th edn (London, 1795).
Perkins, Benjamin Douglas, *Directions for Performing the Metallic Operation with Perkins's Patent Tractors* (London, 1798?).
Perkins, Benjamin Douglas, *The Efficacy of Perkins's Patent Metallic Tractors* (London, 1800).
Quack Doctors Dissected (Gloucester, c1805).
Quincy, John, *Lexicon Physico-Medicum: Or, a Medicinal Dictionary*, 4th edn (London, 1730).
Quincy, John, *Pharmacopoia Officinalis and Extemporanea*, 8th edn (London, 1730).
Report of Dr Benjamin Franklin and other Commissioners, charged by the King of France with the examination of the Animal Magnetism (London: Joseph Johnson, 1785).
Shaw, Peter, *The Reflector* (London: Longman, 1750).

Secondary Sources

Borsay, Anne, 'Falconer, William', *ODNB* (accessed 3 October 2013).
Brewer, Colin, 'The Golden Age of Placebo', *British Medical Journal*, 344 (2012): 47.
Brewer, John, *The Pleasures of the Imagination: English Culture in the Eighteenth Century* (London: Harper Collins, 1997).
Brody, Howard, 'The Doctor as Therapeutic Agent: A Placebo Research Agenda', in *The Placebo Effect: An Interdisciplinary Exploration*, edited by Anne Harrington (London: Harvard University Press, 1997), 77–92.
Darnton, Robert, *Mesmerism and the End of the Enlightenment in France* (Cambridge, Mass: Harvard University Press, 1968).
Davies, Owen, and Francesca Matteoni, '"A Virtue Beyond All Medicine": The Hanged Man's Hand, Gallows Tradition and Healing in Eighteenth- and Nineteenth-Century England', *Social History of Medicine*, 28 (2015): 686–705.
Dixon, Thomas, *From Passions to Emotions: The Creation of a Secular Psychological Category* (Cambridge: Cambridge University Press, 2003).
Ellenberger, Henri F., *The Discovery of the Unconscious: The History and Evolution of Dynamic Psychiatry* (London: Allen Lane, 1970).
Ernst, Edzard, 'Towards a Scientific Understanding of Placebo Effects', in *Understanding the Placebo Effect in Complimentary Medicine: Theory, Practice and Research*, edited by David Peters (Edinburgh: Churchill Livingstone, 2001), 17–29.
Fischer-Homberger, Esther, 'On the Medical History of the Doctrine of Imagination', *Psychological Medicine*, 9 (1979): 619–628.
Helman, Cecil G., 'Placebos and Nocebos: The Cultural Construction of Belief', in *Understanding the Placebo Effect in Complimentary Medicine: Theory, Practice and Research*, edited by David Peters (Edinburgh: Churchill Livingstone, 2001), 3–16.
Hoock, Holger, *Empires of the Imagination: Politics, War and the Arts, 1750–1850* (London: Profile Books, 2010).
Kaptchuk, Ted, 'Powerful Placebo: The Dark Side of the Randomised Controlled Trial', *Lancet*, 351 (1998): 1722–1725.
Kirmayer, Laurence, 'Towards a Medicine of the Imagination', *New Literary History*, 37 (2006): 583–605.
Kradin, Richard, 'The Placebo Response: An Attachment Strategy that Counteracts the Effects of Stress-Related Dysfunction', *Perspectives in Biology and Medicine*, 54 (2011): 438–454.
McMahon, C. E., 'The Role of the Imagination in the Disease Process: Pre-Cartesian History', *Psychological Medicine*, 6 (1976): 179–184.

McMahon, C. E., and J. L. Hastrup, 'The Role of Imagination in the Disease Process: Post-Cartesian History', *Journal of Behavioural Medicine*, 3 (1980): 205–218.

Morris, David B., 'Placebo, Pain and Belief: A Bicultural Model', in *The Placebo Effect: An Interdisciplinary Exploration*, edited by Anne Harrington (London: Harvard University Press, 1997), 187–207.

Pender, Stephen, 'Rhetoric, Grief and the Imagination in Early Modern England', *Philosophy and Rhetoric*, 43 (2010): 54–85.

Raieck, Jacqueline, Bradley Stone and Ted Kaptchuk, 'Placebos in 19th Century Medicine', *British Medical Journal*, 345 (2012): 44–45.

Rance, Caroline, *The Quack Doctor: Historical Remedies for All Your Ills* (Stroud, Gloucestershire: The History Press, 2013).

Shapiro, Arthur K., and Elaine Shapiro, 'The Placebo: Is It Much Ado About Nothing?' in *The Placebo Effect: An Interdisciplinary Exploration*, edited by Anne Harrington (London: Harvard University Press, 1997), 12–36.

Shapiro, Howard, 'Clinical Reflections on the Placebo Phenomenon', in *The Placebo Effect: An Interdisciplinary Exploration*, edited by Anne Harrington (London: Harvard University Press, 1997), 37–55.

Strathern, Paul, *Quacks, Rogues and Charlatans* (London: Royal College of Physicians, 2015).

Wilson, Philip, '"Out of Sight, out of Mind?" The Daniel Turner–James Blondel Dispute over the Power of the Maternal Imagination', *Annals of Science*, 49 (1992): 63–85.

Winter, Alison, *Mesmerized: Powers of Mind in Victorian Britain* (Chicago: Chicago University Press, 1998).

Wolf, Stewart, 'Effects of Suggestion and Conditioning on the Actions of Chemical Agents in Human Subjects—the Pharmacology of Placebos', *Journal of Clinical Investigation*, 29 (1950): 100–109.

CHAPTER 8

Harnessing the Potency of Print

The printed word was the essential facilitator for both the promotion and the effectiveness of patent medicines. It was the vehicle which enabled a few components, perhaps cheap and simple, to be assembled at one site, and then transformed into an expensive medicine which could be sold across England to thousands of purchasers. Chapter 6 investigated the organisation and financing of advertising designed to create and maintain national or regional markets. This chapter explores how print was manipulated in this advertising for two purposes: one was to persuade customers to buy these medicines, which might produce a complete cure, but could also generate major complications or even death; and the other purpose was to increase their effectiveness, a consequence well recognised by contemporaries and attributed by them to the power of the 'imagination'.

As we saw in the last chapter, several contemporary physicians agreed that changes in the imagination could transform the ingredients in the recipe into a potent medicine which was more effective than a mixture of the same ingredients prescribed by a regular practitioner. Many of the elements which were thought to be influential in utilising the imagination (such as 'confident assurances', mystery, familiarity, authority, attitude of the physician and solemnity) also apply to the promotion of patent medicines by the printed word. A partial exception was 'magnificent promises', which, contrary to earlier accounts, is shown in this chapter to feature in the promotion of only a minority of medicines. The methods exploited in print to alter the imagination can be summarised

A. Mackintosh, *The Patent Medicines Industry in Georgian England*, Medicine and Biomedical Sciences in Modern History,
https://doi.org/10.1007/978-3-319-69778-9_8

as increasing the consumer's confidence in the medicine, supported by some mystery, and the newspaper advertisements demonstrate these methods being put to good use.

Medicine advertisements are a prominent feature in preserved provincial newspapers and several historians have commented on them. Up to the 1970s, the common approach was to regard them solely as a source of amusement. In the last quarter of the twentieth century, Georgian irregular medicine became a subject for historical exploration and the medicine advertisements were a useful point of entry into the topic. Porter used them to obtain a great deal of information on 'quackery', but made little attempt to analyse the advertising techniques.[1] Others have reviewed medicine advertisements in a particular locality, relying on impressions rather than any systematic exploration.[2] Cody provided a deeper analysis when she emphasised the importance of advertisements in adding value to patent medicines, but her disapproval of these medicines limited discussion on the mechanisms of doing so.[3] John Strachan's book on the culture of the Romantic Period was not intended to deal with healthcare, but he nevertheless used medicine advertising to establish many of his arguments on advertising in general.[4]

All these publications contained convincing examples of the aims of the advertisements and some of the techniques employed; but they were based on impressions derived from individual examples of the advertisements or a small local sample. Barker provided an exception to this methodology with her systematic approach to medical advertising in newspapers serving large northern towns.[5] Her paper, which studied advertisements for medical services as well as for medicines, was a significant advance, emphasising the importance of creating trust in urban populations in the North of England which were isolated from their traditional family and community support. By tabulating the different marketing methods in the samples, Barker showed that identification of the brand was the commonest technique employed; but it was unclear in the paper how the content of the advertisements, which described multiple aspects of the medicines, was condensed to one or two marketing methods.

The key to both increasing sales and utilising the imagination to enhance efficacy was the confidence of the consumer, accompanied by a degree of mystery, and this chapter shows how this was achieved in practice. The necessary trust in the medicines was generated predominantly by a low-key style rather than the 'hard sell' which has sometimes been considered as an essential component of patent medicine advertising. Once medicine advertising has been explored systematically, many of the earlier assumptions about it are found to be based on a few eye-catching examples which are unrepresentative of the duller majority. In this period, most of the advertisers wanted to position their medicines as close as possible to orthodox therapy: they did not wish to be strident opponents of regular practitioners, or to follow the example of prominent irregulars.

First, however, we need to explore what is meant by confidence, a subject which has not attracted a great deal of attention from the scholars of our period. Some historically informed sociologists, such as Niklas Luhmann, Anthony Giddens and Christel Lane, have pointed out the intimate connection between confidence and trust: trust is the mechanism for dealing with uncertain real-life situations.[6] For Luhmann, trust is imperfect as there is normally insufficient information to give complete assurance; but without trust it would be impossible to deal with the complexities of life. Trust depends either on relationships with individuals, directly or through a form of communication such as the printed word, or on interactions with people grouped in institutions.

Early modernists have emphasised the importance of interpersonal communication in establishing trust and so creating confidence. In his study of credit and commerce in England, Craig Muldrew argued that the necessary trust for the credit-based economy was achieved by interpersonal relations between individuals and was communicated to others by reputation.[7] Steven Shapin felt that trust in seventeenth-century English natural philosophy required the guarantee of a gentleman with the necessary social, economic and genealogical attributes.[8] For the Republic of Letters during the first half of the eighteenth century, Anne Goldgar has also contended that trust depended on personal relationships between the participants.[9] However, the Georgian consumer seeking to place his or her trust in a patent medicine was in a rather different situation, without any direct contact with the person attempting to generate that trust. As Barker has argued, establishing trust in the medicines largely depended on 'thin', impersonal, trust, which was created by social and commercial links, rather than the traditional 'thick' trust formed by the stronger networks of family and friends which had been crucial in the past.[10] The purchasers of medicines advertised in the newspapers could still utilise some 'thick' trust in the form of advice from relatives and friends; but they also had to develop relationships of 'thin' trust with the vendors, and these were dependent on promotional strategies, particularly the advertising content.

The promotional strategies designed to generate the 'thin' trust, can be divided into those which were created by the descriptive words of members of the industry, those which derived from authoritative bodies or individuals outside the industry, and those which resulted from familiarity. Within the industry, the words of the owners, wholesalers and retailers promoted a trust amongst the consumers which was readily produced and flexible, but also potentially fragile as the consumers would know that the descriptions were biased towards the interests of the industry. The authority from outside the industry came from men or women of social or professional distinction, from previous legal decisions, or from the government and the law in the form of the medicine patent or the excise stamp. This external authority had the potential to be more effective in convincing consumers of the benefits of the medicines because the persuasive powers of the external individuals or institutions were based on their expertise, genuine or not, and their position in society, making this authority less obviously promotional in nature. In addition to these methods of generating trust amongst consumers, familiarity with the medicines was a less direct, but particularly important, mechanism in establishing their trust, and the advertisements had a substantial role in achieving it.

The main source for this chapter is another systematic analysis of medicine advertisements in the studied newspapers from Leeds, Birmingham and Salisbury during five years between 1769 and 1822. Of course, as we saw in Chapter 6, print was used to promote patent medicines in other formats, particularly bills but also treatises and puffs in books and periodicals. The content of the bills was broadly similar to that of the newspaper advertisements, though at greater length; but the bills which survive are less numerous and have been chosen for preservation, making them less representative. The newspaper advertisements in complete runs of newspapers have the substantial advantage of being a large and truly primary source, as they have not been subjected to later selection. Advice was also provided to consumers in the printed directions which often accompanied every bottle of a medicine, but few examples of these have been preserved. So this chapter will concentrate on the wording of the newspaper advertisements, the largest and most representative source.

Content of the Newspaper Advertisements

Medicine advertising by print has a long history going back to the earliest periodicals in the seventeenth century. A medicine advertisement can be found in the periodical *Mercurius Politicus* in 1652.[11] In the late seventeenth century, it was possible to achieve a national market for an owned medicine without resorting to substantial newspaper advertising. Anthony Daffy sold 65,000 bottles of his elixir outside London and had 38 overseas agents in the 1670s and 1680s, a time when no provincial newspapers were published.[12] He used printed handbills for point-of-sale advertising and also probably for a wider distribution of information: the survival of at least ten copies of his bills suggests that they were widely distributed.[13] Advertisements of all types were sparse in the early provincial newspapers at the beginning of the eighteenth century, but medicine advertisements were among them.[14] Medicine advertising grew with the expansion of provincial newspapers during the century: for example, soon after its foundation in 1736, the *Salisbury Journal* printed about forty medicine advertisements per year and by the 1770s this number had risen to over 300 per year.[15] By the second half of the eighteenth century, the major wholesalers regarded newspaper advertising as essential, and, as we saw in Chapter 6, some were prepared to spend hundreds of pounds a year on it.

What have historians discovered when they have looked at the persuasive content of a selection of medicine advertisements? Barker has emphasised the role of gaining trust in promoting medicines, whereas others have picked out some less convincing features.[16] Porter felt that hyperbole was essential and suggested that linking the medicine name to a famous physician or a desirable attribute such as 'solar' or 'gold' was beneficial.[17] Strachan also noticed that the medicines were often named after dead, distinguished, practitioners.[18] For Barry, effective local testimonials were important together with approval, especially in the early eighteenth century, by the gentry or aristocracy, or by local practitioners.[19] Attacking the quackery or the patent medicines of others was a prominent feature for Cody, and she also noticed that the proprietors wished to hide behind anonymity and used fraudulent testimonials.[20] All these characteristics can be found in individual advertisements in my systematic analysis of the studied newspapers, but, with the exception of gaining trust, none provides a true picture of the main methods of persuading consumers to buy a patent medicine.

Both the structure and content of medicine advertisements were very flexible, but, in broad terms, medicines could be promoted in the advertisements in three ways. The commonest method was to focus the advertisement on a small number of medicines, usually just one, and to describe their indications, virtues, prices and availability in some detail. This method was favoured by most of the major wholesalers/owners, including Francis Newbery, Spilsbury, Godbold, the Wrays, the Butlers and Shaw and Edwards. A second method, preferred by the Diceys, was to advertise several medicines together, commonly by describing a single medicine, accompanied by a list of other available medicines from the same wholesaler, perhaps with a brief description of each. In a series of advertisements from the same wholesaler, different single medicines could be selected from the medicines mentioned in the list. The Jackson/Barclay firm sometimes used this second method, and sometimes the first. A variation on the format of this second method, which was sometimes adopted by the Hill family or by some newspaper printers, was just to list medicines with a brief description and to avoid portraying any of them in more detail. The third method, normally only employed by local retailers to advertise their own stock, was to list the medicines, often in two columns, with no additional information apart from the price.

In the studied newspapers, the contents of the advertisements were explored systematically in those advertisements which provided enough details for assessment. The numerical analyses used the first advertisement, in each study period, of each of these *featured medicines*, defined as those described with a minimum of four lines of advertisement text. Other advertisements which only briefly mentioned the medicine do not generate enough information to help in the analysis, nor could they have contributed a great deal to establishing the consumers' trust at the time. They would have increased the medicines' familiarity and provided information on availability, but not much else. Fortunately, for our purposes, many of the medicines in printed lists joined the ranks of the featured medicines when they were described at more length in other advertisements. Otherwise they do not contribute to the detailed analyses in this chapter.

The type of information provided varied, with some advertisements containing only a brief description and the name of a retailer, whereas a few were over a hundred lines long and packed with detail. The advertisements could adopt unusual formats: they could be inserted in the editorial matter, appear to be just advertising a book rather than a medicine, be devoted to summarising a legal case, require a £5 advance payment for the medicine to be sent by post, or be partly written in Latin.[21] However, a core of information was common to many, and the large majority of the advertisements were more straightforward, with a variable combination of the indications, the rationale for taking the medicine, the price and a list of retailers. All four of these components were normally present to variable degrees, though one or more could be omitted; for example, an advertisement might just be a reprint of a testimonial letter. As discussed in Chapter 2, the indications were more focussed on a narrow range of conditions than has been suggested, while the patent medicines as a whole could provide something for nearly all common, non-traumatic, conditions at all ages. Chapter 5 explored the different types of retailers, especially the members of the print trades, and the stability of the prices. Much of this chapter will investigate the remaining component: the rationale for taking the medicine.

Creating Trust by the Words of the Industry

Trust in the effectiveness of the medicine had to be built up as part of the negotiating process between the consumer and the retailer. Establishing this trust was not just a question of proclaiming the virtues of a medicine: consumers also had to be reassured that it was not harmful or dangerous, as problems with patent medicines were publicised and commented on. A prominent example was the death of Oliver Goldsmith in 1774, which was generally attributed to taking Dr. James's Fever Powder, though opinions differed as to whether he had taken an excessive dose in spite of medical advice, received a defective counterfeit preparation, or had just been unfortunate.[22] In the following year Jane Butterfield, a housekeeper, was tried for murder and acquitted following her administration of Dodd's Rheumatic Tincture to her employer who died soon afterwards.[23] Thomas Clayton of Hull circulated a handbill in 1804 describing how his three-year-old son had died as a consequence of taking Ching's Worm Lozenges.[24] Several books also reiterated the dangers of patent medicines.[25] So, consumers were provided with good reasons *not* to take a patent medicine, as well as being subjected to advertisements encouraging them to do so.

A major part of generating the 'thin' trust amongst consumers was the words of the owners, wholesalers and retailers within the patent medicines industry, and this is the topic of this section. The next two sections will explore trust engendered by external authority and the benefits of familiarity. Overall, the trust generated by the industry was achieved by a rather low-key advertising style which sought to position the patent medicines as close as possible to regular therapy, largely avoiding the hyperbole and shrill claims which would have been associated with quackery. As the last chapter revealed, 'confident assurances' and solemnity were both regarded by contemporaries as important in generating the beneficial alterations in the imagination, and this low-key approach translates them into the medium of print. Positioning the medicines close to regular therapy also imitates in print another important factor in improving the imagination: the attitude of the physician. The use of testimonials and case reports as part of this low-key style will be considered first. They have been prominent in previous accounts of patent medicines, but we will find that they were used relatively sparingly. The rest of the text and the headings mimicked a bland public announcement at times, and claims were normally restrained. One common aim of the advertisements was reassurance on the accurate composition of the medicine, which could be provided by including the proprietor in the medicine's name and by clarifying that he, or she, was the source of the medicine.

Testimonials and case reports stand out in any inspection of newspaper advertisements for medicines, particularly as they provide revealing vignettes of a medical world which is foreign to twenty-first century readers. A testimonial was a case description sent into the proprietor by the patient or a third party, apparently spontaneously and sometimes with witnesses. A case report was a description of a cure by the proprietor/advertiser, sometimes with the approval, but never at the initiation, of the patient. A case report could be very brief and vague, so, arbitrarily, only those over three lines in length, a sufficient length to generate some detail, have been included here. The use of a testimonial or a case report was far from universal: overall 22% of the first advertisements for a featured medicine contained one, whereas, as we shall see, 40% had a warning about counterfeits. Testimonials or case reports were used in between a quarter and a third of the advertisements in the first four studied years, followed by a sharp fall in 1822 (Table 8.1). Barker also found that the majority of advertisements in her study did not utilise testimonials, and that their appearances declined sharply in the early nineteenth century.[26] This restraint could have been due to no suitable testimonial

Table 8.1 Use of testimonials and case reports in advertisements for the featured medicines (% of all featured medicines in that year)

	1769		*1781*		*1794*		*1807*		*1822*	
Number of featured medicines	**53**	%	**130**	%	**94**	%	**114**	%	**168**	%
Testimonial/case report	**18**	34	**32**	25	**32**	34	**29**	25	**11**	7
Testimonial/case report from an identifiable source	**9**		**23**		**25**		**15**		**5**	
Testimonial/case report stating failure by regulars	**2**	4	**7**	5	**18**	19	**10**	9	**3**	2

being available, but a case report in some form could normally have been inserted by the advertiser without the need for any response from a patient. A more probable explanation, consistent with the sharp reduction in the numbers in 1822, was that many advertisers preferred to omit testimonials or case reports as they were increasingly at odds with their desire to be close to the style of regular practice.

An assertion about the testimonials and case reports which finds little support in the studied newspapers was that their subjects and reporters were made up, and that a testimonial could be easily fabricated.[27] Like Barker, I found that many of them gave full names with an occupation and/or an address, sometimes accompanied by similar information from a witness (Table 8.1).[28] Overall, 63% of the testimonials and case reports came from an apparently identifiable source, and some of these sources, as we shall see in the next section, contained authority when they were derived from the gentry, a magistrate or a senior clergyman. Of course, the advertisers could have been relying on a checkable source not being checked, especially as most of the sources were living outside the newspaper circulation area. But exposure could generate unfavourable publicity, such as when Sarah Adams rejected in print the contents of a testimonial attributed to her in an advertisement for Dr. Lamert's Nervous Balsam, forcing Dr. Lamert, a travelling irregular, to pay for a long advertisement explaining the situation and denying that he intended to deceive.[29] Most of the checkable testimonials probably came from real people, even if they might have been misquoted.

Also, the flamboyant, hard-hitting, testimonial was less common than might be expected in the competitive medical market. One feature of this type of testimonial was the inclusion of medical failure, which could be vague such as a mention that the Faculty was unable to help, or more specific, for example the inability of Worcester Royal Infirmary to treat the patient in comparison to the complete cure provided by three bottles of the medicine.[30] As we can see from Table 8.1, with the exception of 1794, only a minority of the published testimonials or case reports referred to failure by regular medical practice. In Georgian England people often used both orthodox and irregular medicine for the same condition[31]; so a rather higher proportion of these testimonial writers, with a troublesome, often chronic, medical condition and sufficient money to buy patent medicines, would probably have sought unsatisfactory assistance from local practitioners: yet advertisers seem reluctant to mention the favourable comparison for their medicines. The reason for this hesitancy in highlighting regular medical failure is probably that the patent medicine owners in this era positioned their products as close as possible to regular therapy, and so they did not want to appear in opposition to 'the faculty'.

The predominant tone of most testimonials was authenticity, a common format being a detailed description of the illness prior to a full response to the medicine. Many were written as an apparently private correspondence between the patient, a relative, a friend or the retailer and the medicine proprietor. Small details were often supplied, particularly the reason for taking the medicine and its source, which were often described more carefully than would have been necessary just to demonstrate the virtues of the product. Indeed, the precise clinical descriptions, which included cases of venereal disease, seem surprising to our eyes in a newspaper intended for readership by all. The description of the treatment often emphasised two phases, an initial symptomatic response, followed by a more gradual, but usually complete, cure. A wide variation in the timing of the two phases was apparent. For example, Alice Cook had immediate symptomatic relief for her painful, swollen, arm with the first application of Le Coeur's Imperial Oil, and a complete cure after one bottle.[32] By contrast, Mr. Newnham had improvement in his chronic breathlessness with one bottle of Spilsbury's Drops, but he required several months of continuous treatment for a cure.[33] An emphasis on rapid symptomatic relief was probably intended to encourage starting the medicine, and the achievement of a complete cure, even if it took some time, would have fostered persistence. The predominant tone of the testimonials was an undemonstrative factual approach, which had more in common with a private conversation between acquaintances than a public proclamation.

As the testimonials and case reports were used sparingly, how was the rest of the text used to encourage the purchase of the medicines? Contrary to Porter's claim that hyperbole was essential, the major finding was that the claims in the text of the *majority* of the advertisements showed a relative lack of 'hard sell'.[34] Hyperbole such as 'the greatest medical blessing the world has ever received' was sometimes employed, but most advertisements avoided such extravagant claims.[35] Indeed, some seemed to be trying to avoid any substantial praise, perhaps just stating that a fresh supply was available followed by a list of indications. For example, one advertisement for Spilsbury's Antiscorbutic Drops was headed 'To the Public' and started with a declaration that Mr. Meggitt, a Wakefield bookseller, had received a fresh supply, followed by a brief description of the indications and encouragement for 'a trial of its virtue'.[36] It appears that this advertiser was imitating the many bland newspaper advertisements about bankruptcy proceedings, property to let, meetings of societies and other topics. Another form of understated promotion was to assume that the virtues of a medicine were already widely known: for example, the advertiser of Hunt's Aperient Family Pills merely commented that their excellence was long established so there was no need to comment further.[37]

A specific demonstration of the lack of hyperbole was that infallibility or certain success were only claimed in a small and declining number of advertisements (Table 8.2). Some such claims were for complete infallibility, such as 'certain cure' or 'never known to fail'; while for others it was qualified, such as 'certain cure if directions followed' or infallibility only being asserted for one of several indications. Overall, only 15% of

Table 8.2 Incidence of claims of infallibility or best medicine of type in advertisements for the featured medicines (% of all featured medicines in that year)

	1769		*1781*		*1794*		*1807*		*1822*	
Number of featured medicines	**53**	%	**130**	%	**94**	%	**114**	%	**168**	%
Claims for complete or qualified infallibility	**15**	28	**21**	16	**15**	16	**13**	11	**19**	11
Described as best of type	**6**	11	**17**	13	**9**	10	**22**	19	**40**	24

the advertisements claimed complete or qualified infallibility. The more modest claim of being the best medicine within its type rose in the later years, though it was still far from universal (Table 8.2). The text often assumed success or mentioned specific examples, rather than asserting a comprehensive effectiveness.

If the majority of advertisements avoided a hard sell by extravagant claims, what did they contain to promote the medicine? The answer is an enormous variety of material; there were no standard formats for the text of medicine advertisements. Some might concentrate on the indications, while others might employ persuasive arguments why the treatment would be successful. Another technique, as we have seen, was to print little more than a factual understated advertisement and let any evidence speak for itself. Some advertisers might want to prove that they had legal ownership of the medicine, perhaps accompanied by legal threats and warnings of the health risks of counterfeits. Another approach was seemingly to be advertising a treatise on a medical condition and only mention the medicine towards the end. Warnings that the condition might become dangerous if the medicine was not taken could also prove useful. Within this diversity, two specific components, namely the nature of headings and the reassurance on the composition, were used to build up the consumers' trust in the medicines.

How were the headings of the advertisements used to promote the medicines? The headings were partly at the discretion of the newspaper printers, unlike the main text which was usually provided by the wholesalers as we saw in Chapter 6. So the use of headings was less consistent between newspapers, but some common techniques can be seen. One, which was often employed in some runs of the studied newspapers, was not to have a heading at all, so that the advertisement had more resemblance to editorial matter and perhaps seemed closer to a non-promotional medical discussion. In 1807, the *Salisbury and Winchester Journal* used the technique in 38% of its advertisements for featured medicines, *Aris's Birmingham Gazette* used it in 13%, and the two Leeds newspapers in none. In general, the headings became more complex in the later years with multiple lines often being used in 1822. However, even in this last year we do not see the multiple typefaces and variable font sizes which became popular soon afterwards. The headings could be non-specific, such as 'a card' or 'to the public', or could consist of the owner's name and address at the start of a testimonial. The most frequent headings were either a bald statement of the name of the medicine or some of the indications: they appeared in 39 and 37% respectively of the first advertisements which used a heading, and both could be present. Claims for efficacy in a heading were unusual. In the headings, once again, we see a low-key, factual, approach to enhance trust by the consumer. This is supported by the rarity of another aspect of the visual display: pictures or crests only appeared in 10 out of the 559 first advertisements for featured medicines, and they did not increase in frequency over the years.

Considering the advertisements' text itself, the factual style did not mean that a large amount of information was provided about the constituents of the medicines and their mode of action. Advertisers wanted to preserve the 'mystery', a recognised means of altering the imagination. About 20% of the advertisements mentioned that the medicine was safe, gentle or innocent, but most gave no information on its mechanism of action: 82% of the first advertisements provided no details of how the medicine achieved its effects beyond a reiteration of its indications. Only 12% of the advertisements had any description of the composition of the medicine, and many of these were non-specific, such as 'vegetable' or 'not mercury'. An additional explanation for this lack of information is that the owners thought that details could raise concerns about the medicines. This is supported by the almost complete absence of notifications of potential side effects from the featured medicines; only two warnings were given in the 559 advertisements. Overall, the consumers were to be persuaded that the medicines were effective and routinely used, without puncturing the mystery or exposing possible side effects.

In addition to persuading consumers by an appropriate provision of information about the medicines, the advertisers also sought to increase the consumers' trust in their medicines by emphasising the consistent, proven, composition of their products. In Chapter 2 we saw that both medical practitioners and laypeople expressed concern that prescribed, and other locally prepared, medicines were incorrectly made up, either through incompetence or by deliberate fraud with cheaper ingredients. Patent medicines made up by an experienced individual, at one location to an unchanged recipe, could have an advantage over regular or other medicines composited by the local, perhaps poorly trained, apothecaries or druggists. We can see this emphasis on the correct and consistent composition in three aspects of patent medicines and their advertising, namely in the choice of a name for the medicine, in the frequent documentation of the proprietor, and in the reminders about the risks of counterfeits.

First, the owners commonly used either their own name or that of the supposed inventor in the medicine's name (Table 8.3). Some medicines had nobody in their names, such as Beaume de Vie or Medicated Spice Nuts, but 82% were linked to a person, such as Whitehead's Family Cerate or Ching's Patent Worm Destroying Lozenges. A few of the supposed inventors were dead and almost certainly had no real link with the medicine, as in the case of Dr. Sydenham's Family Pills of Health, or the enigmatically named Dr. Boerhaave's Red Pill Number Two. However, Porter's claim that many medicines were named after a famous dead physician seems to be an exaggeration:[38] the name more commonly featured the living inventor. Nor is Nancy Cox's finding that they were often linked to foreign doctors or foreign locations reproduced here.[39] It is possible that the inclusion of a person in the medicine's title could just have been a continuation of the common practice of attaching a person to an advertised book title, and this could be especially true early in our period when London booksellers were often involved in distributing medicines.[40] However, this style for naming medicines did not decrease as medicine wholesaling was increasingly carried out by chemists and medicine specialists, and 85% of the names of the featured medicines in 1822 still incorporated a person. Linking most patent medicines to a specific, living, person reminded consumers about their uniform, and potentially superior, composition.

Second, when we look at the content of the advertisements rather than the names of the medicines, the proprietor or preparer of the medicine was mentioned in about two thirds of the advertisements (Table 8.3).

Table 8.3 Incidence of the inventor or proprietor in the medicine name, of the proprietor being mentioned in the advertisement, and of a counterfeit caution in the advertisement (% of all featured medicines in that year)

	1769		1781		1794		1807		1822	
Number of featured medicines	**53**	%	**130**	%	**94**	%	**114**	%	**168**	%
Inventor or proprietor in medicine name	**40**	75	**93**	72	**82**	87	**101**	89	**142**	85
Proprietor/preparer mentioned in advertisement	**35**	66	**84**	65	70	74	**74**	65	**97**	58
Counterfeit caution in advertisement	**20**	38	**38**	29	**41**	44	**63**	55	**63**	38

Occasionally the proprietor would be an anonymous regular, such as 'a physician' with specified experience, but normally he or she would be named, often with an address. Commonly, no further information about the proprietor was provided: only 53 (15%) of the 360 advertisements which mentioned the owners or preparers recorded any details about their previous experience or expertise outside the development and preparation of their medicine.[41] This lack of emphasis on the experience of both regularly qualified and unqualified proprietors suggests that the primary aim of identifying them was not to impress the consumers with their skills, but it was rather to establish a single person as responsible for the supposedly consistent quality of the medicine. The key point being made was that one person was preparing the patent medicine from the same recipe over the years, and the medicine would be exactly the same whether it was bought in Carlisle or Penzance.

The third finding which indicates the importance of the medicines' consistent composition was the frequent caution about the possibility of counterfeiting and the resulting risks to the consumer (Table 8.3). Dicey and Company often emphasised the problem in their advertisements as they sold long-standing medicines of uncertain ownership which were in competition with other medicines with similar names. One advertisement for their Dr. Bateman's Drops (patented in 1726) contained this warning:

> But great as the good effects are from taking the *true and genuine* Bateman's Drops, the consequences resulting from taking the *Counterfeit Sorts*, are too frequently as much the reverse, the ill effects of which have been often experienced: It is therefore recommended to every one to take particular notice, that the words *Dicey and Co No. 10 Bow Church Yard*, are printed in the stamp affixed to each bottle, and signed at the top of each Bill of Directions—All others are COUNTERFEIT.[42]

Many of the warnings were more limited in scope, but. overall, 40% of all the first advertisements contained some caution on counterfeiting. The warnings had three purposes. Firstly, they promoted branding which encouraged the consumer to identify and buy that particular medicine, and, secondly, they created an impression that a medicine was effective and so worth counterfeiting, increasing the trust in the medicine's value. Thirdly, they provided a guarantee of a medicine's consistency and purity, in contrast to the counterfeits, or indeed other medicines for the condition, which might contain anything. By mentioning the risks of counterfeiting, the advertiser was seeking to convince the consumer that the choice was between the precision and reliability of his product and the random nature of its competitors.

In summary, this systematic exploration of the wording in medicine advertisements shows that testimonials were often not included and that those that were used attempted to be authentic. Exaggerated claims of absolute superiority were largely avoided, and shrill headings were rare. The ownership and preparation by a single living person was often mentioned to emphasise the consistency of the patent medicine in comparison with other, locally prepared, medicines. The advertised indications for the patent medicines, as discussed in Chapter 2, also fit into this pattern of restraint: rather than claiming to be panaceas, most of them were recommended for a comparatively narrow range of conditions. However, this factual approach did not extend to the components and the action of the medicine itself: strikingly little information was provided on the ingredients and how the medicine achieved its benefits. The emphasis was on when the medicine should be used and the overall results, not on the mechanisms involved.

How did the low-key factual approach create the required trust in the medicines? One part of the answer was that this was the almost universal pattern for all newspaper advertisements in this period. This method of promotion was only just becoming respectable for movable goods as opposed to the many advertisements for property and announcements for the benefit of the public.[43] To us, used to the eye-catching illustrations and exaggerations of modern advertising, Georgian advertisements look dull and restrained. But this was the style of the time: anything more strident could have provoked disdain and might have been counter-productive. Contemporaries were well aware of the risks of hyperbole: Hugh Blair, the well-known lecturer on rhetoric and belles lettres, wrote that 'when hyperboles are unreasonable, or too frequent, they render a composition frigid and unaffecting'.[44] The consumers of these advertised medicines could probably see through exaggerated bombast and were more impressed by factual information. Even Porter felt that Georgian opinion was offended by the hyperbole, which he nonetheless regarded as the essence of medicine advertising.[45]

Yet this is not the complete answer. Within the Georgian pattern, advertisers had scope for more vigorous marketing if they wished. This low-key factual, style generated trust amongst consumers by positioning patent medicines as close as possible to orthodox therapy. The advertisers wanted the medicines to correspond to regular treatments, offering a convenient alternative, or providing a replacement when regular therapy had failed: they were not seeking to be in opposition to orthodoxy. Thus, the advertisers were mostly reluctant to criticise regular medicine and often kept the published testimonials clear and detailed, mimicking the writing style commonly employed in orthodox medicine. The implied claim that the patent medicines from a central source were better made than locally obtained medicines was primarily a criticism of the druggists, not the medical practitioners. At the same time, the predominant style distanced patent medicines from the overstated manner of quackery, such as the mountebank on a stage who 'comes once a week to cozen a parcel of poor deluded creatures out of their money', as Thomas Turner wrote in his diary in 1760.[46] Most medicine advertisers did not want to be associated with the extravagant claims of some irregular practitioners.

Creating Trust by External Authority

Trust and confidence in patent medicines amongst the public was also enhanced by the capture of authority external to the industry. Like the words and actions from within the industry, this external authority incorporated ingredients which contemporaries considered as being active in altering the imagination, in this case authority itself, peer pressure and again 'confident assurances'. It could come from the recommendations of social superiors, or from the advice of suitable professionals: it could also be provided by legal decisions or legal threats, and an official authority was bestowed by the owners' adaptation of the patent system and accidentally by the medicine excise stamp. Overall, advertisers were hesitant to use respected recommendations from social superiors or experts, but they were delighted with the legal authority of the patent and the inadvertent official approval of the excise stamp.

One representation of authority to consumers was an endorsement of the medicine by a social superior, especially from royalty or the nobility, or by a reputable medical practitioner. This endorsement could be a specific recommendation or a 'certificate' by the named or unnamed source of authority, or it could be a vaguer association with an individual, who may not have suspected that he or she was involved, or was already dead. Dr. Hadley's Powders were 'patronized' by the political hostess the Duchess of Rutland, the notorious novelist Lady Caroline Lamb, and eight other titled ladies.[47] A wider range of the aristocracy apparently supported Godbold's Vegetable Balsam, which was certified as demonstrating 'superior efficacy' in their own families by the Marquis of Lothian, followed by the names of four other peers, two viscountesses, three other ladies, four baronets and two colonels.[48] An association with royalty, then as now, could also be beneficial: for example, the advertisement for Glass's Magnesia stated that it had been 'given with good effect' to the Prince of Wales and the younger princes, as well as being 'approved and recommended by the most eminent of the faculty'.[49] This particular endorsement must have been regarded as effective as the same text was used in both 1769 and 1781. The authority of the 1794 recommendation by the 'late celebrated physician' Sir Edward Hulse, baronet, in favour of James Inglish's version of Dr. Anderson's Scots Pills may have been diminished by the fact that Sir Edward had died 35 years earlier, and had been known to be out of his mind in his last years.[50]

Such examples stick in the memory, but the detailed analysis shows that these associations with figures of authority were not common. Of the first advertisements for featured medicines, 41 (7%) named a non-medical social superior outside a testimonial, and even if we extend the search to the use or recommendation by unnamed non-specific superiors, such as 'many persons of distinction and rank in society',[51] the total number of recommendations by non-medical social superiors, however vague, only rises to 88 (16%). Even fewer recommendations from reputable medical practitioners, dead or alive, were mentioned: 18 (3%) advertisements named one. Another 47 (8%) referred to support from the faculty or other medical bodies, but this was often very non-specific. So, contrary to the findings of both Barry and Strachan, only a small number of patent medicines were advertised as receiving the named endorsement of social superiors or medical practitioners.

Similarly, *testimonials* from a person with implicit or explicit authority, or witnessed by such a person, were rare. For this type of testimonial, the authority would be provided by naming a social superior, medical practitioner or magistrate as a subject, correspondent or witness. For example, Sir Clement Trafford vouched for the effectiveness of Hill's Pectoral Balsam of Honey on himself, and stated that Hill was 'at liberty to use the contents of this letter as he pleases'.[52] A testimonial could be sworn before a mayor or other magistrate to provide legal authority: one example was the testimonial from Thomas Fortune in favour of Dicey's Daffy's Elixir which was sworn before William Fisher, major of Yarmouth.[53] A testimonial backed by some form of authority would seem to have been particularly effective in boosting confidence. Yet they were used sparingly: only 14 examples (3%) were found amongst the 559 first advertisements.

Rather less reticence was shown in proclaiming the explicit authority of the royal patent or the excise stamp. One of the main benefits of expending both time and money in getting a patent was the apparent government authority, which could then be put to good effect in advertisements and other promotion.[54] For the studied newspapers in 1769 and 1781, about a quarter of the featured medicine advertisements cited a patent in the heading or text (Fig. 8.1). This proportion is larger than it seems, because we should remember that most of the advertised medicines had never received a patent, and so bestowed patents were frequently mentioned in these years. Unfortunately, exactly how often a bestowed patent was mentioned in the advertisements cannot be determined as the names of medicines in patent records were less precise that those in the advertisements.

In the last three studied years, the excise stamp largely replaced the patent in advertisements as the expression of government authority (Fig. 8.1). From 1783, every bottle or box of a patent medicine

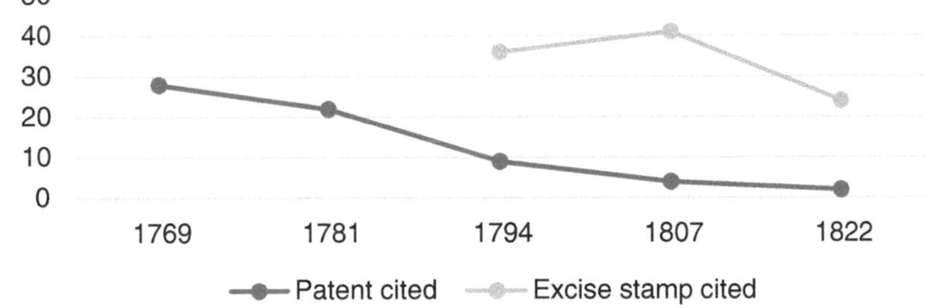

Fig. 8.1 Citations of the patent and excise stamp in the advertisements for featured medicines (% of all featured medicines in that year)

was required to have an attached excise stamp with a royal crown, as explained in Chapter 2, and the stamp could easily be interpreted as a form of government approval for the medicine. Advertisements described the stamp and emphasised its official nature by, for example, explaining that it was attached to each bottle by order of 'His Majesty's Hon. Commissioners'.[55] The excise stamp had several additional advantages over the patent as a form of authority, including the engraving or signing of the name of the owner or wholesaler on the stamp which provided reassurance to the consumer about the authenticity and composition of the medicine. It also could generate a more convincing legal threat against counterfeiters: in 1794 and 1807, about a third of the advertisements which mentioned the stamp also contained a legal warning, often related to forging the stamp. A common phrase was 'imitation is a felony', inserted immediately after the description of the stamp. Attempts to prosecute for the imitation of a *medicine*, even one with a current patent, would be subject to long costly and uncertain legal processes, but any imitation of an *excise stamp*, an official document, was a forgery and clearly a felony: it could, and in at least one case did, attract the death penalty.[56] For all these reasons, the excise stamp largely replaced the patent in advertisements during the last three studied years with a peak incidence in 1807 when it was mentioned in 41% of the first advertisements for featured medicines (Fig. 8.1). By 1822, 24% of the featured advertisements still mentioned the stamp, but none carried a legal threat. The reason for this lack of legal warnings is not clear, but it may have been part of a general trend in society, by the 1820s, to use the criminal law more precisely with less extreme penalties.[57]

Why was the authority of social superiors or medical practitioners not invoked more often, while advertisers were quick to utilise the authority of the patent and the excise stamp? The answer to this question is not clear with no advertiser leaving an explanation of his aims. Of course, the support of the living nobility or gentry, or the endorsement of practising physicians, might have been difficult to obtain for a newspaper advertisement, but anonymous recommendations by individuals or ill-defined groups could have been easily created, and the names of dead physicians would have been available to all. Once again, it seems that many, but not all, advertisers wanted to avoid the flamboyant style of some quackery, and preferred to gain trust by maintaining a relatively undemonstrative factual approach in a style similar to that of orthodox medicine. Indeed, during their public dispute, Thomas Henry mocked the Glass family for continuing to claim over many years that the royal princes were still taking Glass's Magnesia.[58] The use of recommendations and testimonials from social superiors had a flavour of an itinerant irregular and could have been counter-productive. The reason for not creating many non-specific endorsements from regular medicine is less apparent. Advertisers could easily have employed phrases such as 'approved by the faculty' or 'prescribed by the most eminent physicians'. One possible explanation is that although the advertisers wanted to be close to regular therapy, they still wanted their medicines to maintain a distinction from orthodox therapy. Another is that they felt that any disputes with the regulars as to whether any backing had indeed been given would be harmful to sales. However, no evidence has been found to support these possibilities.

By contrast, the patent and the excise stamp carried no association with quackery. The excise stamp was more effective than the patent in adding royal and government authority as well as in reducing the risks of counterfeits, and so the advertisements often mentioned it once it was available. It was not applicable to other forms of irregular therapy, and so it was compatible with the advertisers' overall aim of keeping a clear distance from quackery. We may ask why more advertisements did not mention the excise stamp. Well, they did not need to: it was already on every bottle or box of medicines.

Familiarity of Patent Medicines

Familiarity was recognised by the eighteenth-century experts as an important factor in developing the confidence necessary to alter the imagination: an example from the last chapter was the reduction of the stimulus needed to induce the effects of animal magnetism on a second occasion. Recurrent advertising in a weekly newspaper provided ample opportunity to increase the familiarity of a patent medicine amongst potential purchasers, helping to establish their trust and confidence. Well-known medicines would have been used for years, often decades, and many consumers might already have confidence in their benefits and in a paucity of harmful effects. Examples would include Daffy's Elixir and Anderson's Scots Pills, which were sold in the seventeenth century, as we have seen, and Dr. Bateman's Pectoral Drops and Dr. James's Fever Powder, which had been patented in the first half of the eighteenth century and were still popular at the beginning of the next. Familiarity was well established for these four medicines, but this section will consider how the advertisements could use repetition to enhance the familiarity of less well-known medicines, improving their chances of being selected when a consumer needed to come to a decision on which medicine to take.

The nature of the decision-making process for patent medicines could be prolonged, perhaps with the assistance of relatives and friends, and this length increased the importance of familiarity. Customers had to weigh up the potential advantages and potential hazards of several medicines, and this was more likely to be a protracted process than a snap decision. Testimonials often provided an illustration of this process when they recorded that the medical problem had been present for a considerable time before the patent medicine was started, and that help from friends, relatives or even a bookseller was required. Amongst the advertisements in the *Salisbury and Winchester Journal* for just two months at the beginning of 1807, several examples of the prolonged decision making and the friends' advice can be seen in the testimonials. Mr Davies was urged by an unnamed person to try Dr. Bateman's Pectoral Drops after 15 weeks of severe rheumatic pain; and a recommendation from his medical practitioner persuaded Mr. Thorn to try the Cordial Cephalic Snuff after 29 years of giddiness in the head.[59] The intervention of friends was needed for a lady to take the Cordial Balm of Gilead for her long-standing low and weak state, which had not responded to prescriptions from the regulars; and a friend advised Mr. Stiell to use Brodum's Botanical Syrup to fix skin problems, which had been troublesome for two years.[60] Dixon's Antibilious Pills were recommended by a friend for the chronic bilious complaints of Mr. Lucas, 'chief officer' of an East India Company's ship, who in turn recommended it to his brother.[61] Familiarity with a medicine within a community would give it a significant advantage when the progression of the disease in an individual and the lack of response to other therapies made the time right for selection.

The branding of the medicines was a substantial part of establishing and maintaining their familiarity amongst the public. 'To brand' in the sense of creating an awareness by consumers of a specific product, rather than affixing a recognised surface marking, was a meaning introduced in the mid-nineteenth century: so the term does not feature in contemporary accounts of the patent medicines industry.[62] Nevertheless, patent medicines are early examples of successful branding, and many of the methods of promotion described in this and earlier chapters contributed to it, including the name of the medicine, the reputation of the owner, the packaging, the advertising description, warnings about counterfeits, any associated treatises, the directions, and perhaps the price.

Erica Storm and Jennifer Basford have both investigated the importance to branding of the physical appearance of the medicines and their packaging, with Storm also exploring the role of taste and touch.[63] The techniques used were diverse and complex, but branding of medicines was well established in Georgian England.

The branding was supported by the repetition of newspaper advertisements which could be striking. Spilsbury's Drops were advertised 54 times in the *Leeds Intelligencer* during the whole of 1794, while Hill and Berry's Medicine for the Bite of a Mad Dog was not far behind with 45 advertisements in the same newspaper in the same year. Such a frequency, approaching or even exceeding one per week, was exceptional, but many medicines were advertised five to twelve times during the six-month studied periods, and a single advertisement was uncommon. It might be suggested that inertia was responsible for advertisements continuing unchecked, with the printer not having sufficient alternative copy to fill up the space and the advertiser not being unduly concerned about the number of times the advertisements were repeated. However, as we saw in Chapter 6, all medicine advertisements were normally charged to the advertiser, and so the repetition would have had a significant cost. The *Hampshire Chronicle* records also show up to 12 insertions for a single advertisement being ordered in advance. So the repetition was rarely, if ever, due to inertia: it was intended, and it was considered by the advertisers to be effective, or they would not have spent hundreds of pounds a year on it.

How did the advertisers use repetition to increase the familiarity of their medicines? As usual in the patent medicines industry, a variety of approaches can be seen, but two extremes can be picked out. One was to repeat the same advertisement content over and over again, retelling the main reasons for taking the medicine without providing fresh ones: the main aim was probably to keep the name of the medicine in public view. As demonstrated earlier, the same text linking the medicine to the royal princes was used for Glass's Magnesia for at least twelve years. Over a comparable period, similar texts for Masson's Medicine for the Itch exhorted readers that families were 'liable to catch it [the itch] from connections in business, fresh servants, exc.'[64] The other extreme was to vary the content of the advertisement as much as possible, probably with the aim of attracting the readers' attention, and offering them additional motives to go out and buy the medicine. Solomon's Cordial Balm of Gilead was advertised in this way across all four of the studied newspapers in 1822. The advertisements were printed approximately monthly in each newspaper, with the indications for the medicine and the justification for its use constantly varying, and with their order juggled around to ensure that consecutive advertisements looked different while sometimes containing the same overall message.[65]

The methods of most of the major advertisers were between these two extremes, using both reiteration and fresh material to ensure that their medicines were familiar to the readers. As we saw in Chapter 6, during 1781, Francis Newbery advertised each of his medicines in bursts of three or six insertions of the same text in the *Hampshire Chronicle*, before moving on to a fresh advertisement. In the same newspaper, the Wrays preferred 12 insertions of the same text. The Diceys took a different approach in this and other newspapers. They often listed much of their stock without additional information, except a single medicine which was described in detail in the initial text and in the heading: as the advertisements were repeated, a different described medicine was selected from their advertised stock. All the principal London wholesalers, except Bacon whose provincial advertising was limited, used substantial repetition of advertisements for at least some of their stock.

So the expensive repetition of advertisements aimed to ensure that a consumer, and perhaps relatives and friends, would read about a medicine on several occasions and so become familiar with it. When the time came to decide on selecting a medicine, the consumer would already know about it and would respond positively to any advice given by friends, or to recommendations in a newspaper advertisement or printed bill. In this way, repetition ensured that the consumers' trust in a medicine was enhanced by their familiarity with its name and its indications. However, familiarity did not mean that consumers should know everything, because mystery encouraged the favourable effects of the imagination. As we have seen, advertisers were reluctant to reveal any details about the composition of their medicines and their mode of action: the aim was that the medicine should be familiar as a brand, but that its action should remain mysterious.

Conclusion

This analysis of the newspaper advertisements has shown that the advertisers were designing the content of their advertisements to create the confidence necessary for altering the imagination. Largely, they were doing this not by the hard sell described by some historians, but rather by a low-key factual, approach which attempted to position the patent medicines close to orthodox therapy and as far as possible from irregular medicine. The emphasis on the purity and consistency of the medicine manufacture was part of this style, and the consumers' confidence was enhanced by the authority of the patent when available, and later by the excise stamp. Familiarity also augmented the confidence, but the mystery of the medicine's composition and mode of action had to be preserved.

In their use of the printed word, the advertisers of patent medicines put to practical use the findings of their contemporary physicians on the methods of altering the imagination. The advertisements were well suited to boosting the imagination by promoting trust amongst consumers which provided the essential confidence. As a result, the medicines became more effective, not as assessed by twenty-first century medical science, but in terms of improving the overall health and wellbeing of the consumer. For the advertiser, this benefit should translate into increased sales. However, it would be a mistake to assume that the contemporary expert opinions of physicians concerning the imagination had a direct effect on the content of the advertisements. In the first place, the understanding of the mechanisms involved in the medical powers of the imagination only emerged towards the end of the eighteenth century, which is after the main mid-century development of the patent medicines industry. Also, no direct evidence links the advertisement content to the physicians' opinions on the imagination. It is more probable that the advertisers had learnt from experience which style of content resulted in the best sales: more medicines would be sold if the consumers were finding them to be effective. In other words, the advertisements were stimulating the imagination even though the advertisers had not intended that the printed words should have this effect. The medical experts and the advertisers came to broadly similar conclusions by different methods, with the experts using observations, experience and experiments to explore the powers of the imagination, whereas the promoters of the patent medicines were reacting to the feedback from the sales.

Notes

1. Particularly in Porter, *Health*.
2. Brown, 'Venders'; Barry, 'Publicity', 32–36.
3. Cody, 'No Cure', 104–108.

4. Strachan, *Advertising*, 4–68.
5. Barker, 'Medical Advertising'.
6. Luhman, *Trust and Power*, 4–32; Giddens, *Modernity*, 30–36; Lane, 'Introduction'.
7. Muldrew, *Economy of Obligation*, 5.
8. Shapin, *Social History of Truth*, 42–49.
9. Goldgar, *Impolite Learning*, 10.
10. Barker, 'Medical Advertising', 381.
11. Isaac, 'Pills and Print', 29.
12. Hancock and Wallis, 'Quacking', 14–18.
13. *Early English Books Online.*
14. For example, Dr. Guider's Ague and Fever Plaister, *Stamford Mercury*, 21 January 1720.
15. Ferdinand, 'Selling', 398–399.
16. Barker, 'Medical Advertising', 397.
17. Porter, *Health*, 100–104.
18. Strachan, *Advertising*, 28.
19. Barry, 'Publicity', 34–35.
20. Cody, 'No Cure', 106, 110 and 123.
21. Twenty-four lines about Rowland's Alsana Extract in the Parliamentary Report, *LM*, 11 May 1822; Solomon's Guide to Health, *SWJ*, 17 June 1822; Jackson's Asthmatic Candy, *ABG*, 17 March 1794; Antwerp Medicine, *SWJ*, 25 March 1822; Alternative Powder, *LM*, 1 May 1781.
22. Spilsbury, *Free Thoughts*, xxx and xxxvi; Hawes, *Dr. Goldsmith's Illness* (London, 1774).
23. Gurney and Blanchard, *Trial of Jane Butterfield.*
24. *MPJ*, 17 (1807), 173–175.
25. Two vigorous examples were Adair, *Essays* and *Quack Doctors.*
26. Barker, 'Medical Advertising', 389.
27. Cody, 'No Cure', 123.
28. Barker, 'Medical Advertising', 389.
29. *LM*, 27 July and 17 August 1793.
30. Spilsbury's Antiscorbutic Drops, *SWJ*, 10 February 1794: Parker's Rheumatic Drops, *ABG*, 1 January 1781.
31. Porter and Porter, *Patient's Progress*, 108.
32. *LI*, 17 April 1781.
33. *ABG*, 13 January 1794.
34. Porter, *Health*, 100.
35. Dr. Norris's Drops, *LI*, 9 January 1781.
36. *LI*, 17 March 1794.
37. *SWJ*, 4 February 1822.
38. Porter, 'Language', 85.
39. Cox, *Complete Tradesman*, 201.

40. For example, *Hume's History of England*, *Gifford's History of the French Revolution*, and *Shaw's Methodical English Grammar* which were all advertised in the *Leeds Intelligencer* during 1794. In the first half of 1794, 42% of the books and pamphlets advertised in the studied newspapers were linked to a person.
41. Even when provided, the details could be brief, for example 'Royal Chemist'.
42. *LI*, 16 February 1807. A stamp from Dicey and Co. is pictured in Fig. 2.1.
43. Cranfield, *Press and Society*, 184; Styles, 'Product Innovation', 158.
44. Blair, *Lectures on Rhetoric*, vol. 1, 370.
45. Porter, 'Language', 76.
46. Vaisey, *Thomas Turner*, 208.
47. *SWJ*, 10 June 1822.
48. *LM*, 22 February 1794.
49. *SJ*, 23 January 1769; *SWJ*, 1 January 1781.
50. *LM*, 4 January 1794; Creighton, 'Hulse, Sir Edward'.
51. Ching's Patent Worm Lozenges, *SWJ*, 15 April 1822.
52. *LI*, 23 January 1781.
53. *SWJ*, 13 January 1794.
54. Mackintosh, 'Authority', 546.
55. Robberd's Balsamic Elixir, *LI*, 4 January 1807.
56. As described in Chapter 2, Thomas Collicott was sentenced to death in 1812 for forging medicine excise stamps.
57. O'Gorman, *Eighteenth Century*, 289–290.
58. Henry, *Letter*, 21.
59. *SWJ*, 12 January 1807.
60. *SWJ*, 26 January, 2 February 1807.
61. *SWJ*, 16 February 1807.
62. 'Brand, v.', *OED* (accessed 27 March 2017).
63. Storm, 'Gilding the Pill'; Basford, 'Commodity of Good Names'.
64. *ABG*, 8 January 1781 and 6 January 1794.
65. Five times in *LI*, seven in *LM*, five in *ABG*, and five *SWJ* in the first half of 1822. Solomon had died in 1819, but his anonymous successor(s) continued to promote his medicines for several years.

Bibliography

Newspapers and Journals

Aris's Birmingham Gazette (ABG).
Leeds Intelligencer (LI).

Leeds Mercury (LM).
Medical and Physical Journal (MPJ).
Salisbury and Winchester Journal (SWJ).
Stamford Mercury.

Printed Primary Sources

Adair, James, *Essays on Fashionable Diseases* (London: T. P. Bateman, 1790?).
Blair, Hugh, *Lectures on Rhetoric and Belles Lettres*, 6th edn, 3 vols. (London: Strahan and Cadell, 1796).
Early English Books Online, http://0-eebo.chadwyck.com (accessed 7 June 2016).
Gurney, Joseph, and William Blanchard, *Trial of Jane Butterfield* (London: Owen & Kearsley, 1775).
Hawes, William, *An Account of the Late Dr. Goldsmith's Illness*, 3rd edn (London, 1774).
Henry, Thomas, *A Letter to Dr. Glass* (London: Joseph Johnson, 1774).
Quack Doctors Dissected (Gloucester, c.1805).
Spilsbury, Francis, *Free Thoughts on Quacks and Their Medicines* (London, 1776).

Secondary Sources

Barker, Hannah, 'Medical Advertising and Trust in Late Georgian England', *Urban History*, 36 (2009): 379–398.
Barry, Jonathan, 'Publicity and the Public Good: Presenting Medicine in Eighteenth-Century Bristol', in *Medical Fringe and Medical Orthodoxy 1750–1850*, edited by W. F. Bynum and Roy Porter (London: Croom Helm, 1987), 29–39.
Basford, Jennifer, 'A Commodity of Good Names: The Branding of Products, c.1650–1900' (PhD Thesis, University of York, 2012).
Brown, P. S., 'The Venders of Medicines Advertised in Eighteenth-Century Bath Newspapers', *Medical History*, 19 (1975): 352–369.
Cody, Lisa Forman, '"No Cure, No Money," or the Invisible Hand of Quackery: The Language of Commerce, Credit, and Cash in Eighteenth-Century British Medical Advertisements', *Studies in Eighteenth Century Culture*, 28 (1999): 103–130.
Cox, Nancy, *The Complete Tradesman: A Study of Retailing, 1550–1820* (Aldershot: Ashgate, 2000).
Cranfield, Geoffrey, *The Press and Society: From Caxton to Northcliffe* (London: Longman, 1978).
Creighton, Charles, 'Hulse, Sir Edward', rev. Patrick Wallis, *ODNB* (accessed 10 June 2017).

Ferdinand, C. Y., 'Selling It to the Provinces: News and Commerce Round Eighteenth-Century Salisbury', in *Consumption and the World of Goods*, edited by John Brewer and Roy Porter (London: Routledge, 1993), 393–411.

Giddens, Anthony, *The Consequences of Modernity* (Cambridge: Polity Press, 1990).

Goldgar, Anne, *Impolite Learning: Conduct and Community in the Republic of Letters 1680–1750* (London: Yale University Press, 1995).

Hancock, David, and Patrick Wallis, 'Quacking and Commerce in Seventeenth-Century London: The Proprietary Medicine Business of Anthony Daffy', *Medical History*, Supplement 25 (2005).

Isaac, Peter, 'Pills and Print', in *Medicine, Mortality and the Book Trade*, edited by Robin Myers and Michael Harris (Folkestone, Kent: St. Paul's Bibliographies, 1998), 25–47.

Lane, Christel, 'Introduction: Theories and Issues in the Study of Trust', in *Trust Within and Between Organisations: Conceptual Issues and Empirical Applications*, edited by Christel Lane and Reinhard Bachmann (Oxford: Oxford University Press, 1998), 1–30.

Luhman, Niklas, *Trust and Power* (Chichester: Wiley, 1979).

Mackintosh, Alan, 'Authority and Ownership: The Growth and Wilting of Medicine Patenting in Georgian England', *British Journal for the History of Science*, 49 (2016), 541–559.

Muldrew, Craig, *The Economy of Obligation: The Culture of Credit and Social Relations in Early Modern England* (Basingstoke: Palgrave Macmillan, 1998).

O'Gorman, Frank, *The Long Eighteenth Century: British Political and Social History 1688–1832* (London: Hodder Arnold, 1997).

Porter, Dorothy, and Roy Porter, *Patient's Progress: Doctors and Doctoring in Eighteenth-Century England* (Stanford: Stanford University Press, 1989).

Porter, Roy, 'The Language of Quackery in England, 1660–1800', in *The Social History of Language*, edited by Peter Burke and Roy Porter (Cambridge: Cambridge University Press, 1987), 73–103.

Porter, Roy, *Health for Sale: Quackery in England 1660–1850* (Manchester: Manchester University Press, 1989).

Shapin, Steven, *A Social History of Truth: Civility and Science in Seventeenth-Century England* (London: University of Chicago Press, 1995).

Storm, Erica M., 'Gilding the Pill: The Sensuous Consumption of Patent Medicines, 1815–1841', *Social History of Medicine*, advance publication online, https://doi.org/10.1093/shm/hkw133.

Strachan, John, *Advertising and Satirical Culture in the Romantic Period* (Cambridge: Cambridge University Press, 2007).

Styles, John, 'Product Innovation in Early Modern London', *Past and Present*, 168 (2000): 124–169.

Vaisey, David, ed., *The Diary of Thomas Turner 1754–1765* (Oxford: Oxford University Press, 1984).

CHAPTER 9

The Legacy of the Patent Medicines Industry

The exposure of the Georgian patent industry provides fresh perspectives for historians of commerce, print and medicine. In commerce, the uncovering of the detailed techniques for the national wholesaling and retailing of patent medicines rebalances the debate on when 'modern' retailing began in Britain. In the second half of the eighteenth century, the patent medicines industry employed a centrally led national distribution system for a class of branded fixed-price products, accompanied by retailing in shops with a degree of specialisation. In his authoritative account of 60 years ago, James Jefferys described modern retailing as being specialised and entirely separate from production, operating from fixed premises throughout the year and using fixed openly displayed prices.[1] This description is mostly applicable to the retailing of patent medicines in the eighteenth century. But Jefferys felt that this modern retailing largely dated from the second half of the nineteenth century, though he did recognise that patent medicines were one of the first pre-packed advertised and fixed-price goods.[2] In a recent book, Ian Mitchell has revealed a more fractured development of retailing, but he has still placed the major developments in the mid-nineteenth century.[3] In contrast, others have attributed the early growth of modern retailing to the eighteenth century. Hoh-Cheung and Lorna Mui believed that most of the required changes were well underway by the second half of the eighteenth century, while Nancy Cox thought that her 'complete tradesman',

A. Mackintosh, *The Patent Medicines Industry in Georgian England*, Medicine and Biomedical Sciences in Modern History,
https://doi.org/10.1007/978-3-319-69778-9_9

who provided the full retailing environment with supportive supply networks, was established by the same period.[4]

The demonstration of organised specialised distribution and vending for medicines across the country from at least the mid-eighteenth century supports those who favour the eighteenth century as the key period in the growth of modern retailing. It proves nothing about the marketing of other goods, but it does indicate the available techniques for wholesaling and retailing: in principle, other commodities *could* have been promoted, distributed and sold in the same way as medicines. Books were also heavily advertised in newspapers, and the provincial retailing of books would almost certainly repay a more comprehensive study than has hitherto been carried out. Unfortunately, advertisements for other branded consumer goods were few in number during the eighteenth century; but studies on branded teas and spirits might extend our knowledge of the origins of modern retailing. In general, more investigation is required on specific consumer goods using multiple, perhaps imperfect, sources, rather than just information from a small number of well-preserved archives.

Turning to print history, the printed word was indispensable for the patent medicines industry. The study of print in all its forms has developed in recent years as a specialised field of history, and the exploration of print in this Georgian industry expands this field by demonstrating that print was both a vehicle for, and a component of, that industry. As a vehicle, the printed word was more than just the method of communication amongst the participants in the industry and their consumers: it provided power for the industry. Promotion and instruction by the printed word in bills, newspaper advertisements, treatises, directions and puffs enabled relatively cheap, perhaps simple, ingredients to be transformed into sought-after, expensive, patent medicines. Then the printed word facilitated their distribution and retailing across the country, generating substantial profits for some participants. The printers and booksellers, who were in the strongest position to harness this power, could do well in the business of medicines. Without the power from print, patent medicines would still have been made and sold to some extent, but the substantial national industry would not have existed.

In addition to this role in providing power to ensure the commercial success of the industry, the printed word had a more direct healthcare function as a necessary component of the patent medicines. Thus, the printed word can be considered as a raw material for the patent medicines. Patent medicines were rendered effective by the confidence created by the printed word, and so print had the therapeutic potency to improve consumers' well-being: it was an essential raw material for the patent medicines industry alongside the pharmaceuticals. Indeed, it was the one material which was present, albeit to a varying degree, in *all* the successful patent medicines. The printed word was a crucial and universal material for the industry as well as being a supplier of power.

The revelation of print as an essential contributor to the efficacy of patent medicines has wider implications for Georgian, and later, healthcare. Is the concept of the printed word as a therapeutic agent applicable to a broader range of Georgian therapies, such as spas, sea bathing, electricity or homeopathy? Did the effects extend to other countries in Europe with their different medical cultures and systems of regulation? The printed word may also have improved the efficacy of many later alternative therapies, and some orthodox ones as well. In general, the finding that the printed word could be an active agent in Georgian medicines should be reproducible for other types of medical therapy.

However, the most important historical legacy of the Georgian patent medicines industry is a new approach to the overall structure of Georgian commercial healthcare. Porter envisaged it as having two components: irregular medicine/quackery and orthodox medicine, with no hard division between them.[5] This position needs to be altered to accommodate a third component: the patent medicines industry. The industry overlapped with regular medicine and irregular medicine/quackery, while remaining distinct from both of them. Patent medicines were largely separated from irregular medicine by their ownership and distribution often being in the hands of reputable people who did not undertake irregular practice, and who operated as a coherent industry with its own methods. These methods included a range of cooperative business practices between producers, wholesalers, advertisers and retailers which contrast sharply with the individuality of most irregular medical practitioners. Correspondingly, the patent medicines industry was distinguished from regular therapy by its different methods of production and sale, its dependency on the printed word, and the lack of medical training for most of the medicine proprietors and retailers. Again, regular medical practitioners largely operated as individuals, unlike the collaborative skills of participants in the patent medicines industry.

Despite these clear differences, the pluralistic and unregulated nature of the Georgian medical market meant that the industry also overlapped with both regular and irregular therapy. Some regular practitioners owned patent medicines, either as a separate business or as part of their practice, and some patent medicines owned by others were prescribed by regulars. Irregular practitioners could also own and distribute patent medicines: many of the local owners described in Chapter 3 were probably irregulars though details are sparse, and the well-publicised itinerant irregulars who owned medicines featured in the satire of the time. Regular medicine and quackery also overlapped outside the patent medicines industry due to the lack of generally recognised qualifications for many practitioners in this period and the extensive medical knowledge of some laypeople. Finally, a few controversial medicine owners who aspired to be regular practitioners despite seemingly inadequate training, such as William Brodum, Samuel Solomon and John Lignum, would be active in all three components.

So, from the mid-eighteenth century, the patent medicines industry became the third force of Georgian commercial healthcare, a distinctive stable and successful alternative to both orthodox medicine and irregular medicine/quackery. But what happened afterwards?

Patent Medicines After the Georgian Era

During the nineteenth century, the largely unregulated medical market of the Georgian period evolved towards a controlled medical profession, accompanied by a legally circumscribed pharmaceutical industry. The status of patent medicines altered as changes in the market from the second quarter of the nineteenth century destabilised their industry and pushed patent medicines closer to irregular practice. Controlling or eliminating patent medicines were often an ambition of the reformers, but proved difficult to achieve in the free market climate of the time. On the contrary, sales of patent medicines increased substantially during the nineteenth century, much to the disgust of the leaders of the medical profession, though less so to the rank and file. Eventually, patent medicines were prohibited by Act of Parliament in 1941—or were they?

Friedrich Engels recorded the extensive use of patent medicines in Manchester in the 1840s, which he felt contributed to ill health in the working class:

> Large numbers of patent medicines are sold as cures for all sorts of actual and imaginary complaints, Morrison's Pills, Parr's Life Pills, Dr Mainwaring's Pills and thousands of other pills, medicines and ointments which are all capable of curing all the illnesses under the sun.[6]

Engels noted that Parr's Life Pills were taken to relieve an astonishing variety of different complaints, such as both constipation and diarrhoea, and that 20,000–25,000 boxes were sold each week. Sales of patent medicines crept up in the early Victorian period, followed by a boom in the half century from around 1860 to the First World War. Subject to few legal restrictions and recommended for both trivial and serious diseases, they were heavily advertised and available in a wide variety of shops.[7] The net receipts from the patent medicine stamp duty in 1860 were £43,366, only a modest increase from £35,858 in 1810.[8] But after nearly another half century, the receipts in the year 1907/1908 were £334,142 with about 80% of the revenue coming from sales at the unchanged lowest rate of duty of 1½d, indicating that the increase was due to improved sales not to higher rates of duty.[9] Nearly 42 million patent medicine articles were stamped in 1907/1908, which was more than one article per head of the population. Reflecting a surge in sales in the 1860s, the Newbery family business moved from 45 St. Paul's Churchyard to larger premises nearby in 1869, and again to even larger premises three years later.[10]

Some of the techniques of the Georgian patent medicines industry continued, particularly secrecy, national distribution, owners who did not practise irregular medicine, and the heavy use of advertising in periodicals to create a consumer demand and 'pull' the medicines out across the country. Thomas Holloway (1800–1883) was an unsuccessful merchant with no apparent medical experience when he started to sell his Universal Family Ointment in 1837, followed by his digestive pills in 1839.[11] Based in large premises on the Strand, he poured his money into newspaper advertising, spending thousands of pounds a year where his Georgian predecessors had spent hundreds. His skill as a financier meant that profits from the medicines were multiplied into a large fortune, which enabled him to fund and name a mental sanatorium as well as the Royal Holloway College. Another owner who would have felt at home in the Georgian industry was Thomas Beecham (1820–1907), originally a self-educated shepherd and then a postman, who created his pills in Lancashire in the late 1840s.[12] With the assistance of his son Joseph and over £100,000 per year spent on advertising, his pills became the best-selling patent medicine in the country in the 1870s. James Eno (1827–1915) was a chemist and dentist in Newcastle-upon-Tyne whose Fruit Salts achieved a national market with the help of prosecutions of imitators and a huge amount of innovative advertising with literary and philosophical quotations.[13]

However, starting at the end of the Georgian period, the medical market was changing with patent medicines moving closer to irregular practice/quackery and no longer constituting a distinct third force. This was due to several factors including a crumbling of the principles behind irregular practice, increased specialisation within irregular practice, unification of the medical profession, and alterations in both wholesaling and retailing arrangements. Before discussing these factors, a caution is needed. At several points, this book has emphasised the necessity to obtain systematic data from across the country, and to avoid conclusions solely derived from simple examples or earlier opinions. But this rule now has to be suspended: such systematic data does not yet exist for Victorian and later patent medicines, and so the conclusions on the developing medical market will necessarily be provisional until such time as more detailed information has been gathered in.

Until the 1830s, irregular practice had largely followed the theories, and sometimes the style, of orthodox practice. Its practitioners might have commented on the ineffectiveness of the regulars and sometimes might have questioned the regulars' motives, but they rarely questioned the regulars' medical theories. The 1830s and 1840s saw a flourishing of medical movements which provided radical alternatives to the theory and practice of orthodox medicine.[14] Some, such as Animal Magnetism, Homeopathy and Hydropathy, were on the fringe of orthodox medicine, proposing radical theories but still sometimes practised by regularly trained practitioners as well as many irregulars. Others, such as Hygeism, as advocated by James Morison, and Medical Botany were in total opposition to orthodox medicine, advising the public to avoid it completely. In the eighteenth century, most irregular practitioners provided healthcare and made money without espousing any radical system. Now, as Porter and Loudon suggested, the irregulars in the medical movements were reducing the space in the market for the less committed irregulars who were just providing a service and making money.[15]

In the same period, the rise of irregulars with specific skills limited the access to the market for 'general irregulars', as they could be termed, paradoxically just at the time when general practitioners were being recognised as an essential element of orthodox medicine. Irregulars with specific skills, but as yet no formal qualifications, included dentists, chiropodists ('corn-cutters'), bonesetters, oculists and aurists. Together, the medical movements and the specialisation of other irregulars limited the scope for general irregulars to earn their living by seeing patients. The limited available data reflects the magnitude of this change, even when it is collected in different ways for separate purposes. For example, surveys of the town and county of Nottingham for Edward Harrison in 1806 revealed 149 irregulars of both sexes, excluding druggists and midwives; while the 1851 census for the populous West Riding of Yorkshire recorded 28 dentists, 9 medical botanists and only 4 irregulars, described as 'quack doctors'.[16] Faced with this alteration in the medical market, many irregulars, and those who might previously have considered becoming one, concentrated on producing medicines for others to sell rather than dealing directly with patients.

The unification of the medical profession also linked patent medicines more closely with irregular practice/quackery. As we saw in Chapter 2, strong opposition to patent medicines from some regulars had been apparent in the previous century, but the growth in formal medical qualifications and the unification of the profession during the nineteenth century increased the force of these criticisms by allowing leading regulars to present themselves as guardians of the public interest. A clearer division, in both qualifications and reputations, between orthodox medicine and everything else provided less room for a third force, pushing patent medicines further away from regular practitioners and the influential middle classes. Another tool used by the regulars was the law in coroner's inquests and the criminal courts. Their main targets in the second quarter of the nineteenth century were the radical medical movements in opposition to orthodox medicine, with the regulars acting as witnesses to persuade the coroners to give verdicts critical of the radical treatments which had been provided for the deceased. Hygeism in the form of Morison's Pills received unfavourable verdicts in nine inquests between 1834 and 1850, and manslaughter was the verdict in at least six inquests involving the use of Medical Botany between 1847 and 1855.[17] These legal actions did not directly affect the sales of patent medicines: but three subsequent convictions for manslaughter following the use of Morison's Pills confirmed the legal principle that all medical practitioners, regular or irregular, must be competent to provide their treatments.[18] This encouraged irregulars to provide medicines for other retailers, rather than running the risk of dealing directly with patients.

In addition, the withdrawal of the print trades from the wholesaling and retailing weakened patent medicines as a third force. No longer was a substantial section of the distribution system clearly separated from colourful irregulars and doubtful medical practices. As we saw in Chapter 4, the London publishers had withdrawn from medicine wholesaling by the beginning of the nineteenth century, but the control of the national market by a small number of London wholesalers employing similar methods, especially the ex-publishing businesses of the Newberys and Diceys, persisted until at least the 1820s. Also, much of the retailing was shared between the print trades and the druggists in that decade. Exactly what happened afterwards awaits further investigation, but it is clear that wholesaling and retailing, without the stabilising influence of the print trades, had fragmented by later in the century. The boom in patent medicines from the 1860s was facilitated by a much wider range of wholesalers and retailers than existed in the Georgian patent medicine industry. The combined bookseller and patent medicine vendor's shop pictured in Fig. 5.4 was mentioned in a book on Birmingham booksellers because this amalgamation had become rare by the end of the century.[19] Successful owners such as Thomas Beecham and James Eno needed to build up national markets over time by their own efforts, rather than by approaching an existing national wholesaler as John Hunter had suggested to Edward Jenner for Jenner's Tartar Emetic in 1784.[20] In the 1880s, the pharmacists complained that patent medicines were being sold by a wide variety of outlets at discounted prices, and they wondered whether the reduced profits still justified stocking these medicines.[21] For the public and the medical profession, little separation now existed between patent medicines and irregular practice.

For all these reasons, patent medicines ceased to be a third force in healthcare soon after the end of the Georgian era, and they moved closer to Porter's conception of patent medicines as the visible manifestation of quackery. The censorship of any association with patent medicines in biographies by descendants and others dates from this period. The alterations in the creation and supply of nationally available medicines require more study, but for the public in the late nineteenth century, patent medicines were the main exhibition of irregular practice. During the Victorian era, patent medicines, increasingly referred to as 'secret medicines', faced strong opposition from leaders of the medical profession and the pharmacists wanted to confine sales to their own shops. However, despite a plethora of fierce editorials, resolutions at meetings, draft Parliamentary bills, and attempted legal actions, neither the doctors nor the pharmacists created any significant restrictions on the production and selling of patent medicines until the end of the century.

Strident opposition to patent medicines from some, though not all, regular practitioners was not new; but it became more prevalent in the early nineteenth century, and some of the medical journals sought to organise this opposition into campaigns. In 1808, the first three issues of the *Medical Observer*, edited by an anonymous group of 'practical physicians', devoted over 250 pages between them to exposing the inadequacies of patent medicines and their advocates, and the frequent criticisms in the text and editorials of the *Lancet* started with the first issue in 1823.[22] The attacks in the *Lancet* were unsparing: for example, in a chain of correspondence in 1846, the Queen's obstetrician Charles Locock was forced to explain that Da Silva, a corn-cutter, had named Dr. Locock's Pulmonic Wafers after Locock's deceased father, a regular practitioner, without any consent, and there was little he could do about it.[23] The *Lancet* felt that this response was inadequate: Locock should have tried to obtain legal redress, and if unsuccessful should have used his public prominence to campaign to alter the law. The *Provincial Medical and Surgical Journal*, predecessor of the *British Medical Journal*, joined in after its foundation in 1840, advising that the 'nostrum of the quack' should be 'absolutely prohibited'.[24] The leaders of the medical profession were certainly trying to diminish, and ultimately eliminate, patent medicines in the medical market. At times, they were abetted by the leaders of the Pharmaceutical Society who wished to confine sales to fully qualified members.[25]

Yet these campaigns achieved nothing of practical significance until the end of the century, and not much until well into the next one. Multiple attempts from the 1850s onwards to impose some statutory control over patent medicines were unsuccessful or were emasculated. Partly, this can be explained by a prevailing belief in free markets, and resulting profits, which made any form of monopoly unattractive to legislators. The financial beneficiaries of patent medicines were spread wider than their owners: for example, influential publications such as the national newspapers derived increasing revenues from patent medicine advertisements.[26] In addition, the leaders of the medical profession did not always reflect the views of the rank and file who often wished to recommend secret medicines, and might be involved in their promotion.[27]

A small success for the campaigning doctors was achieved in 1885 when the words 'this stamp implies no government guarantee' were added to all medicine stamps, a hundred years after the stamp's introduction.[28] The next limited achievement was to bring some control to the supply of patent medicines containing powerful drugs. One such medicine was the various versions of chlorodyne which contained chloroform and morphine. With increasing awareness of their addictive properties, the British Medical Association (BMA) prompted a series of legal actions in 1892–1894 aimed at including chlorodyne and similar medicines under the 1868 Pharmacy Act.[29] This Act limited the sale of poisons, such as chloroform, morphine and belladonna to registered medical practitioners and registered pharmacists: but the Act included a specific exemption for 'the making and dealing in patent medicines', which had been taken to include all secret advertised remedies. In these legal actions, the BMA persuaded the courts that this exemption only applied to medicines which held a current legal patent, not to patent medicines in general, and the supply of powerful patent medicines was restricted to registered pharmacists.

The BMA continued its campaign and it published the constituents of many patent medicines, hoping that the blandness of the recipes, particularly for the higher priced remedies, would reduce demand.[30] The pressure resulted in a Parliamentary Select Committee on patent medicines which in 1914 confirmed that British law still allowed any secret medicine to be made as long as it did not contain a scheduled poison, to be advertised as a cure for anything, and to be sold at any price.[31] The Select Committee's recommendations were restrained, a confidential register of patent medicine producers and their products; but their publication on the first day of the First World War ensured inaction. In the following years, small victories for the doctors were achieved with the 1917 Venereal Diseases Act prohibiting the sale of patent medicines for these conditions, more controls on an increased number of scheduled poisons during the 1920s, and the 1939 Act confining the treatment of cancer to registered medical practitioners.[32] But a century-and-a-half of campaigning by at least some members of the medical profession had accomplished only a modest control over these 'abominable impositions'.

Then, suddenly, their aims were achieved. In the entirely different wartime medical market when government direction was the norm, the 1941 Pharmacy and Medicines Act prohibited patent medicines by requiring full disclosure of all active ingredients on the medicine label.[33] Patent medicines as defined in this book, and as the term was commonly used, disappeared. In addition, the Act forbade advertisements claiming cures for many common, chronic diseases, limiting the role of other proprietary medicines.

However, consumers in Britain today still buy advertised branded medicines in shops other than pharmacies, without any medical advice. A quick inspection of a supermarket usually reveals medicines on sale for a variety of aches and pains, respiratory disorders, stomach and bowel problems, sleeplessness, poor memory and lack of energy, amongst other problems, and the same medicines are available all over the country. They are in addition to health-preserving medicines such as vitamin pills and other dietary supplements, and beauty-preserving topical products. Some of these medicines and products are extensively, and expensively, advertised to the general public: though unlike the patent medicines of the past, today's proprietary medicines are intended for unpleasant, but not dangerous, problems. The medicines are not secret, but many consumers do not understand, or pay little attention to, their listed contents, concentrating more on the brand names. To use an anatomical analogy, the flesh of the Georgian patent medicines industry has long since disappeared, but its skeleton is still with us today.

Postscript

This book started with the grave of John Newbery, revealing the description of Dr. James's Fever Powder as 'the most powerful discovery in the annals of medicine'. Let us end on a more buoyant note with a vision of three successful medicine owners sitting in their country houses. Their owners were all well-publicised participants in the patent medicines industry, who did not indulge in medical practice, and we have met them at several points in this book. Francis Newbery owned Heathfield Park, Sussex, previously the home of Lord Heathfield, the victor of the Siege of Gibraltar, from 1795 till his death in 1818; Thomas Dicey owned Claybrooke Hall, Leicestershire, together with other land in neighbouring parishes, from 1775 till his death in 1807; and Nathaniel Godbold owned Westbrook Place, Surrey, formerly the family home of General Oglethorpe, the founder of the colony of Georgia, from 1790 till his death in 1799. Surveying their extensive parklands from the windows of their large houses in the 1790s, all three of these patent medicine owners, wholesalers and country gentlemen would have known that they were part of a stable and very profitable industry which was separate from both regular medicine and quackery.

Notes

1. Jefferys, *Retail Trading*, 1–5.
2. Jefferys, *Retail Trading*, 1 and 381.
3. Mitchell, *Tradition*, 11.
4. Mui and Mui, *Shops*, 289–291; Cox, *Complete Tradesman*, 14.
5. Porter, *Health*, 4.
6. Engels, *Condition of the Working Class*, 117.
7. Loeb, 'Doctors and Patent Medicines', 409.
8. Vaughan, 'Secret Remedies', 102; *House of Commons Papers (Accounts and Papers)*, ix, Finance Accounts of Great Britain, 20–21. Per head of the population, this is a reduction in receipts.
9. *Secret Remedies*, 183.
10. 'The Messrs. Newbery', 116.
11. Corley, 'Holloway, Thomas'.
12. Corley, 'Beecham, Thomas'; 'Thomas Beecham'.
13. Corley, 'Eno, James'; 'Pharmacy's Offspring', 739.
14. Porter, *Health*, 231; Brown, *Performing Medicine*, 204; Miley and Pickstone, 'Medical Botany', 144.
15. Porter, *Health*, 234; Loudon, *Medical Care*, 210.
16. *MCR*, 13 (1806), ci–cii; Marland, *Medicine and Society*, 209.
17. Brown, *Performing Medicine*, 210; Miley and Pickstone, *Medical Botany*, 147.
18. Brown, *Performing Medicine*, 211.
19. Hill, *Bookmakers*, 101.
20. Corley, 'Beecham, Thomas'; 'Pharmacy's Offspring', 739; Paget, *John Hunter*, 165. John Hunter's suggested wholesaler was Francis Newbery.
21. 'The Patent Medicine Trade'; Holloway, *Pharmaceutical Society*, 309.
22. The initial 99 pages of the first issue of the *Medical Observer* criticised individual medicines in turn, while the *Lancet* published 24 recipes of 'quack medicines' in its first four issues (*Medical Observer*, 1 (1808): 1–99; *Lancet*, 1 (1823): 30, 62, 89 and 138).
23. *Lancet*, 47 (1846): 251–252, 307–308 and 311. Charles Locock attended all Victoria's births and was created a baronet in 1857.
24. *Provincial Medical and Surgical Journal*, 2 (1841): 333.
25. Holloway, *Pharmaceutical Society*, 221–247.
26. Vaughan, 'Secret Remedies', 102.
27. Burney, 'Medicine in the Age of Reform', 178; Loeb, 'Doctors and Patent Medicines', 415–418.
28. Booth, *Revenue Stamps*, A147.
29. Holloway, *Pharmaceutical Society*, 247–248; Aronson, 'Patent Medicines'.
30. *Secret Remedies*; *More Secret Remedies*.

31. Vaughan, 'Secret Remedies', 109.
32. Holloway, *Pharmaceutical Society*, 394–396; Vaughan, 'Secret Remedies', 110.
33. Holloway, *Pharmaceutical Society*, 397.

Bibliography

Journals

Lancet.
Medical and Chirurgical Review (MCR).
Medical Observer.
Provincial Medical and Surgical Journal.

Printed Primary Sources

House of Commons Papers (Accounts and Papers).

Engels, Friedrich, *The Condition of the Working Class in England*, translated by W. O. Henderson and W. H. Chaloner, 2nd edition (Oxford: Basil Blackwell, 1971).

Secondary Sources

Aronson, Jeffrey K., 'Patent Medicines and Secret Remedies', *British Medical Journal*, 339 (2009): 1394–1396.

Booth, R. G., *A Catalogue of the Revenue Stamps of the U K, Isle of Man, Channel Islands and Eire, vol. 2* (Hexham, Northumberland: Tom Clutterbuck, 1982).

Brown, Michael, *Performing Medicine: Medical Culture and Identity in Provincial England, c.1760–1850* (Manchester: Manchester University Press, 2014).

Burney, Ian, 'Medicine in the Age of Reform', in *Rethinking the Age of Reform*, edited by Arthur Burns and Joanna Innes (Cambridge: Cambridge University Press, 2003), 163–181.

Corley, T. A. B., 'Beecham, Thomas', *ODNB* (accessed 30 March 2017).

Corley, T. A. B., 'Holloway, Thomas', *ODNB* (accessed 30 March 2017).

Corley, T. A. B., 'Eno, James Crossley', *ODNB* (accessed 6 April 2017).

Cox, Nancy, *The Complete Tradesman: A Study of Retailing, 1550–1820* (Aldershot: Ashgate, 2000).

Hill, Joseph, *The Bookmakers of Old Birmingham: Authors, Printers and Booksellers* (Birmingham: Cornish Brothers, 1907).

Holloway, S. W. F., *Royal Pharmaceutical Society of Great Britain 1841–1991: A Political and Social History* (London: The Pharmaceutical Press, 1991).

Jefferys, James B., *Retail Trading in Britain 1850–1950* (Cambridge: Cambridge University Press, 1954).

Loeb, Lori, 'Doctors and Patent Medicines in Modern Britain: Professionalism and Consumerism', *Albion*, 33 (2001): 404–425.

Loudon, Irvine, *Medical Care and the General Practitioner 1750–1850* (Oxford: Clarendon Press, 1986).

Marland, Hilary, *Medicine and Society in Wakefield and Huddersfield, 1780–1870* (Cambridge: Cambridge University Press, 1987).

Miley, Ursula, and John V. Pickstone, 'Medical Botany Around 1850: American Medicine in Industrial Britain', in *Studies in the History of Alternative Medicine*, edited by Roger Cooter (Basingstoke: Macmillan Press, 1988), 140–154.

Mitchell, Ian, *Tradition and Innovation in English Retailing, 1700–1850: Narratives of Consumption* (Farnham, Surrey: Ashgate, 2014).

More Secret Remedies, What They Cost and What They Contain (London: British Medical Association, 1912).

Mui, Hoh-Cheung, and Lorna Mui, *Shops and Shopkeeping in Eighteenth-Century England* (London: Routledge, 1989).

Paget, Stephen, *John Hunter: Man of Science and Surgeon (1728–1793)* (London: Fisher Unwin, 1897).

'Pharmacy's Offspring', *Chemist and Druggist*, 124 (1936): 737–743.

Porter, Roy, *Health for Sale: Quackery in England 1660–1850* (Manchester: Manchester University Press, 1989).

Secret Remedies, What They Cost and What They Contain (London: British Medical Association, 1909).

'The Messrs. Newbery', *Chemist and Druggist*, 15 (1874): 112–116.

'The Patent Medicine Trade', *Chemist and Druggist*, 34 (1889): 17–18.

'Thomas Beecham', *Chemist and Druggist*, 70 (1907): 544–545.

Vaughan, Paul, '"Secret Remedies" in the Late Nineteenth and Early Twentieth Centuries', in *Alternative Medicine in Britain*, edited by Mike Saks (Oxford: Clarendon Press, 1992), 101–111.

Appendix A: Investigated Patent Medicines

Each of these medicines has a part in this book, sometimes directly through explicit mention, sometimes indirectly in the analysis of the newspaper advertisements, and in many cases in both of these ways. A few of the medicines were, or could have been, rival versions with similar names. The information derives from the studied advertisements in 1769, 1781, 1794, 1807 and 1822, augmented by other sources described in the book.

Section A.1 lists the 'national' medicines; that is, those medicines which were evidently intended for a wide distribution as they were advertised in more than one location, or in one location in more than one year with a London wholesaler, or in both the Leeds newspapers in the same year with a wholesaler outside Yorkshire. Section A.2, 'other medicines', records the remaining medicines which were advertised in the

A. Mackintosh, *The Patent Medicines Industry in Georgian England*, Medicine and Biomedical Sciences in Modern History,
https://doi.org/10.1007/978-3-319-69778-9

studied newspapers. Some of these were distributed nationally, but we do not have evidence in the studied newspapers; even if they had a London wholesaler, their promotion and distribution might have been organised by a local retailer, particularly the newspaper printer. Section A.3 lists some patent medicines of the period which are named in the book without appearing in the advertisements in the studied newspapers.

Much of the information in this appendix derives from promotional and imperfect material: some of it was out-of-date, and it is unlikely that all of it was accurate, particularly the details of the owner. 'Owner' refers to the person or company advertised as the source of the medicine, and it includes people described as 'preparer', 'proprietor' and 'inventor', as well as anybody who was designated as licensing or approving the wholesaler. Prices were for the smallest advertised quantity, and from 1794 they included the excise duty. Apparent sharp changes in price may reflect the advertisement of rival versions of the same medicine.

A.1 National Medicines

Inclusion criteria: Advertised in the studied newspapers in more than one location, or in one location in more than one year with a London wholesaler, or in both the Leeds newspapers in the same year with a wholesaler outside Yorkshire. Addresses were in London unless specified otherwise.

Name	Price	Year	Newspapers	Owner	Address	Wholesaler	Address
Adam's Solvent	5s 6d	1781	ABG SWJ	'Proprietor'	Argyle St.	'Proprietor'	Argyle St.
	4s 6d	1807	SWJ	Perry (surgeon)	Southampton St., Bloomsbury	Perry (surgeon)	Southampton St., Bloomsbury
American Soothing Syrup	2s 9d	1822	LI, ABG, SWJ	Johnson & Burgess	28 York Place, City Road	Johnson & Burgess	28 York Place, City Road
Amboyna Lotion	4s 6d	1807	LI, LM	–		Shaw & Edwards	66 St. Paul's Churchyard
Appleby's Balsam of Health	1s 6d	1769	ABG	Dr. Appleby	–	Dr. Appleby	–
	1s 6d	1781	ABG	Andrew Ledbroke (druggist)	Leicester	Andrew Ledbroke	Leicester
						Mary Woodward	–
	1s 6d	1781	LM	–		J. Bowling	Leeds
(Dr.) Arnold's Pills	5s 5d	1794	ABG	Dr. Arnold	–	–	
	2s 9d	1807	ABG	–		? Mr. Axtell	1 Finch Lane, Cornhill
Barclay's Antibilious Pills	5s 6d	1807	LI, LM, SWJ	Rev. Dr. Barclay	London	R. Butler	4 Cheapside
Barclay's Ointment for the Itch	1s 9d	1807	LM, ABG	Barclay & Son	95 Fleet Market	Barclay & Son	95 Fleet Market
	1s 9d	1822	LI, LM, SWJ	Barclay & Sons	95 Fleet Market	Barclay & Sons	95 Fleet Market
(Dr.) Bateman's Pectoral Drops	1s	1769	LI	Cluer Dicey & Co.	London	Cluer Dicey & Co.	London
	1s	1781	LM	? (probably a rival version)		Jackson, Warter & Co.	95 Fleet Market
	–	1781	ABG	Dicey & Co.	10 Bow Churchyard	Dicey & Co.	10 Bow Churchyard
	–	1794	LI, ABG, SWJ	Dicey & Co.	10 Bow Churchyard	Dicey & Co.	10 Bow Churchyard
	1s 1½d	1807	LI, LM, SWJ	Dicey & Co.	10 Bow Churchyard	Dicey & Co.	10 Bow Churchyard
	1s 1½d	1822	LI, ABG, SWJ	Sutton & Co.	10 Bow Churchyard	Sutton & Co.	10 Bow Churchyard
Beaume de Vie	3s	1769	LI, ABG, SJ	? (3 patentees in 1767, see Appendix B)		W. Nicoll	51 St. Paul's Churchyard
	3s	1781	ABG, SWJ	T. Becket	Adelphi	W. Nicoll	51 St. Paul's Churchyard
	3s 6d	1794	ABG	–		Dicey & Co.	10 Bow Churchyard
	3s 6d	1807	LI	T. Becket	–	Dicey & Co.	10 Bow Churchyard
Betton's British Oil	1s 1½d	1794	SWJ	Dicey & Co.	10 Bow Churchyard	Dicey & Co.	10 Bow Churchyard
	1s 9d	1807	LI, LM	Dicey & Co.	10 Bow Churchyard	Dicey & Co.	10 Bow Churchyard
	1s 9d	1822	LI, ABG, SWJ	Sutton & Co.	10 Bow Churchyard	Sutton & Co.	10 Bow Churchyard
(Dr.) Boerhaave's Red Pill No. 2	4s 6d	1822	LI, LM, ABG, SWJ	–		–	
Bolderson's Worm Cakes	1s	1769	LI, ABG	–		Cluer Dicey & Co.	10 Bow Churchyard
Bott's Corn Salve	1s	1781	ABG	–		Pearson & Rollason	High St., Birmingham
	1s	1781	SWJ	George Bott	Nottingham	–	

(Continued)

(Continued)

Name	*Price*	*Year*	*Newspapers*	*Owner*	*Address*	*Wholesaler*	*Address*
British Powder for Teeth and Gums	1s	1781	LM, SWJ	–		Thomas Jackson, Warter & Co.	95 Fleet Market
(Dr.) Brodum's Botanical Syrup	5s 5d	1794	LI, LM	Dr. Brodum	9 Albion St., Blackfriars	Dr. Brodum	9 Albion St., Blackfriars
	6s	1807	SWJ	Dr. Brodum	London	–	
	7s	1822	SWJ	–		–	47 Salisbury Square
(Dr.) Brodum's Nervous Cordial	5s 5d	1794	LI, LM	Dr. Brodum	9 Albion St., Blackfriars	Dr. Brodum	9 Albion St., Blackfriars
	6s	1807	SWJ	Dr. Brodum	London	–	
Butler's Restorative Tooth Powder	2s 9d	1807	ABG, SWJ	Mr. Butler	4 Cheapside	Butler	4 Cheapside
Carrington's Life Pills	1s 1½d	1822	LM, SWJ	Barry & Sons	Bristol	Sutton & Co.	10 Bow Churchyard
						Barclay & Sons	95 Fleet Market
						Butlers (chemists)	4 Cheapside
						Sanger	150 Oxford St.
Ching's Patent Worm Lozenges	1s 1½d	1807	LI, ABG	Mr. Ching (apothecary)	–	Ching & Butler	4 Cheapside
	1s 1½d	1822	LI, LM, SWJ	Butlers (chemists)	4 Cheapside	Butlers (chemists)	4 Cheapside
Church's Cough Drops	2s 6d	1807	LI, LM	Shaw & Edwards	66 St. Paul's Churchyard	Shaw & Edwards	66 St. Paul's Churchyard
	2s 9d	1822	LI, SWJ	–		Shaw & Edwards	66 St. Paul's Churchyard
Cockle's Compound Antibilious Pills	1s 1½d	1822	LM, SWJ	Cockle (apothecary)	6 Speldhurst St., Burton Crescent	Barclay & Sons	95 Fleet Market
Cordial Balsam of Rakasiri	11s	1822	LI, LM, SWJ	Drs. C. & J. Jordan	9 Gt Surrey St. & 28 Berwick St.	Drs. C. & J. Jordan	
						Barclay & Sons	95 Fleet Market
Cordial Cephalic Snuff	6d	1769	ABG	Benjamin Collins	Salisbury	Newbery & Carnan	65 St. Paul's Churchyard
	–	1781	ABG, SWJ	Collins & Johnson?	Salisbury?	F. Newbery	45 St. Paul's Churchyard
						Dicey, Beynon & Co.	10 Bow Churchyard
						Collins and Johnson	Salisbury
	–	1807	SWJ	–		F. Newbery & Sons	45 St. Paul's Churchyard
	1s 1½d	1822	SWJ	F. Newbery & Sons	45 St. Paul's Churchyard	F. Newbery & Sons	45 St. Paul's Churchyard
Dalby's Carminative	1s 6d	1781	ABG	J. Dalby (apothecary)	–	F. Newbery	45 St. Paul's Churchyard
	1s 6d	1794	ABG	Mrs. Frances Gill (daughter of the late Joseph Dalby)	–	F. Newbery	45 St. Paul's Churchyard
	1s 9d	1807	LI, ABG	Mrs. Frances Gill	North St., Westminster	F Newbery & Sons	45 St. Paul's Churchyard

(Continued)

(Continued)

Name	*Price*	*Year*	*Newspapers*	*Owner*	*Address*	*Wholesaler*	*Address*
De Velno's Vegetable Syrup	11s 6d	1794	LI, ABG	Isaac Swainson	Frith St., Soho	–	
	–	1807	SWJ	Swainson	Frith St., Soho	Swainson	Frith St., Soho
	13s	1822	SWJ	Thomas Canham	52 Berners St., Oxford St.	Thomas Canham	52 Berners St., Oxford St.
Dicey's Anderson's Scots Pills	1s	1781	SWJ	Cluer Dicey	10 Bow Churchyard	Cluer Dicey	10 Bow Churchyard
	1s 1½d	1794	ABG, SWJ	Dicey & Co.	10 Bow Churchyard	Dicey & Co.	10 Bow Churchyard
	1s 1½d	1807	LI	Dicey & Co.	10 Bow Churchyard	Dicey & Co.	10 Bow Churchyard
	1s 1½d	1822	LI, ABG, SWJ	Sutton & Co.	10 Bow Churchyard	Sutton & Co.	10 Bow Churchyard
Dicey's Daffy's Elixir	–	1781	ABG	Dicey & Co.	10 Bow Churchyard	Dicey & Co.	10 Bow Churchyard
	1s 1½d	1794	LI, ABG, SWJ	Dicey & Co.	10 Bow Churchyard	Dicey & Co.	10 Bow Churchyard
	1s 8d	1807	LI, LM	Dicey & Co.	10 Bow Churchyard	Dicey & Co.	10 Bow Churchyard
	2s	1822	LI, ABG, SWJ	Sutton & Co.	10 Bow Churchyard	Sutton & Co.	10 Bow Churchyard
Dickinson's Red and White Drops	5s	1769	SJ	–		Roach, Holborn & Fox	Westminster Hall
	5s	1781	SWJ	–		–	
Dickinson's Gowland's Lotion	5s 3d	1794	LI, LM	T. Vincent & R. Dickinson	Long Acre	T. Vincent & R. Dickinson	Long Acre
	5s 6d	1807	LI	Robert Dickinson	Long Acre	–	
Dixon's Antibilious Pills	2s 9d	1807	ABG, SWJ	Dixon (apothecary)	–	Butler	4 Cheapside
	2s 9d	1822	LI, LM, SWJ	–		Butlers (chemists)	4 Cheapside
English Coffee	2s 6d	1781	ABG, SWJ	Lee Roe	9 Silver St., Fleet St.	Lee Roe	9 Silver St., Fleet St.
Essence of Coltsfoot	3s 6d	1794	ABG	James Ryan (surgeon)	Bristol	F. Newbery	45 St. Paul's Churchyard
	3s 6d	1807	LI	James Ryan (surgeon)	Bristol	F. Newbery & Sons	45 St. Paul's Churchyard
	3s 6d	1822	LI, ABG, SWJ	James Ryan (surgeon)	Bristol	F. Newbery & Sons	45 St. Paul's Churchyard
Evil Salve	–	1807	LI, LM	–		Barclay & Son	95 Fleet Market
Ford's Pectoral Balsam of Horehound	1s 1½d	1807	LM	R. Ford (chemist)	Barbican	R. Ford (chemist)	Barbican
	1s 9d	1822	SWJ	Robert Ford	–	Robert Ford	–
(Dr.) Freeman's Gutta Salutaris	3s	1781	ABG	Dr. S. Freeman MD	1 Staple's Inn, Holborn	Dr. S. Freeman MD	1 Staple's Inn, Holborn
	2s 9d	1807	LI, LM, ABG	Dr. Freeman	–	Butler	4 Cheapside
(Dr.) Freeman's Ointment for Itch	1s 1½d	1822	LI, LM, SWJ	S. Freeman	–	–	
Fruit Lozenges	1s 1½d	1822	LI, LM, ABG	Phillips & Scholefield	Pitt St., Liverpool	Phillips & Scholefield	Pitt St., Liverpool

(Continued)

(Continued)

Name	Price	Year	Newspapers	Owner	Address	Wholesaler	Address
Glass's Magnesia	6s	1769	ABG, SJ	S. Glass (surgeon)	Oxford	R. Davies	Piccadilly
						J. Fletcher	St. Paul's Churchyard
	6s	1781	ABG, SWJ	Samuel Glass	Oxford	W. Davis	Piccadilly
						W. Nicoll	St. Paul's Churchyard
Godbold's Vegetable Balsam	–	1794	LI, LM, ABG, SWJ	N. Godbold	3 Bloomsbury Square	N. Godbold	3 Bloomsbury Square
	18s per pint	1807	LM, SWJ	N. & S. Godbold	3 Bloomsbury Square	N. & S. Godbold	3 Bloomsbury Square
	9s	1822	SWJ	Messrs Godbold	3 Bloomsbury Square	Messrs Godbold	3 Bloomsbury Square
Greenough's Pectoral Lozenges of Tolu	1s	1781	LI, ABG, SWJ	T. Greenough (apothecary)	10 Ludgate Hill	W. Barley	Cockspot St.
						F. Newbery	45 St. Paul's Churchyard
						Dicey & Co.	10 Bow Churchyard
	1s 1½d	1794	LI, ABG	R. Hayward	10 Ludgate Hill	R. Hayward	10 Ludgate Hill
	1s 1½d	1807	SWJ	R. Hayward (chemist)	10 Ludgate Hill	R. Hayward	10 Ludgate Hill
	1s 1½d	1822	SWJ	R. Hayward (chemist)	10 Ludgate Hill	R. Hayward	10 Ludgate Hill
Greenough's Tooth Tinctures	–	1781	LI, SWJ	T. Greenough	–	T. Greenough	–
	–	1794	LI, ABG	R. Hayward	10 Ludgate Hill	R. Hayward	10 Ludgate Hill
	1s 1½d	1807	SWJ	R Hayward (chemist)	10 Ludgate Hill	R. Hayward	10 Ludgate Hill
	2s 9d	1822	SWJ	R. Hayward (chemist)	10 Ludgate Hill	R. Hayward	10 Ludgate Hill
(Dr.) Green's Specific Drops	2s 6d	1781	ABG, SWJ	Dr. Green	1 Little Hoe Lane, Plymouth	Dr. Green	1 Little Hoe Lane, Plymouth
						Thomas Wilson	8 Slaney St., Birmingham
Hallam's Bilious Pills	2s 9d	1807	LI, LM	Edward Hallam (surgeon)	Bury St. Edmonds	Shaw & Edwards	66 St. Paul's Churchyard
Hamilton's Tincture for the Teeth	2s 6d	1781	ABG, SWJ	Hilton Wray	14 Birchin Lane	Martha & Hilton Wray	14 Birchin Lane
(Dr.) Hammond's Specific Pill	6s	1769	ABG, SJ	Dr. Hammond	–	–	
Henry's Calcined Magnesia	3s 6d	1781	LI, ABG	Thomas Henry (apothecary)	Manchester	J. Johnson	St. Paul's Churchyard
	2s 6d	1794	LI	Thomas Henry	Manchester	J. Johnson	72 St. Paul's Churchyard
	2s 9d	1822	ABG	T&W Henry	Manchester	T. & W. Henry	Manchester

(Continued)

(Continued)

Name	*Price*	*Year*	*Newspapers*	*Owner*	*Address*	*Wholesaler*	*Address*
(Dr.) Henry's Chemical	7s	1769	SJ	–		–	
Nervous Medicine	7s	1781	ABG, SWJ	Dr. Peter Henry	4 Hatton St.	W. Nicoll	51 St. Paul's Churchyard
						Mrs. Randalls	Royal Exchange
						Davis & Co.	Piccadilly
Hickman's Pills	–	1781	LI, ABG	Mr. Hickman	–	M. & H. Wray	14 Birchin Lane
	2s 9d	1822	LI, LM, SWJ	–		Butlers (chemists)	4 Cheapside
Hill's Canada Balsam	3s	1781	ABG, SWJ	Lady Hill	29 St. James's Place	Lady Hill	29 St. James's Place
	3s 6d	1794	SWJ	Lady Hill	–	W. Bacon	150 Oxford St.
Hill's Elixir of Bardana	3s	1769	LI, ABG, SJ	John Hill	London	John Hill	London
	3s	1781	ABG, SWJ	Lady Hill	29 St. James's Place	Lady Hill	29 St. James's Place
Hill's Essence of Waterdock	3s	1769	LI, ABG, SJ	John Hill	London	John Hill	London
	3s	1781	ABG, SWJ	Lady Hill	29 St. James's Place	Lady Hill	29 St. James's Place
Hill's Pectoral Balsam of	3s	1769	LI, ABG, SJ	John Hill	London	John Hill	London
Honey	3s	1781	LI, ABG, SWJ	Lady Hill	29 St. James's Place	Lady Hill	29 St. James's Place
	3s 6d	1794	SWJ	Lady Hill	–	W. Bacon	150 Oxford St.
Hill's Tincture of Centaury	3s	1769	LI, ABG, SJ	John Hill	London	John Hill	London
	3s	1781	ABG, SWJ	Lady Hill	29 St. James's Place	Lady Hill	29 St. James's Place
Hill's Tincture of Sage	3s	1769	LI, ABG, SJ	John Hill	London	John Hill	London
Hill's Tincture of Spleenwort	3s	1769	LI, ABG, SJ	John Hill	London	John Hill	London
Hill's Tincture of Valerian	2s 6d	1769	LI, ABG, SJ	John Hill	London	John Hill	London
	2s 6d	1781	ABG, SWJ	Lady Hill	29 St. James's Place	Lady Hill	29 St. James's Place
(Dr.) Hodson's Vegetable Syrup	10s 6d	1794	LI, LM, ABG	Dr. Hodson	24 Hatton Gardens	Dr. Hodson	24 Hatton Gardens
Hudson's Botanic Toothpowder	2s 9d	1822	ABG, SWJ	–		Mr. Atkinson	44 Gerard St., Soho
Inglish's Anderson's Scots Pills	1s	1781	LM, SWJ	James Inglish	165 The Strand	James Inglish	165 The Strand
	–	1794	LM	James Inglish	165 The Strand	James Inglish	165 The Strand
	1s 1½d	1807	LI	B. H. Inglish	165 The Strand	B. H. Inglish	165 The Strand
	1s 1½d	1822	LM, SWJ	H. Inglish	165 The Strand	H. Inglish	165 The Strand
Jackson's Asthmatic Candy	–	1794	LM	Jackson & Co.	95 Fleet Market	Jackson & Co.	95 Fleet Market
	1s 1½d	1794	ABG	J. Barclay	95 Fleet Market	J. Barclay	95 Fleet Market
Jackson's British Powder	1s	1781	ABG	Thomas Jackson	95 Fleet Market	Thomas Jackson	95 Fleet Market
for Teeth	–	1794	ABG	J. Barclay	95 Fleet Market	J. Barclay	95 Fleet Market

(Continued)

(Continued)

Name	*Price*	*Year*	*Newspapers*	*Owner*	*Address*	*Wholesaler*	*Address*
Jackson's Corn Salve	1s 6d	1781	SWJ	Thomas Jackson	95 Fleet Market	Thomas Jackson, Warter & Co.	95 Fleet Market
	–	1794	ABG	J. Barclay	95 Fleet Market	J. Barclay	95 Fleet Market
Jackson's Ointment for the Itch	1s 6d	1781	LM, ABG, SWJ	Thomas Jackson	95 Fleet Market	Thomas Jackson, Warter & Co.	95 Fleet Market
	1s 9d	1794	LM	Jackson & Co.	95 Fleet Market	Jackson & Co.	95 Fleet Market
	–	1794	ABG	J. Barclay	95 Fleet Market	J. Barclay	95 Fleet Market
Jackson's Tincture	1s	1769	SJ	Jackson & Co.	Fleet Market	Jackson & Co.	Fleet Market
	1s	1781	LM, SWJ	Thomas Jackson	95 Fleet Market	Thomas Jackson, Warter & Co.	95 Fleet Market
	–	1794	LM	Jackson & Co.	95 Fleet Market	Jackson & Co.	95 Fleet Market
	–	1794	ABG	J. Barclay	95 Fleet Market	J. Barclay	95 Fleet Market
(Dr.) James's Analeptic Pills	4s	1781	LI, ABG, SWJ	Francis Newbery	45 St. Paul's Churchyard	Francis Newbery	45 St. Paul's Churchyard
	4s 6d	1794	ABG	Francis Newbery	45 St. Paul's Churchyard	Francis Newbery	45 St. Paul's Churchyard
(Dr.) James's Fever Powder	–	1769	LI, SJ	F. Newbery jnr	65 St. Paul's Churchyard	Newbery & Carnan	65 St. Paul's Churchyard
	2s 6d	1794	ABG	F. Newbery	45 St. Paul's Churchyard	F. Newbery	45 St. Paul's Churchyard
	2s 9d	1807	LI	F. Newbery & Sons	45 St. Paul's Churchyard	F. Newbery & Sons	45 St. Paul's Churchyard
	–	1822	LI, ABG, SWJ	Messrs Newbery	45 St. Paul's Churchyard	Messrs Newbery	45 St. Paul's Churchyard
Keighley's Tincture for Rheumatism	2s 6d	1781	LM	D. Keighley	York	Mr. Shute	Leeds
	4s	1794	LM, ABG	–		–	
Lane's Antiphlegmonic Ointment	–	1807	LI, LM	–		Barclay & Son	95 Fleet Market
Lane's Haemacathartic Tincture	4s 6d	1807	LI, LM	–		Barclay & Son	95 Fleet Market
Le Coeur's Imperial Oil	2s 6d	1781	LI, ABG, SWJ	Macklin	Southampton	W. Hodson	Bridgewater Square
Leake's Pills	2s 6d	1781	LI, LM, ABG, SWJ	Walter Leake	13 Bride Lane, Fleet St.	Walter Leake	13 Bride Lane, Fleet St.
	2s 9d	1807	LI	Thomas Taylor (surgeon)	London	Thomas Taylor	9 New Bridge St.

(Continued)

(Continued)

Name	*Price*	*Year*	*Newspapers*	*Owner*	*Address*	*Wholesaler*	*Address*
Lignum's Antiscorbutic Drops	4s 6d	1794	LI, LM	Mr. Lignum	Thomas St., Manchester	Mr. Lignum	Thomas St., Manchester
	4s 6d	1807	LI, LM	Mr. Lignum	Manchester	Mr. Lignum	Manchester
	2s 9d	1822	LI, LM, ABG	Mr. Lignum	63 Bridge St., Manchester	Mr. Lignum	63 Bridge St., Manchester
Lignum's Lotion	1s 1½d	1794	LM	Mr. Lignum	Thomas St., Manchester	Mr. Lignum	Thomas St., Manchester
	2s 9d	1807	LI	Mr. Lignum	Manchester	Mr. Lignum	Manchester
	2s 9d	1822	LI, LM, ABG	Mr. Lignum	63 Bridge St., Manchester	Mr. Lignum	63 Bridge St., Manchester
Lignum's Pills	2s 9d	1794	LM	Mr. Lignum	Thomas St., Manchester	Mr. Lignum	Thomas St., Manchester
	2s 9d	1807	LI	Mr. Lignum	Manchester	Mr. Lignum	Manchester
	2s 6d	1822	LI, LM, ABG	Mr. Lignum	Manchester	Simpkin & Marshall C. Evans	London London
(Dr.) Lowther's Nervous Medicine	3s	1769	SJ	Dr. Lowther	Golden Lamp, Hatton Gardens	Dr. Lowther Cooke (bookseller) Paillet	Hatton Gardens Paternoster Row Princes St., Leicester Fields
	3s	1781	SWJ	–		–	
Marshall's Cerate	1s 1½d	1807	LM	–		Shaw & Edwards	66 St. Paul's Churchyard
	1s 1½d	1822	LI, ABG	E. Marshall	London	Shaw & Edwards	66 St. Paul's Churchyard
Masson's Medicine for the Itch	1s	1781	ABG	Mr. Masson (surgeon)	London	–	
	1s 1½d	1794	ABG	Thomas Appleby	–	Thomas Appleby	–
Morris's Royal Brunswick Corn Plaster	1s 1½d	1822	LI, LM, SWJ	G. Morris (chemist)	Kensington	Butlers (chemists)	4 Cheapside
(Dr.) Norris's Drops	5s 3d	1769	ABG, SJ	Dr. T. Norris	Duke St., Westminster	Dr. T. Norris Isaac Fell	Duke St., Westminster 14 Paternoster Row
	2s 6d	1781	LI, ABG, SWJ	Dr. Norris	New Bridge St., Blackfriars	Dr. Norris Fieldhouse & Walker	New Bridge St., Blackfriars Paternoster Row
	2s 9d	1822	SWJ	–		Thomas Moore	–

(Continued)

(Continued)

Name	*Price*	*Year*	*Newspapers*	*Owner*	*Address*	*Wholesaler*	*Address*
Norton's Maredant's Drops	6s	1769	LI, ABG, SJ	John Norton (surgeon)	Golden Square, Piccadilly	John Norton (surgeon)	Golden Square, Piccadilly
	6s	1781	LI, ABG, SWJ	John Norton (surgeon)	Golden Square, Piccadilly	John Norton (surgeon)	Golden Square, Piccadilly
Oxley's Essence of Jamaica Ginger	2s 9d	1807	LI, LM	Samuel Oxley	21 Tavistock St.	Samuel Oxley	21 Tavistock St.
	2s 9d	1822	SWJ	Mrs. S. Oxley	Pontefract	Mrs. S. Oxley	Pontefract
Paregorick Lozenges	1s 1½d	1794	ABG, SWJ	H. Steers	Old Bond St.	H. Steers	Old Bond St.
						F. Newbery	45 St. Paul's Churchyard
Perry's Essence	1s 1½d	1822	LI, LM, SWJ	–		Butlers (chemists)	4 Cheapside
Pike's Grand Antidote/ Ointment	1s 6d	1769	ABG	Ann Pike	–	Cluer Dicey & Co.	10 Bow Churchyard
	1s 6d	1781	SWJ	Cluer Dicey	10 Bow Churchyard	Cluer Dicey	10 Bow Churchyard
	1s 9d	1807	LI	Dicey & Co.	10 Bow Churchyard	Dicey & Co.	10 Bow Churchyard
Pullin's Antiscorbutic Pills	2s 6d	1781	LM, ABG	Swinfen & Sons (surgeons and apothecaries)	Hinckley	Mrs. Newbery	St. Paul's Churchyard
	2s 9d	1794	ABG	Edmund Swinfen (surgeon and apothecary)	Leicester	Edmund Swinfen	Leicester
						W. Bacon	Oxford St.
						John Wye	59 Coleman St.
						Tutt	Royal Exchange
						Mrs. Newbery	St. Paul's Churchyard
Pullin's Female Pills	1s	1781	LM, ABG	Swinfen & Sons (surgeons and apothecaries)	Hinckley	Mrs. Newbery	St. Paul's Churchyard
	1s 1½d	1794	ABG	Edmund Swinfen (surgeon and apothecary)	Leicester	Edmund Swinfen	Leicester
						W. Bacon	Oxford St.
						John Wye	59 Coleman St.
						Tutt	Royal Exchange
						Mrs. Newbery	St. Paul's Churchyard

(Continued)

(Continued)

Name	*Price*	*Year*	*Newspapers*	*Owner*	*Address*	*Wholesaler*	*Address*
Pullin's Purging Pills	1s	1781	LM, ABG	Swinfen & Sons (surgeons and apothecaries)	Hinckley	Mrs. Newbery	St. Paul's Churchyard
	1s 1½d	1794	ABG	Edmund Swinfen (surgeon and apothecary)	Leicester	Edmund Swinfen	Leicester
						W. Bacon	Oxford St.
						John Wye	59 Coleman St.
						Tutt	Royal Exchange
						Mrs. Newbery	St. Paul's Churchyard
(Dr.) Radcliffe's Elixir	–	1769	LI, ABG	Cluer Dicey & Co.	10 Bow Churchyard	Cluer Dicey & Co.	10 Bow Churchyard
	–	1781	SWJ	Cluer Dicey	10 Bow Churchyard	Cluer Dicey	10 Bow Churchyard
	1s 1½d	1794	SWJ	Dicey & Co.	10 Bow Churchyard	Dicey & Co.	10 Bow Churchyard
	1s 1½d	1807	LI	Dicey & Co.	10 Bow Churchyard	Dicey & Co.	10 Bow Churchyard
	1s 1½d	1822	LI, ABG, SWJ	Sutton & Co.	10 Bow Churchyard	Sutton & Co.	10 Bow Churchyard
Ramsay's Anti-odontalgia	1s 1½d	1807	LI, LM	G. Ramsay (apothecary)	Penrith	Barclay & Son	95 Fleet Market
	1s 1½d	1822	LI, LM	G. Ramsay (apothecary)	Penrith	Barclay & Sons	95 Fleet Market
Ramsay's Caledonian Cream	1s 1½d	1807	LI, LM	G. Ramsay (apothecary)	Penrith	Barclay & Son	95 Fleet Market
	1s 1½d	1822	LI, LM	G. Ramsay (apothecary)	Penrith	Barclay & Sons	95 Fleet Market
Ramsay's Cumberland Bituminous Fluid & Pills	2s 9d	1807	LM	G. Ramsay (apothecary)	Penrith	Barclay & Son	95 Fleet Market
	2s 9d	1822	LI, LM, SWJ	G. Ramsay (apothecary)	Penrith	Barclay & Sons	95 Fleet Market
Ramsay's Medicated Spice Nuts	9d	1807	LM	G. Ramsay (apothecary)	Penrith	Barclay & Son	95 Fleet Market
	9d	1822	LI, LM, SWJ	G. Ramsay (apothecary)	Penrith	Barclay & Sons	95 Fleet Market
Ramsay's Pectoral Balsam	1s 1½d	1807	LI, LM	G. Ramsay (apothecary)	Penrith	Barclay & Son	95 Fleet Market
	2s 9d	1822	LI, LM	G. Ramsay (apothecary)	Penrith	Barclay & Sons	95 Fleet Market
Remedy for Deafness	11s	1822	LI, LM, SWJ	–		The Hermitage	21 Edgeware Road
Restorative Salo Pills	22s	1794	LI, SWJ	Mr. & Mrs. White	St. Paul's Churchyard	Mr. & Mrs. White	St. Paul's Churchyard
Robberd's Balsamic Elixir	2s 9d	1807	LI, LM, ABG, SWJ	–		Butler	4 Cheapside
Robert's Medicated Vegetable Water	5s 5d	1794	ABG, SWJ	Daniel Roberts	Painswick, Gloucestershire	Daniel Roberts	Painswick, Gloucestershire
Rowland's Alsana Extract	2s 9d	1822	LM, SWJ	A. Rowland & Son	Kirkby St., Hatton Garden	A. Rowland & Son	Kirkby St., Hatton Garden
Rowley's British Herb Snuff & Tobacco	–	1769	ABG, SJ	–		James Rowley	St. Paul's Coffee House
	6d	1781	LI, LM, ABG, SWJ	–		–	London Coffee House, Ludgate Hill

(Continued)

(Continued)

Name	*Price*	*Year*	*Newspapers*	*Owner*	*Address*	*Wholesaler*	*Address*
Ruspini's Tincture	–	1781	ABG, SWJ	Ruspini (surgeon/ dentist)	Pall Mall	Ruspini	Pall Mall
						Ball (perfumer)	78 New Bond St.
Samaritan Water	2s 6d	1781	ABG, SWJ	W. Bayley (perfumer)	Cockspur St.	F. Newbery	45 St. Paul's Chyd
						Dicey & Co.	10 Bow Churchyard
(Dr.) Sibly's Lunar Tincture	10s 6d	1807	SWJ	Dr. Sibly	40 New Bridge St.	Dr. Sibly	40 New Bridge St.
	10s 6d	1807	LM	C. W. Saffell	–	'Doctor's House'	40 New Bridge St.
	4s 6d	1822	LI, LM, SWJ	J. R. Saffell	35 Gloucester St.	Barclay & Sons	95 Fleet Market
(Dr.) Sibly's Solar Tincture	7s 6d	1794	LM	Dr. Sibly	–	Dr. Sibly	–
	7s 6d	1807	SWJ	Dr. Sibly	40 New Bridge St.	Dr. Sibly	40 New Bridge St.
	6s	1807	LM	C. W. Saffell	–	'Doctor's House'	40 New Bridge St.
	6s	1822	LI, LM, SWJ	J. R. Saffell	35 Gloucester St.	Barclay & Sons	95 Fleet Market
Simson's Aethereal Tincture	1s 1½d	1794	LI, LM, ABG	John Wye	59 Coleman St.	John Wye	59 Coleman St.
Smith's Pectoral Stomachic Tincture	2s	1781	LM, ABG	–		–	
(Dr.) Smith's Restorative Nervous Drops	11s	1822	LI, SWJ	–		The Hermitage	21 Edgeware Road
(Dr.) Smyth's Nature's Restorative	10s 6d	1781	LI, ABG, SWJ	Dr. Smyth	14 Suffolk St., Charing Cross	Dr. Smyth	14 Suffolk St., Charing Cross
(Dr.) Smyth's Specific Drops	2s 6d	1781	LI, ABG, SWJ	Dr. Smyth	14 Suffolk St., Charing Cross	Dr. Smyth	14 Suffolk St., Charing Cross
	2s 8d	1794	ABG	Dr. Smyth	Great Suffolk St., Charing Cross	Dr. Smyth	Great Suffolk St.
						W. Bacon	150 Oxford St.
						Fridden	100 Fleet St.
						Axtel	1 Finch Lane
						Cawill	Holborn
(Dr.) Solomon's Abstergent Lotion	4s 6d	1807	LI, LM	Dr. Solomon	Liverpool	–	
	2s 9d	1822	LI, ABG	Samuel Solomon	Liverpool	–	
(Dr.) Solomon's Cordial Balm of Gilead	10s 6d	1807	LM, ABG, SWJ	Dr. Solomon	Liverpool	–	
	10s 6d	1807	LI	Dr. Solomon	Liverpool	Dicey & Co.	10 Bow Churchyard
	11s	1822	LI, LM, ABG, SWJ	Samuel Solomon	Liverpool	–	
(Dr.) Solomon's Drops (Anti-Impetigines)	10s 6d	1807	LI, LM, SWJ	Dr. Solomon	Gilead House, Liverpool	–	
	11s	1822	LI, LM, SWJ	Samuel Solomon	Liverpool	–	

(Continued)

(Continued)

Name	*Price*	*Year*	*Newspapers*	*Owner*	*Address*	*Wholesaler*	*Address*
Spilsbury's Antiscorbutic Drops	4s	1781	LI, LM, ABG, SWJ	Francis Spilsbury	Mount Row, Westminster Bridge	Francis Spilsbury	Mount Row, Westminster Bridge
	5s	1794	LI, ABG, SWJ	Mrs. Spilsbury	Soho Square	Mrs. Spilsbury	Soho Square
	5s 6d	1807	LI, LM, ABG, SWJ	–	15 Soho Square	The Dispensary	15 Soho Square
	–	1822	SWJ	–		–	
Squire's Grand Elixir	–	1822	ABG, SWJ	Sutton & Co.	10 Bow Churchyard	Sutton & Co.	10 Bow Churchyard
(Dr.) Steers's Calomile Drops	1s 1½d	1794	ABG, SWJ	H. Steers	Old Bond St.	H. Steers F. Newbery	Old Bond St. 45 St. Paul's Churchyard
(Dr.) Steers's Opodeldoc	1s 6d	1781	LI, ABG, SWJ	H. Steers	9 below Northumberland House, Charing Cross	H. Steers F. Newbery	9 below Northumberland House, Charing Cross 45 St. Paul's Churchyard
	2s	1794	LI, ABG, SWJ	H. Steers	10 Old Bond St.	H. Steers F. Newbery	10 Old Bond St. 45 St. Paul's Churchyard
	2s 6d	1807	LI, ABG, SWJ	F. Newbery & Sons	45 St. Paul's Churchyard	F. Newbery & Sons	45 St. Paul's Churchyard
	2s 6d	1822	LI, ABG, SWJ	F. Newbery & Sons	45 St. Paul's Churchyard	F. Newbery & Sons	45 St. Paul's Churchyard
Swinfen's Electuary	2s 6d	1781	LM, ABG	R. Swinfen (surgeon)	Hinckley	Mrs. Newbery	St. Paul's Churchyard
	1s 9d	1794	ABG	Edmund Swinfen (surgeon/ apothecary)	Leicester	W. Bacon John Wye Tutt Mrs. Newbery	150 Oxford St. 59 Coleman St. Royal Exchange St. Paul's Churchyard
(Dr.) Sydenham's Family Pills of Health	1s 1½d	1822	LI, LM, SWJ	–		Butlers (chemists)	4 Cheapside
Tasteless Ague & Fever Drops	–	1781	LI, ABG, SWJ	–		–	
(Dr.) Taylor's Remedy for Deafness	8s 6d	1807	LI	Mrs. Matthews (bookseller)	Strand	Dicey & Sutton Shaw & Edwards Barclay & Son Butler	10 Bow Churchyard 66 St. Paul's Churchyard 95 Fleet Market 4 Cheapside
	8s 6d	1822	LM, SWJ	–		Barclay & Sons	95 Fleet Market
Thomas's Tolu Essence	2s 9d	1794	LI, LM, ABG	–		John Wye	59 Coleman St.

(Continued)

(Continued)

Name	*Price*	*Year*	*Newspapers*	*Owner*	*Address*	*Wholesaler*	*Address*
Towers's Stomachic Essence	4s 6d	1822	LM, SWJ	John Towers	–	'All the wholesale medicine venders in London'	
Towers's Tonic Pills	2s 9d	1822	LM, ABG, SWJ	John Towers	–	'All the wholesale medicine venders in London'	
Trinder's Specific for Piles	4s 6d	1822	LI, SWJ	J. F. Trinder	–	–	
Tumor Plaister	–	1807	LI, LM	–		Barclay & Son	95 Fleet Market
Turlington's Balsam of Life	1s 9d	1769	LI, SJ	Thomas Jackson & Son	95 Fleet Market	Thomas Jackson & Son	95 Fleet Market
	1s 9d	1781	LI, ABG	Hilton Wray	14 Birchin Lane	Martha & Hilton Wray	14 Birchin Lane
	1s 10d	1794	ABG	H. Wray & Co.	14 Birchin Lane	H. Wray & Co	14 Birchin Lane
Tyce's Ointment for the Itch	1s 9d	1794	ABG	Mr. Tyce	–	–	
	1s 9d	1807	ABG	–		–	
	1s 9d	1822	ABG	C. W. Turner	20 Hatton Gardens	C. W. Turner	20 Hatton Gardens
Universal Cerate	1s	1781	LI, LM, ABG, SWJ	N. Falck MD	47 Jewin St.	N. Falck MD	47 Jewin St.
Vegetable Tooth Powder	2s 9d	1822	LI, LM, SWJ	–		Butlers (chemists)	4 Cheapside
Velno's Vegetable Syrup	10s 6d	1781	ABG	–		Dr. Mercier	Frith St., Soho
	10s 6d	1781	SWJ	W. Hodson	Bridgewater Square	W. Hodson	Bridgewater Square
(Mrs.) Vincent's Gowland's Lotion	2s 9d	1807	SWJ	Mrs. Vincent	4 Davies St., Grosvenor Square	Bacon & Co.	150 Oxford St.
	2s 9d	1822	SWJ	M. E. Vincent	–	–	
Waite's Worm Medicine	1s 1½d	1794	LI, LM, ABG, SWJ	W. Howard	Reading	W. Howard	Reading
				J. Evans	41 Long Lane, West Smithfield	J. Evans	41 Long Lane, West Smithfield
(Dr.) Walker's Jesuits' Drops	2s 6d	1769	LI, ABG, SJ	J. Wessels & Co.	45 Old Bailey	J. Wessels & Co	45 Old Bailey
	2s 6d	1781	LI	Temple Ravenscroft	London	Temple Ravenscroft	London
	2s 6d	1781	ABG	J. Wright	45 Old Bailey	J. Wright	45 Old Bailey
	2s 6d	1781	SWJ	J. Wessels	45 Fleet St.	J. Wessels	45 Fleet St.
	2s 9d	1807	LI, LM	Shaw & Edwards	66 St. Paul's Churchyard	Shaw & Edwards	66 St. Paul's Churchyard

(Continued)

(Continued)

Name	*Price*	*Year*	*Newspapers*	*Owner*	*Address*	*Wholesaler*	*Address*
(Dr.) Walker's Specific Purging Remedy	2s 6d	1769	LI, ABG, SJ	J. Wessels & Co.	45 Old Bailey	J. Wessels & Co.	45 Old Bailey
	–	1781	LI	Temple Ravenscroft	London	Temple Ravenscroft	London
	–	1781	ABG	J. Wright	45 Old Bailey	J. Wright	45 Old Bailey
	2s 6d	1781	SWJ	J. Wessels	45 Fleet St.	J. Wessels	45 Fleet St.
Whitehead's Essence of Mustard	2s 9d	1807	LI, LM, ABG, SWJ	R. Johnson (apothecary)	15 Greek St., Soho	R. Johnson (apothecary)	15 Greek St., Soho
	2s 9d	1822	SWJ	R. Johnson (apothecary)	15 Greek St., Soho	R. Johnson (apothecary)	15 Greek St., Soho
Whitehead's Family Cerate	1s 1½d	1807	LI, LM, ABG, SWJ	R. Johnson (apothecary)	15 Greek St., Soho	R. Johnson (apothecary)	15 Greek St., Soho
	1s 1½d	1822	SWJ	R. Johnson (apothecary)	15 Greek St., Soho	R. Johnson (apothecary)	15 Greek St., Soho
Wood's Laxative Pills	1s 1½d	1822	LI, LM, SWJ	James Wood	32 High St., Bristol	Shaw & Edwards	66 St. Paul's Churchyard

A.2 Other Patent Medicines in the Studied Newspapers

Inclusion criterion: Patent medicines other than 'national medicines' advertised in a studied newspaper.

Addresses were in London unless specified otherwise.

Name	Price	Year	News-paper	Owner	Address	Wholesaler	Address
Alternative Powder	–	1807	LM	–		–	
Amboyna tooth powder	2s 6d	1807	ABG	–		Shaw & Edwards	66 St. Paul's Churchyard
Anderson's Scots pills	1s	1781	ABG	–		–	London Coffee House
Anodyne Linament	2(?)s 6d	1781	ABG	Ward (surgeon)	Henley-in-Arden	Pearson & Rollason	High Street, Birmingham
Antwerp Medicine	£5	1822	SWJ	Francis Mapleton & Co.	Aberystwyth	–	'forwarded from London'
Arnold's Pectoral Balsam of Coltsfoot	1s 9d	1822	ABG	J. & J. Arnold	59 Barbican	J. & J. Arnold	59 Barbican
Atkinson & Barker's Infant Preservative	1s 1½d	1822	LM	Atkinson & Barker	1 Market Place Manchester	Barclay & Sons	95 Fleet Market
Balm of Quito	10s 6d	1807	LM	–		W. Withers	229 Strand
Balsam of Liquorice	2s 9d	1807	LI	J. Pidding (surgeon)	–	Barclay & Son	95 Fleet Market
Bannister's Chilblain Lotion	1s 1½d	1822	SWJ	J. Bannister	Rowde, nr Devizes	J. Bannister	Rowde, nr Devizes
Barclay's Asthmatic Candy	4s 6d	1807	LI	Barclay & Son	95 Fleet Market	Barclay & Son	95 Fleet Market
Beaume de Sante	2s 6d	1781	ABG	–		J. Bew	Paternoster Row
Beddome's Cough Pills	1s 1½d	1822	ABG	Josephus Beddome	Gloucestershire	–	
Beddome's Desideratum for Gout	4s 6d	1822	ABG	Josephus Beddome	Gloucestershire	–	
Bennett's Cough Drops	2s 9d	1807	LI	–		Barclay & Son	95 Fleet Market
Bezoardic Pill	2s 6d	1781	ABG	–		–	
Birthdom's Bilious Pills	1s 1½d	1807	SWJ	–		–	
Blackwood's Cordial Elixir	2s	1769	SJ	–		J. Wilkie (bookseller)	71 St. Paul's Churchyard
(Dr.) Blatz's German Corn Plaister	1s	1781	SWJ	'Proprietor'	10 Birchin Lane	–	10 Birchin Lane
Blenkinsop's Bilious Pills	2s 6d	1807	SWJ	'Proprietor'	29 Queen St., Edgeware Road	–	29 Queen St., Edgeware Road
(Dr.) Boerhaave's Antiscorbutic Pills	–	1781	ABG	–		–	
British Herb Tobacco & Snuff	2s 6d	1769	LI	Rev. J. Jones	–	Evans (goldbeater)	Long Acre
British Ointment for Corns	1s 6d	1794	LM	W. Naylor (colour maker)	Bishopgate within	Naylor & Bailey John Wye	Bishopgate within 59 Coleman St.
Brussels Tincture	2s 9d	1822	SWJ	S. Tozer (chemist)	Bridge Parade, Bristol	S. Tozer (chemist)	Bridge Parade, Bristol

(Continued)

(Continued)

Name	*Price*	*Year*	*News-paper*	*Owner*	*Address*	*Wholesaler*	*Address*
Chalybeate Aperient	–	1822	SWJ	Savory, Moore & Davidson (chemists)	136 New Bond St.	Savory, Moore & Davidson (chemists)	136 New Bond St.
Chalybeate Elixir	5s	1781	SWJ	–		Satchell (milliner)	John St., Berkeley Square
Ching's Prepared Charcoal	2s 6d	1807	ABG	–		Mrs. Crisp Gattie & Lee W. Bacon Pearsall	Spring Gardens Bond St. 150 Oxford St. Holborn
Chymical Drops	1s	1769	ABG	–		–	
Clarke's Florida Balsam	–	1822	SWJ	–		Barclay & Sons	95 Fleet Market
Clarke's Florida Electuary	–	1822	SWJ	–		Barclay & Sons	95 Fleet Market
Clarke's Sudorific Pills	2s 9d	1822	SWJ	–		Barclay & Sons	95 Fleet Market
Concrete Acidulated Kali	–	1822	SWJ	Savory, Moore & Davidson (chemists)	136 New Bond St.	Savory, Moore & Davidson (chemists)	136 New Bond St.
Cornwell's Oriental Cordial	5s 6d	1807	LM	–		Shaw & Edwards	66 St. Paul's Churchyard
Court Sticking Plaister	–	1769	SJ	–		–	
Cuff's Digesting Pills	1s 1½d	1822	SWJ	J. Cuff	15 Milsom Road, Bath	J. Cuff	15 Milsom Road, Bath
Cundell's Balsam of Honey	2s 9d	1807	LI	–		Howard & Evans	42 Long Lane
Cyprian Preventative	10s 6d	1781	ABG	–		Yates & Robinson	Panton St.
De Velno's Vegetable Pills	4s 6d	1807	ABG	C. Staples	–	Shaw & Edwards	66 St. Paul's Churchyard
Deering's Drops	1s 6d	1794	ABG	–		W. Bacon	150 Oxford St.
Dennis's Family Pills	–	1822	LM	W. Dennis (surgeon)	Stillington, York	Barclay & Sons	95 Fleet Market
Devolen's Antiscorbutic Drops	3s 6d	1781	ABG	–		Pearson & Rollason	High St., Birmingham
Edward's Tincture for Ague	–	1769	ABG	Mr. Edwards	–	–	
Elliott's Ceylonian Powder	1s 9d	1807	LM	Elliott (druggist)	Huddersfield	Shaw & Edwards	66 St. Paul's Churchyard
Elliott's Family Cordial	2s 9d	1807	LM	Elliott (druggist)	Huddersfield	Shaw & Edwards	66 St. Paul's Churchyard
Elliott's Restorative Tincture	2s 6d	1807	LM	Elliott (druggist)	Huddersfield	Shaw & Edwards	66 St. Paul's Churchyard
Essence of Peppermint	1s	1769	SJ	J. Juniper (apothecary)	Maxfield St., Soho	J. Juniper (apothecary)	Maxfield St., Soho
Ethereal Anodyne Opodeldoc	–	1822	SWJ	Savory, Moore & Davidson (chemists)	136 New Bond St.	Savory, Moore & Davidson (chemists)	136 New Bond St.

(Continued)

(Continued)

Name	*Price*	*Year*	*News-paper*	*Owner*	*Address*	*Wholesaler*	*Address*
Foot's Cathartic Mixture (version 1)	5s	1822	SWJ	Amelia Foot	Fovant, Salisbury	Amelia Foot	Fovant, Salisbury
Foot's Cathartic Mixture (version 2)	–	1822	SWJ	Mary & Stephen Foot	Donhead St. Mary, Wiltshire	Mary & Stephen Foot	Donhead St. Mary, Wiltshire
Fothergel's Chymical Nervous Drops	10s 6d	1781	ABG	Dr. S. Freeman MD	1 Staple's Inn, Holborn	Dr. S. Freeman MD	1 Staple's Inn, Holborn
Fraunces's Female Elixir	1s 6d	1781	ABG	–		Dicey & Co.	10 Bow Churchyard
Freeman's Antiscorbutic Drops	–	1781	ABG	–		–	
Friar's Balsam	1s	1781	SWJ	–		–	
Friar's Drops	3s	1781	ABG	–		–	
Gapper's Lozenges for Heartburn	–	1794	SWJ	Mr. Gapper (surgeon)	Mere, Wiltshire	Baldwin, Hernon & Langton (druggists)	Giltspur St.
Gardner's Worm Medicine	–	1807	ABG	Mr. Gardner (surgeon)	High St., Birmingham	Mr. Gardner (surgeon)	High St., Birmingham
German Corn Plaister	1s 1½d	1794	SWJ	W. & R. T. Axtell	1 Finch Lane	W. & R. T. Axtell	1 Finch Lane
Godden's Eye Ointment	2s	1822	SWJ	–		Sanger	150 Oxford St.
Godfrey's Cordial	6d	1781	SWJ	–		–	
Grant's Chymical Drops	1s	1781	SWJ	–		–	
Green's Royal Antiscorbutic Drops	5s 6d	1822	SWJ	John Green	Newton Bushel, Devon	John Green	Newton Bushel, Devon
Grimble's Scald & Burn Salve	2s 6d	1822	ABG	–		–	
Hadley's Convulsion Powders	4s 6d	1822	SWJ	'proprietor'	1 Kirkby St., Hatton Gardens	'proprietor'	1 Kirkby St., Hatton Gardens
Hardy's Itch Ointment	1s 1½d	1822	ABG	–		Evans (apothecary)	Birmingham
Hardy's Worm Cakes	1s 1½d	1822	ABG	–		Evans (apothecary)	Birmingham
Hemet's Pearl Dentrifice	2s 9d	1822	SWJ	J. Hemet	London	Bayley & Blew	Cockspur St.
Henry's Aromatic Spirit of Vinegar	–	1822	ABG	T. & W. Henry	Manchester	T. & W. Henry	Manchester
Henry's Uncalcined Magnesia	2s	1781	LI	Thomas Henry (apothecary)	Manchester	J. Johnson	St. Paul's Churchyard
Hoare's Cathartic Family Pills	1s 1½d	1822	SWJ	Hoare (surgeon & apothecary)	Warminster	Hoare (surgeon & apothecary)	Warminster

(Continued)

(Continued)

Name	*Price*	*Year*	*News-paper*	*Owner*	*Address*	*Wholesaler*	*Address*
Hodgson's Antiscorbutic Tincture	2s 6d	1781	SWJ	Robert Hodgson (apothecary)	68 Snow Hill	Robert Hodgson (apothecary)	68 Snow Hill
Hodgson's Pectoral Lozenges	1s	1781	SWJ	Robert Hodgson (apothecary)	68 Snow Hill	Robert Hodgson (apothecary)	68 Snow Hill
Hodgson's Tincture for Teeth & Gums	1s	1781	SWJ	Robert Hodgson (apothecary)	68 Snow Hill	Robert Hodgson (apothecary)	68 Snow Hill
Hodgson's Tincture for Toothache	1s	1781	SWJ	Robert Hodgson (apothecary)	68 Snow Hill	Robert Hodgson (apothecary)	68 Snow Hill
(Dr.) Hodson's Persian Restorative	10s 6d	1794	ABG	Dr. Hodson	29 Hatton Gardens	Dr. Hodson	29 Hatton Gardens
(Dr.) Hooper's Female Pills	1s	1781	SWJ	–		–	
Hooping Cough Medicine	3s	1769	SJ	'gentleman of the faculty'	–	W. Harris (bookseller)	70 St. Paul's Churchyard
Howland's Specific Drops	5s 3d	1769	ABG	Mr. Howland	North Audley St.	–	
Hudson's Botanic Tincture	2s 9d	1822	ABG	–		Atkinson	44 Gerard St., Soho
Hunter's Powder for Wens	11s 6d	1794	LM	Dr. Brodum	9 Albion St., Blackfriars	Dr. Brodum	9 Albion St., Blackfriars
Hunt's Aperient Family Pills	1s 1½d	1822	SWJ	–		–	
Huxham's Tincture of Bark	3s 6d	1781	SWJ	–		–	
Imperial Asthmatic & Consumptive Balsam	5s 3d	1781	ABG	–	20 Birchin Lane, Cornhill	–	
Ipecacuanha Lozenges	–	1822	SWJ	Savory, Moore & Davidson (chemists)	136 New Bond St.	Savory, Moore & Davidson (chemists)	136 New Bond St.
Jackson's Essence of Life	6s	1769	SJ	Jackson & Co.	Fleet Market	Jackson & Co.	Fleet Market
Jackson's Hepatic Pills	1s 1½d	1822	SWJ	S. Jackson (druggist)	Market Place, Romsey	S. Jackson (druggist)	Market Place, Romsey
Jackson's Restorative Electuary	6s	1769	SJ	Jackson & Co.	Fleet Market	Jackson & Co.	Fleet Market
Jackson's Tussis Remedy	1s 1½d	1822	SWJ	S. Jackson (druggist)	Market Place, Romsey	S. Jackson (druggist)	Market Place, Romsey
(Dr.) James's Mild Fever Powder	–	1769	SJ	Newbery & Carnan	65 St. Paul's Churchyard	Newbery & Carnan	65 St. Paul's Churchyard
John Lord's Corn Salve	2s 6d	1769	SJ	–		Warren & Co. (perfumers)	Golden Fleece, Marylebone St.
(Dr.) John Scot's Pills	10s 6d	1794	SWJ	Dr. John Scot	–	W. Bacon	150 Oxford St.

(Continued)

(Continued)

Name	Price	Year	News-paper	Owner	Address	Wholesaler	Address
Kempson's Fever Julep	–	1807	ABG	Scott Kempson (surgeon)	Cleobury Mortimer, Shropshire	Scott Kempson (surgeon)	Cleobury Mortimer, Shropshire
(Dr.) Lamert's Nervous Balsam	4s 6d	1822	SWJ	Dr. Lamert	10 Church St., Spitalfields	Dr. Lamert	10 Church St., Spitalfields
Lancaster Black Drop	2s 9d	1822	SWJ	M. Braithwaite	–	Savory, Moore & Davidson	136 New Bond St.
(Dr.) Lowther's Antiscorbutic Powders	–	1769	SJ	Dr. Lowther	Golden Lamp, Hatton Gardens	Dr. Lowther	Golden Lamp, Hatton Gardens
						Cooke (bookseller)	Paternoster Row
						Paillet	Princes St., Leicester Fields
Lymington Marine Epsom Salts	2s 3d	1822	SWJ	West (salt proprietor)	Lymington	West (salt proprietor)	Lymington
Mann's Approved Remedy	2s 6d	1807	SWJ	Mann	–	–	
Mason's Worm Nuts	1s 1½d	1794	LI	Robert Mason	–	John Scott	417 The Strand
Medicine for bite of a mad animal	–	1794	LI	Hill & J. Berry	–	Hill & J. Berry	–
Medicine for bite of a mad dog (Hey)	–	1769	LM	George Hey	Colne	George Alderson (surgeon)	Ilkley, Yorkshire
Medicine for bite of a mad dog (Hill)	5s 3d	1781	LI	William Hill	Ormskirk	James Berry	London
Medicine for bite of a mad dog (Johnson)	2s	1781	LI	Edward Johnson	Bedale	–	
Medicine for bite of a mad dog (Barton)	3s	1781	LM	Miles Barton (surgeon)	Ormskirk	–	
Mevac's Turkey Rhubarb & Jamaican Ginger	1s 6d	1822	ABG	E. H. Mevac (chemist)	Bull Ring, Birmingham	E. H. Mevac (chemist)	Bull Ring, Birmingham
Minster's Remedy for Coughs & Consumptions	3s	1781	ABG	Thomas Minster (surgeon)	Gloucestershire	–	
Molineux's Smelling Medicine	1s	1769	LI	Molineux	Newcastle (upon-Tyne?)	T. Slack	Newcastle (upon-Tyne?)
Montpellier Pectoral Drops	1s 6d	1781	SWJ	–		Swinney & Evetts	Birmingham
Morris's Cough Drops	1s 1½d	1822	SWJ	Geoffrey Morris (chemist)	High St., Kensington	Geoffrey Morris (chemist)	High St., Kensington
Morris's Golden Antibilious Pills	1s 1½d	1822	SWJ	Geoffrey Morris (chemist)	High St., Kensington	Geoffrey Morris (chemist)	High St., Kensington

(Continued)

(Continued)

Name	Price	Year	News-paper	Owner	Address	Wholesaler	Address
Naylor's British Ointment for Corns	1s 6d	1794	ABG	William Naylor	London	John Wye	59 Coleman St.
						F. Newbery	45 St. Paul's Churchyard
Nendick's Popular Pills	1s 6d	1781	ABG	–		W. Bailey (perfumer)	Cockspur St.
						F. Newbery	45 St. Paul's Churchyard
						Dicey & Co.	10 Bow Churchyard
Orange Peas for Issues	1s	1781	SWJ	–		–	
Oxford Eye Water	1s 1½d	1807	SWJ	–		–	
Parker's Rheumatic Drops	1s 6d	1781	ABG	–		–	
Patirosa Lozenges	2s 9d	1794	SWJ	–		J. Fuller	8 South side of Covent Garden
Patton's Pills	10s 6d	1781	SWJ	–		–	
Pilulae Antiscrophulae	1s 1½d	1822	SWJ	Dr. Roberts	Bridport	Sutton & Co.	10 Bow Churchyard
						Barclay & Sons	95 Fleet Market
						Butlers	4 Cheapside
						F. Newbery & Sons	45 St. Paul's Churchyard
						Shaw & Edwards	66 St. Paul's Churchyard
Poor Man's Friend	1s 1½d	1822	SWJ	Dr. Roberts	Bridport	Sutton & Co.	10 Bow Churchyard
						Barclay & Sons	95 Fleet Market
						Butlers	4 Cheapside
						F. Newbery & Sons	45 St. Paul's Churchyard
						Shaw & Edwards	66 St. Paul's Churchyard
Powell's Cough Electuary	9d	1822	LM	Powell & Co.	Loughborough	Barclays & Sons	95 Fleet Market
(Dr.) Rambin's Quintessence for Stone & Gravel	5s 3d	1769	ABG	Dr. Rambin	King St., Golden Square	Dr. Rambin	King St., Golden Square
Randall's Domestic Medicine	1s 1½d	1822	SWJ	J. M. Randall (apothecary)	Poole, Dorset	J. M. Randall (apothecary)	Poole, Dorset
Randall's Elixir	2s 9d	1822	SWJ	J. M. Randall (apothecary)	Poole, Dorset	J. M. Randall (apothecary)	Poole, Dorset
Ranibin's Quiet Pills	5s 3d	1781	ABG	–		–	London Coffee House
Renlivoglio's Dentrifical Tincture	2s	1781	ABG	'imported by London physician'		Pearson & Rollason	High St., Birmingham
Reynolds's Specific	4s 6d	1822	SWJ	S. Reynolds	Enfield, Middlesex	S. Reynolds	Enfield, Middlesex
Rooke's Matchless Balsam	2s 9d	1794	LI, LM	Mr. Rooke	Kirkby Lonsdale	–	

(Continued)

(Continued)

Name	*Price*	*Year*	*News-paper*	*Owner*	*Address*	*Wholesaler*	*Address*
Royal Cinnamon Drops	3s 6d	1781	ABG	Hilton Wray	14 Birchin Lane	Martha & Hilton Wray	14 Birchin Lane
Royal Stomachic Cordial Bitters	3s 6d	1781	ABG	–		–	16 Carey St.
Ruspini's Styptic	7s 6d	1794	SWJ	Chevalier Ruspini	–	Swinney	20 Pall Mall
Rymer's Cardiac and Nervous Tincture	2s 9d	1794	ABG	James Rymer (surgeon & apothecary)	36 Gerard St., Soho	–	
(Dr.) Sandwell's Issue Plaisters	1s	1781	SWJ	–		–	
(Dr.) Saul's Golden Eye Drops	–	1807	ABG	Dr. Saul	Rathbone Place	Dr. Saul	Rathbone Place
(Dr.) Saul's Pills	–	1807	ABG	Dr. Saul	Rathbone Place	Dr. Saul	Rathbone Place
Seidlitz Powders	–	1822	SWJ	Savory, Moore & Davidson	136 New Bond St.	Savory, Moore & Davidson	136 New Bond St.
Sern's Balsamic Aether	–	1781	ABG	N. D. Falck MD	47 Jewin St.	–	
Skene's Attenuating Tincture	3s 6d	1781	SWJ	J. Skene (surgeon)	Cursitor St.	J. Skene (surgeon)	Cursitor St.
Smart's Tincture of Cascarilla Bark	2s 3d	1794	LM	W. Smart	87 Fleet St.	W. Smart John Wye	87 Fleet St. 59 Coleman St.
(Dr.) Smellome's Eye Salve	2s 6d	1781	SWJ	–		–	
(Dr.) Smith's Tincture	11s	1822	LM	–		The Hermitage	21 Edgeware Road
Snook's Aperient Family Pills	1s 1½d	1822	SWJ	J. Snook (chemist)	Bridgewater, Somerset	J. Snook (chemist)	Bridgewater, Somerset
Speediman's Stomach Pills	1s 6d	1781	ABG	Mr. Speediman	186 The Strand	Mr. Speediman	186 The Strand
Stearne's Aether	6s	1781	SWJ	–		–	
Stoughton's Drops	1s	1781	SWJ	–		–	
Strickland's Antiseptic Toothpaste	1s 1½d	1822	ABG	E. R. Strickland (chemist)	Coventry	E. R. Strickland (chemist)	Coventry
Strickland's Balsam of Liquorice	1s 1½d	1822	ABG	E. R. Strickland (chemist)	Coventry	E. R. Strickland (chemist)	Coventry
Strickland's Antibilious Pills	1s 1½d	1822	ABG	E. R. Strickland (chemist)	Coventry	E. R. Strickland (chemist)	Coventry
Sumatra Tooth Powder	2s 9d	1807	ABG	Howard & Evans	42 Long Lane	Howard & Evans	42 Long Lane
Swinfen's Antiacid Lozenges	1s 1½d	1794	ABG	Edmund Swinfen (surgeon & apothecary)	Leicester	W. Bacon John Wye Tutt Mrs. Newbery	150 Oxford St. 59 Coleman St. Royal Exchange St. Paul's Churchyard
Swinfen's Hydropic Elixir	5s	1781	LM	R. Swinfen (surgeon)	Hinckley	–	

(Continued)

(Continued)

Name	*Price*	*Year*	*News-paper*	*Owner*	*Address*	*Wholesaler*	*Address*
Swinfen's Jesuits' Drops	2s 6d	1781	LM	R. Swinfen (surgeon)	Hinckley	–	
Swinfen's Ointment	1s 1½d	1794	ABG	Edmund Swinfen (surgeon & apothecary)	Leicester	W Bacon John Wye Tutt Mrs. Newbery	150 Oxford St. 59 Coleman St. Royal Exchange St. Paul's Churchyard
Swinfen's Worm Cakes	1s 1½d	1794	ABG	Edmund Swinfen (surgeon & apothecary)	Leicester	W. Bacon John Wye Tutt Mrs. Newbery	150 Oxford St. 59 Coleman St. Royal Exchange St. Paul's Churchyard
Switzerland Arquebusade Water	3s	1781	SWJ	–		–	
Taylor's Essence of Jamaican Ginger	2s 9d	1807	LM	Taylor (chemist)	York	Taylor (chemist) R. Johnson	York 15 Greek St., Soho
Thompson's Restorative Toothpaste	–	1807	SWJ	Mrs. Thompson	73 High St., Southampton	Cobb (chemist)	2 Haymarket
Trowbridge Pills	1s	1781	SWJ	Mrs. Wynne	Trowbridge	–	
Turner's Aether	2s	1781	SWJ	–		–	
Tyce's Wash	2s 9d	1807	ABG	–		–	
Venable's Indian Syrup	4s 6d	1822	LM	Hugh Venables	Greenwich	Butlers (chemists)	4 Cheapside
Volatile Vegetable Essence	2s	1781	ABG	–		–	London Coffee House
(Mrs.) Walter's Recipe for Pulmonary Complaints	–	1822	SWJ	Miss Martha Hall	Oldbury on the Hill, Tetbury, Gloucestershire	Miss Martha Hall	Oldbury on the Hill, Tetbury, Gloucestershire
Ward's Celebrated Venereal Specific	2s 6d	1781	ABG	Mr. Ward (surgeon)	Henley-in-Arden, Warwickshire	Pearson & Rollason	High St., Birmingham
Ward's Dropsy Powders	–	1769	SJ	John Fielding & Robert Dingley	–	John Fielding & Robert Dingley	–
Ward's Essence	–	1769	SJ	John Fielding & Robert Dingley	–	John Fielding & Robert Dingley	–
Ward's Liquid Sweat	–	1769	SJ	John Fielding & Robert Dingley	–	John Fielding & Robert Dingley	–
Ward's Paste	–	1769	SJ	John Fielding & Robert Dingley	–	John Fielding & Robert Dingley	–
Ward's Pill and Sweating Powders	–	1769	SJ	John Fielding & Robert Dingley	–	John Fielding & Robert Dingley	–

(Continued)

(Continued)

Name	*Price*	*Year*	*News-paper*	*Owner*	*Address*	*Wholesaler*	*Address*
Ward's Sack or Emetic Drops	–	1769	SJ	John Fielding & Robert Dingley	–	John Fielding & Robert Dingley	–
Ward's White Drops	–	1769	SJ	John Fielding & Robert Dingley	–	John Fielding & Robert Dingley	–
Wash for Itch	1s	1769	ABG	–		–	
Wastell's Mercury Preparation	7s 6d	1781	ABG	Henry Wastell (surgeon)	–	–	
(Dr.) Waugh's Cuckoo Plaister	1s	1794	LI	Dr. Waugh	–	–	Worm Shop, Knaresborough
Welch's Female Pills	2s 9d	1807	LM	C. Kearsley	Fleet St.	C. Kearsley	Fleet St.
						Dicey & Co.	10 Bow Churchyard
						Bacon & Co.	150 Oxford St.
						F. Newbery & Sons	45 St. Paul's Churchyard
Wheatley's Remedies for Itch	1s 9d	1794	LM	–		John Wye	59 Coleman St.
William's Aperient Pills	1s 6d	1794	SWJ	T. Williams (apothecary)	–	W. Bacon	150 Oxford St.
Wilson's Anderson's Pills	1s	1769	ABG	Thomas Wilson	Slaney St., Birmingham	Thomas Wilson	Slaney St., Birmingham
						W. Jemeson & Co. (linen drapers)	Oswestry
Wilson's Antiscorbutic Drops	5s 6d	1807	ABG	Thomas Wilson	9 Worcester St., Birmingham	Thomas Wilson	9 Worcester St., Birmingham
Wilson's Cure for Jaundice	2s 6d	1769	ABG	Thomas Wilson	Slaney St., Birmingham	Thomas Wilson	Slaney St., Birmingham
						W. Jemeson & Co. (linen drapers)	Oswestry
Wilson's Imperial Antiscorbutic Drops	4s	1794	ABG	Thomas Wilson	32 Edgbaston Road, Birmingham	Thomas Wilson	32 Edgbaston Road, Birmingham
Wilson's Specific Pills for Venereal Disease	2s 9d	1807	ABG	Thomas Wilson	9 Worcester St., Birmingham	Thomas Wilson	9 Worcester St., Birmingham
Wilson's Sugar (Worm) Cakes	1s 1½d	1794	ABG	Thomas Wilson	32 Edgbaston Road, Birmingham	Thomas Wilson	32 Edgbaston Road, Birmingham
	1s 1½d	1807	ABG	Thomas Wilson	9 Worcester St., Birmingham	Thomas Wilson	9 Worcester St., Birmingham
Wood's Corn Dissolvent & Specific for Cancers	–	1822	SWJ	Mr. & Mrs. Wood	111 Queen St., Portsea	Mr. & Mrs. Wood	111 Queen St., Portsea
Woodward's Worm Powder	–	1781	ABG	Andrew Ledbroke (druggist)	Leicester	Andrew Ledbroke	Leicester
						Mary Woodward	?

(Continued)

(Continued)

Name	*Price*	*Year*	*News-paper*	*Owner*	*Address*	*Wholesaler*	*Address*
(Mrs.) Wyles's Medicine for Toothache	7½d	1794	SWJ	Mrs. Wyles	Chelmsford	Riley	33 Ludgate St.
Wyman's Antibilious Pills	2s 9d	1807	ABG	W. Wyman (surgeon)	Kettering	Dicey & Sutton	10 Bow Churchyard
Yooll's Anderson's Scotch Pills	1s	1769	ABG	James Yooll (merchant)	Newcastle upon Tyne	James Yooll	Newcastle upon Tyne
						Parker	Birchin Lane

A.3 Additional Patent Medicines

Inclusion criteria: Patent medicines intended for sale in Georgian England which are mentioned in this book, but not in advertisements in the studied newspapers. Addresses were in London unless specified otherwise.

Name	*Price*	*Owner*	*Address*	*Wholesaler*	*Address*
Coghlan's Medicated Snuff	1s	James Coghlan	37 Duke St., Grosvenor Square	James Coghlan	37 Duke St., Grosvenor Square
Dodd's Rheumatic Tincture	–	John Dodd	London	W. Harris	70 St. Paul's Churchyard
Dover's Powder	–	Multiple			
Edinburgh Febrifuge Powder	3s (4 powders)	Edward Galliard	Edinburgh	John Murray	32 Fleet St.
(Dr.) Jebb's Antibilious Pills	–	–		Thomas Collicott	Oxford St.
Jenner's Tartar Emetic	–	Edward Jenner	Gloucestershire	–	
Jesuits' Balsamic Cordial	1s	James Coghlan	37 Duke St., Grosvenor Square	James Coghlan	37 Duke St., Grosvenor Square
Jesuits' Nervous Pills	1s (10 pills)	James Coghlan	37 Duke St., Grosvenor Square	James Coghlan	37 Duke St., Grosvenor Square
(Dr.) Johnson's Yellow Ointment	2s	William Singleton	Lambeth Butts	William Singleton	Lambeth Butts
Jones's Tincture of Peruvian Bark	3s 6d	William Jones	24 Gt Russell St.	William Jones	24 Gt Russell St.
Medicine for the Bite of a Mad Dog	1s or more	Elizabeth Shackleton	Colne, Lancashire	Elizabeth Shackleton	Colne, Lancashire
Mucilage of Marshmallows	7s	Thomas Curtis	21 Tavistock St. Covent Garden	Thomas Curtis	21 Tavistock St. Covent Garden
(Dr.) Priestley's Antibilious Powders	21s (21 powders)	Dr. Robert Priestley	Kirkgate, Leeds	Dr. Robert Priestley	Kirkgate, Leeds
St. Ignatius, or Jesuits' Bean	5s	James Coghlan	37 Duke St., Grosvenor Square	James Coghlan	37 Duke St., Grosvenor Square
Sulphurated Laxative Pill	2s 8d	James Coghlan	37 Duke St., Grosvenor Square	James Coghlan	37 Duke St., Grosvenor Square

Appendix B: Patentees of Medicines Up to 1830

Derived from Bennet Woodcroft, ed., *Titles of Patents of Invention, Chronologically Arranged* (London: Patent Office, 1854), and Bennet Woodcroft, ed., *Abridgements of Specifications Relating to Medicine, Surgery and Dentistry, 1620–1866* (London: Commissioners of Patents for Inventions, 1872).

All 109 patentees of the 118 medicines patented in England up to 1830 are described, including five patents with no enrolled specification which were probably not legally valid (marked as 'no specification') and Samuel Hannay's patent application which the Lord Chancellor refused to let pass the Great Seal, the final stage of granting a patent, on the grounds of public decency. Four medicines had more than one patentee (marked as 'joint').

The description of the medicine in the patent may differ from the description in advertisements. Declared occupations are as described and abbreviated in *Titles*. The names of saints are the parish: a precise address was unusual.

A. Mackintosh, *The Patent Medicines Industry in Georgian England*, Medicine and Biomedical Sciences in Modern History,
https://doi.org/10.1007/978-3-319-69778-9

Surname	First Name	Year	Declared address	Declared occupation	Abbreviated Description of Medicine
Appleby	Thomas	1768	Knutsford, Cheshire	surgeon	balsam for sand & gravel
Astley	Joseph	1807	Borrowstounness, Scotland	chemist	improved sal. ammoniac
Bacon	John	1779	Covent Garden, Westminster	chymist	medicine for fevers & consumptions
Baker	Walter	1748	–	chymist	Schwanberg's liquid shell
Barclay	William	1802	Manchester Buildings, St. Margaret's Westminster	clerk	Revd Mr. Barclay's antibilious deobstruent pills
Barton	Joseph	1799	Old St., St. Luke, Middlesex.	chymist	aereated, preventative fluid & balsam
Beckett	Thomas	1767	City of London	merchant	Beaume de Vie (joint)
Beer	William	1802	Ely Place, City of London	medical professor and dealer in medicine	Dr. Beer's reanimating vital fluid
Betton	Michael	1742	Wellington, Salop.	gent.	oil for rheumatism & scorbutic complaints (joint)
Betton	Thomas	1742	Shrewsbury, Salop	gent.	oil for rheumatism & scorbutic complaints (joint)
Brandon	Richard (elder)	1805	Lucas St., St. Mary, Rotherhithe, Surrey	–	Brandon's British constitutional pills
Brodum	William	1799	Parish of Christchurch, Surrey	doctor of physick	Dr. Brodum's botanical syrup
		1799	Parish of Christchurch, Surrey	doctor of physick	Dr. Brodum's nervous cordial
Browne	Henry	1799	Derby	chemist	extract of zinc
Burrows	John	1772	St. James, Westminster	doctor in physick	Velno's vegetable syrup
Byfield	Timothy	1711	–	doctor in physick	sal. oleosum volatile
Calvert	Edward	1760	Norton Falgate, Middlesex	druggist	violet cordial
Cerreti	Nicholas	1744	St. Martin within, Ludgate, London	dealer in medicine	Greek water for venereal disease
Chase	Samuel	1772	Luton, Bedfordshire	surgeon	antiscorbutic electuary
		1786	Luton, Bedfordshire	surgeon & apothecary	stomach drops
Ching	John	1796	Launceston, Cornwall	chemist & apothecary	worm medicine, two kinds of lozenges
Ching	Rebecca	1808	Rush Common, St. Mary, Lambeth	widow of John Ching patentee	improved Ching's worm destroying lozenges
Collett	Joseph	1744	St. Clement Danes, Middlesex	dealer in medicine	elixir for dropsy, jaundice, stone & gravel
		1752	–	practitioner in physick	British balsam of health (joint)
		1754	City of London	practitioner in physick	ladies' nervous & cordial drops
		1758	St. Martin in the Fields, Middlesex	–	wine & powder for gout
Collins	Benjamin	1773	Salisbury, Wiltshire	–	cephalic snuff
Conwell	William	1822	Ratcliffe Highway, St. George in the East, Middlesex	surgeon	improved purgative vegetable oil
Cornwell	Brian	1783	St. Dunstan, London	gent.	vegetable cordial
Derbishire	Philip	1828	Ely Place, Holborn, Middlesex	esq.	embrocation for sea sickness

(Continued)

(Continued)

Surname	*First Name*	*Year*	*Declared address*	*Declared occupation*	*Abbreviated Description of Medicine*
Eaton	Robert	1722	–	doctor in physick	chymical preparation to stop external & internal bleeding
Faynard	James	1773	St. Margaret, Westminster	gent.	powder to stop bleeding (no specification)
Felton	Samuel	1809	Berwick St., Soho	botanist	botanical preparation for gravel & stone
Ford	Robert	1816	Crouch End, Hornsey, Middlesex	chemist	medicine for coughs, colds, asthmas called Ford's Balsam of Horehound
Ford	Thomas	1830	Canonbury Square, Islington	chemist	improved Ford's Balsam of Horehound
Fordyce	William	1763	St. James, Westminster	surgeon	stomach pill
Foster	Abraham	1766	Seething Lane, All Hallows, Barking, Essex	peruke maker	medicine for ague
Fraunces	Joseph	1751	Daventry, Northamptonshire	apothecary	female strengthening elixir
Gale	Thomas	1782	New Bridge St., London	chymist	spa elixir
Godbold	Nathaniel	1785	Bloomsbury Square, Middlesex	gent.	Godbold's vegetable balsam
		1798	Bloomsbury Square, Middlesex	gent.	Godbold's vegetable balsam, ointment & pill
Greenough	Thomas	1744	St. Sepulchre, London	apothecary	tooth tincture
		1757	St. Martin, Ludgate, London	apothecary	stomatick lozenges for stomach and bowels
		1757	St. Martin, Ludgate, London	apothecary	volatile balsam for pains of stomach and bowels
		1779	Ludgate Hill, London	apothecary	samaritan water
Grew	Nehemiah	1698	–	doctor of physick	salt of purging waters
Grubb	Robert	1777	St. Martin, Ludgate, London	gent.	Friar's drops for venereal disease, scurvy and rheumatism
		1793	Old Bailey, St. Martin, Ludgate, London	gent.	restorative drops
Hannay	Samuel	1774	Philpott Lane, London	chymist	genital wash to prevent venereal disease
Hayward	Robert	1742	Bristol	apothecary	powder for rheumatism and gout
Hemet	Jacob	1773	St. Pancras, Middlesex	dentist	pearl dentrifice
Henderson	Christopher	1767	City of London	merchant	Beaume de Vie (joint)
Henry	Peter	1744	St. Andrew, Holborn, Middlesex	doctor of physick	nervous medicine
Henry	William	1816	Manchester	doctor of physick	improved sulphate of magnesia
Hooper	John	1743	Reading, Berkshire	apothecary & man-midwife	medicine called 'female pills'
		1752	Reading, Berkshire	–	strengthening balsam & pills for children
Hopkins	John	1767	City of London	druggist	Beaume de Vie (joint)
Howe	Thomas	1788	Bath	chymist & druggist	Howe's pectoral lozenges of horehound
Irwin	James	1773	St. George, Hanover Square, Middlesex	confectioner	new method to make medicine lozenges
Jackson	Humphry	1753	East Smithfield, Middlesex	chemist	cordial bitter stomach tincture
Jackson	Thomas	1747	Shropshire	yeoman	medicine for burns, scalds, bruises, sprains

(Continued)

(Continued)

Surname	*First Name*	*Year*	*Declared address*	*Declared occupation*	*Abbreviated Description of Medicine*
Jackson	Thomas	1761	City of London	chymist	imperial lotion for infections and venereal disease
Jackson	James	1752	–	chymist	British balsam of health (joint)
James	Robert	1747	St. James, Westminster	doctor in physick	fever powder
		1774	Bruton St., Middlesex	doctor in physick	analeptic pills
Jewell	Joseph	1807	Stratford, Essex	chymist	improved calomel for medicinal use
Johnston	Robert	1798	Greek St., Soho, Westminster	chymist & apothecary	Whitehead's essence of mustard
Juniper	John	1762	St. Anne, Soho, Westminster	chymist & apothecary	essence of peppermint
Lovell	Edward	1731	–	–	styptick for internal & external bleeding
Langley	James	1751	–	surgeon	medicine from English vegetables
Leake	Walter	1753	City of London	practitioner in physick	Pilula salutaria, health restoring pills
Lena	Innocenzo della	1800	Piccadilly, Middlesex	surgeon	'Powder of Mars' urine preparation
Lerat	Charles	1769	St. Mary-le-Bone, Middlesex	surgeon	powder to purify the blood
Lobb	Theophilus	1762	–	doctor of physick	tincture for appetite, strength, rheumatism, gout, etc.
La Blache	Louis Goy	1757	St. George, Bloomsbury	surgeon	Royal Military Drops for venereal disease
Lowther	William	1755	City of London	gent.	anti-epileptic powders
		1757	Hatton Gardens, Middlesex	esq.	powders & drops
Martin	Benjamin (senior)	1784	Maidstone, Kent	–	'Antipertussis' for hooping cough
Mason	Robert	1792	Chipping Sodbury, Gloucestershire	surgeon & apothecary	worm medicine
Mettemberg	Joseph de	1825	Foley Place, St. Mary-le-Bone, Middlesex	physician	Mettemberg's Water
Mushet	John	1829	York Square, Regent's Park	gent.	medicine for gouty afflictions (no specification)
Necler	Edmund	1746	Hammersmith, Middlesex	gent.	medicinal paste worn on a belt
Norris	Thomas	1768	Duke St., Westminster	chemist	cure for fevers & inflammatory disorders
Norton	John	1764	St. James, Westminster	surgeon	Maredant's Drops
Okell	Benjamin	1726	–	chymist	Dr. Bateman's Pectoral Drops
Pike	Ann	1760	Peckham, Surrey	wife of Thomas Pike	ointment for itch & scorbutic humours
Radley	William	1776	St. Andrew, Holborn, Middlesex	druggist & chymist	purging carminative tincture
Roche	James	1803	King St., Holborn, Middlesex	gent.	external application for hooping cough
Rock	Richard	1751	St. Bridget, London	licentiate in medicines	medicine for venereal disease
Ryan	John	1758	St. Andrew, Holborn	physician & surgeon	Peru Drops for venereal disease
		1762	St. Andrew, Holborn	physician	white drops

(Continued)

(Continued)

Surname	*First Name*	*Year*	*Declared address*	*Declared occupation*	*Abbreviated Description of Medicine*
Rymer	James	1792	Reigate, Surrey	surgeon & apothecary	cardiac & nervous tincture
Savory	Thomas	1815	New Bond St., Middlesex	chymist	new formulation of Seidlitz powder
Sedgwick	William	1749	Newcastle-upon-Tyne	surgeon	purer sal. ammoniack
Severne	Joseph	1785	Bromyard, Herefordshire	surgeon	aromatic ague cake
Sibly	Ebenezer	1795	Upper Titchfield St., St. Mary-le-Bone, Middlesex	doctor in physic	reanimating solar tincture
Sigmond	Joseph	1800	Bath, Somerset	surgeon dentist	British Imperial Lotion for preserving teeth
Sinclair	George	1722	–	–	medicines from American plants
Smith	Thomas	1749	Spittlefields, London	gent.	medicinal snuff for hypochondria & melancholy
Spilsbury	Francis	1792	Soho Square, Westminster	chymist	antiscorbutic drops
Story	Edward	1759	St. James, Clerkenwell	apothecary	worm destroying cakes
Stoughton	Richard	1712	–	apothecary	restorative cordial, Elixir Magnum Stomachicum
Stringer	Richard	1791	The Strand, Middlesex	chemist & druggist	Stringer's Essence of Myrrh for gum scurvy
Stuart	Ferdinand	1809	Billiricay, Essex	esq.	improved Peruvian bark
Sutton	Daniel	1766	Ingatestone, Essex	surgeon	smallpox treatment (joint) (no specification)
Sutton	Robert	1766	Framingham Earl, Norfolk	surgeon	smallpox treatment (joint) (no specification)
Tanner	Francis	1744	St. George, Bloomsbury, Middlesex	gent.	local sudorific, in one joint or limb
Taylor	Jeremiah	1755	Bristol	gent.	draught for colic
Thompson	John	1786	–	–	concentrated balsam of arquebusade (no specification)
Tickell	William	1786	Walcot, Bath	apothecary	anodyne oethereal spirit
Towers	John	1816	Little Warner St., Cold Bath Fields, Middlesex	chemist	Towers's New London Cough Tincture
Turlington	Robert	1744	London	merchant	balsam of life for stone, gravel & cholic
Wakefield	Robert	1776	St. Paul, Covent Garden	regular bred surgeon, member of Corporation of Surgeons	medicine for children's gripes and convulsions
Walker	Robert	1755	St. Sepulchre, London	dealer in medicines	Jesuits' Drops for Venereal Disease
Warren	Richard	1772	St. James, Westminster	–	volatile essence of lavender
Warren	William	1826	Crown St., Finsbury Square	gent.	improved Peruvian Bark (no specification)
Watt	John James	1828	Stray St., Stepney, Middlesex	surgeon	gas or lotion of chlorine to prevent venereal disease
Wessels	Hart	1759	St. Mary Axe, London	doctor of physick	Tinctura Embryonum
West	George	1752	City of London	surgeon	pectoral elixir for all diseases of the breast
Williams	Thomas	1765	St. James, Westminster	apothecary	essence of flowers of benzoin or pulmonic drops
Wilson	Thomas	1781	St. Sepulchre, London	chymist	medicine for agues & fevers
Wright	Henry	1760	St. Michael, City of London	chymist	Royal Clove Drops
Wright	William	1753	Baldock, Hertfordshire	surgeon	cordial mixture to facilitate childbirth

Index

A. Mackintosh, *The Patent Medicines Industry in Georgian England*, Medicine and Biomedical Sciences in Modern History,
https://doi.org/10.1007/978-3-319-69778-9

N

Q

FSC
www.fsc.org
MIX
Papier | Fördert gute Waldnutzung
FSC® C083411

Kolibri 360 GmbH
Ferdinand-Jühlke-Straße 7
99095 Erfurt, Deutschland
produktsicherheit@kolibri360.de